God, Evolution & Science

God, Evolution & Science

How Our World Evolved from God

William Stolzman

WIPF & STOCK · Eugene, Oregon

GOD, EVOLUTION & SCIENCE
How Our World Evolved from God

Copyright © 2020 William Stolzman. All rights reserved. Except for brief quotations in critical publications or reviews, no part of this book may be reproduced in any manner without prior written permission from the publisher. Write: Permissions, Wipf and Stock Publishers, 199 W. 8th Ave., Suite 3, Eugene, OR 97401.

Wipf & Stock
An Imprint of Wipf and Stock Publishers
199 W. 8th Ave., Suite 3
Eugene, OR 97401

www.wipfandstock.com

PAPERBACK ISBN: 978-1-7252-5787-0
HARDCOVER ISBN: 978-1-7252-5788-7
EBOOK ISBN: 978-1-7252-5789-4

Cover and book illustrations by Nicole Tanner

Scripture text in this work are taken from the *New American Bible, revised edition* © 2010, 1991, 1986, 1970 Confraternity of Christian Doctrine, Washington, D.C. and are used by permission of the copyright owner. All Rights Reserved. No part of the New American Bible may be reproduced in any form without permission in writing from the copyright owner.

Manufactured in the U.S.A.   04/20/20

Contents

Introduction & Summary | 1

Part I: Foundations

1. Evolution of the Notion of Evolution | 13
2. Logical Evolutionary Theory | 21
3. Initial Consequences of Logical Evolutionary Theory | 29
4. Interrelational Logistic | 36
5. Interrelational Philosophy (IPhil) | 51
6. Resolving the Problem of Evil | 69
    I. The Classical Philosophical Problem of Evil | 69
    II. The Biblical Problem of Evil | 77
    III. Dealing with Personal & Material Evil | 88
    IV. Humor and Evil | 93

Part II: Beginnings

7. The Almighty as Primal Absolute Potential Energy | 103
    I. Primal Absolute Potential Energy (PAPE) | 107
    II. Primal Absolute Energetic Reality (PAER) | 116
    III. Primal Absolute Interrelational System (PAIS) | 117
    IV. Primal Absolute Interrelational System as Panentheistic | 126
8. Logical Evolution of Space and Mathematics | 132
9. Logical Evolution of Gravitation | 150
10. Logical Evolution of Electromagnetism | 164
11. Logical Evolution of Strong Force | 174
12. Logical Evolution of Weak Force and Summary Reflections | 179

Part III: Developments

13. Evolution and History | 193
14. Human Development and Evolution | 207
15. Logistic of Knowledge | 230
16. Stages of Human Development | 255
17. Societal Matters | 273
18. Who Is a Person? Is the Almighty a Person? | 299
19. Religions | 319

Part IV: End

20. Soul and Afterlife | 341
21. The Finite-Absolute Progenitor | 351
22. The Big Collapse & New Beginning | 361

Glossary | 373
Bibliography | 381
Index | 387

Introduction & Summary

For whom is this book written? In today's world, this question refers to a very precise demographic, whose primary interest is only in a very narrow slice of reality. Modern scientific disciplines and popular interests are being divided into smaller and smaller pieces. Today's scientists look at the things in their field of interest as through deeply penetrating microscopes. However, how can a person appreciate and understand holistically an entire painting by looking at small pieces of it through a microscope? Specialized approaches to reality have lost sight of the forest for the trees. Although the scientific method is a tool I use throughout this book, this book is not a scientific treatise on a particular topic. Rather, it is a work of philosophy in the classical, universal sense of that word—a love of wisdom. As a book of wisdom, it draws together in a comprehensive way all current knowledge to "make sense of the whole thing." Consequently, this holistic book is countercultural to our specialized world today. It is written for "renaissance people" who are multidisciplinary and interested in making sense of our world in general. Such intellectual explorers are eager to go beyond the narrow fields of modern academic disciplines to gain a holistic, cross-disciplinary, comprehensive vision of our awesome world.

Limited by the academic boundaries of their scientific disciplines, most physicists, cosmologists, or anthropologists could not write this book. Similarly, most theologians are bound by the restrictions of the canonized revelations and traditional texts of their respective denominations. In contrast, IPhil's discussion of the Almighty is not founded on any religious tradition but on the nature of energy. Nonetheless, throughout this book I will make many comparative references to religions worldwide and show their relationship to the interrelational logistic of energy. The Almighty is the same foundation of all spiritual and religious experiences and all scientific knowledge in our world. As a result, IPhil provides a philosophic foundation for the interpretation of different religious traditions and texts, bringing them closer to each other and to the physical sciences, because they are all founded on the pre-big bang, potential energy of the Almighty.

Some professionals have accumulated and ordered many facts in a comparative way, but they usually refuse to make any evaluative judgments about these materials because they currently do not have an independent, common philosophic foundation on which to base any moral judgments objectively. In this book, the objective foundation of judgment is the interrelational character of the potential energy of our evolving world.

Unlike the antithetical judgments currently made within our hypercritical, skeptical world, IPhil's view is synthetic and optimistic, oriented toward interrelationally establishing a holistically better world. Based on progressive potential energy and its physical expressions, IPhil pushes toward more interrelational insights and possibilities.

The material in this book is progressively presented and ordered as the skeletal framework of a skyscraper. It is first deeply grounded in the material earth. A great amount of effort and care is taken to establish the deep, foundational principles upon which this interrelational philosophy (IPhil) is built. In today's world, most people focus on the one floor they are most interested in, letting the people upstairs and downstairs do their own thing. In contrast, this book shows how the different floors of reality build successively one on the next. Like energy, this book is progressive in an orderly way, each chapter building upon the previous in a logical evolutionary way.

Unfortunately, my experience is that many readers go immediately to the chapter they are most interested in. They then reject the materials presented in their pet chapter because they have not taken the time to grasp the logical, evolutionary foundations provided in the preceding chapters. Because this book is evolutionary in its logical development, it is important that the reader progress in a sequential manner through this multidisciplinary study. Each chapter will give the reader intellectual surprises, just as most advances in evolution in our physical world have their surprises.

Most academic research in psychology today focuses on the narrow, in-depth study of just one behavior. In contrast, longitudinal psychological studies lay out the progressive development of different traits through an entire lifetime. Technical studies that focus on in-depth studies of one area of life are important and of greatest interest to professionals. However, longitudinal studies provide ordinary people with general insights into the way life develops generally.

For me, philosophy embodies wisdom in the broad, classical sense. Wisdom is learned from dealing with many problems of daily life, and a good and true philosophy should be capable of answering the "big questions of life" in clear, logically consistent, understandable ways. For example, "What is the meaning of life?" IPhil provides a simple answer: The meaning and purpose of life is to be more interrelational. The more interrelational one's life is, the more meaningful it is. The "more interrelational" thrust of physical reality and life comes from the primal potential energy which was the origin of the big bang. The forward thrust of potential energy remains the progressive foundation of our daily lives and everything in our world today.

Throughout this text I raise many difficult philosophical questions: Why is there something rather than nothing? Why is our physical world bidirectionally 3-dimensional, while the dimension of time is only unidirectional? Why are there four and only four fundamental forces in our world? What is the relationship of mind to brain? Where did freedom, consciousness, and deliberateness come from? What is the soul, and is it eternal? These questions and others are answered in clear and concise ways, using the four premises about energy in IPhil. I am confident that by using the philosophic tools of interrelational philosophy (IPhil), readers will be able to answer their own "big questions" easily and concisely.

Because interrelational philosophy (IPhil) presents a new progressive and evolutionary way of looking at the Almighty and our world, all standard philosophic terms undergo a paradigm shift. Consequently, I have prepared for the readers a glossary that describes how in IPhil all major philosophic terms undergo a paradigm shift, providing a logically consistent, comprehensive description of how our world energetically evolved from the Almighty.

# # # #

In this book, I present a new approach for linking God and evolution based on the nature of energy. Energy is conservative and exists in two forms: potential and kinetic. Scientists have well established that our physical world began as a singular, kinetic big bang. Since energy is conservative, it must have existed before that. By the principle of conservation of energy, there had to exist a *singular, potential, energetic* reality from which came the *singular kinetic energy* of the big bang. I call this singular, antecedent potential reality Primal Absolute Potential Energy (PAPE). Many people would call this primal reality "God"; I prefer to name it the "Almighty." I describe this primal potential energy as absolute insofar as it must have contained all the energy kinetically in the big bang and in its subsequently developed universe.

I begin by expanding the standard tool kit of logic by showing how all logically consistent systems can evolve logically. In this way, I establish that evolution is not only biological, it is also logical.

By the principle of conservation of energy, before the kinetic energy of the big bang, that energy existed in a potential state. Potential energy is forward leaning and therefore changes through time. Because the kinetic energy of the big bang is efficiently subsequent to the Primal Absolute Potential Energy before the big bang, the primal potential energy must inherently be self-initiating and free. Furthermore, I will show that in this transformation of potential energy into kinetic energy, that self-initiating is both evolutionary, quantized, and spatial. Then, from this interrelationally expanding proto-spatial field, I will show by logical evolution how the four fundamental forces of physics were subsequently evolved.

# # # #

INTRODUCTION & SUMMARY

Philosophy begins with physical evidence and makes metaphysical inferences from that evidence. In this book, experiential observations provide evidence for an orderly philosophic worldview which I call "interrelational philosophy" (IPhil). The major tenets of this philosophy are:

1. *Potential Energy.* The founding ur-stuff of reality is potential energy, namely, the singular Primal Absolute Potential Energy (PAPE) antecedent to the singular kinetic big bang.

2. *Intrarelated Individuals.* The singularity of the kinetic energy of the big bang indicates that the potential energy from which the big bang came was individually One. From this comes the insight that potential energy has the interrelational capacity to form other enduring, integral, energized individuals.

3. *Interrelational Systems.* Just as our bodies are integrally one, our bodies consist of many interrelational parts. Evidence indicates that energy is capable of forming reciprocal interrelational systems. Through a study of enduring physical systems, I will show that the process of forming and sustaining energized systems is interrelational, reciprocal, logically consistently, well-ordered, and fourfold in its advancing phases.

4. *Free Self-Initiatives.* Besides being forward leaning in time, the potential energy before the big bang had to have the capacity to freely self-initiate the kinetic, physical big bang. Consequently, potential energy must have the capacity to freely self-initiate its potential energy into kinetic, evolutionary advancements.

The above summary of the major tenets of IPhil is not assumed *a priori*. A major challenge of this book is to establish all the above statements from observed physical evidence, and then to show how our world was evolved and generated from these fourfold tenets of potential energy.

# # # #

Knowledge of evolution did not begin with Darwin. He recognized that, long before his time, humans practiced artificial evolution in their breeding and agrarian practices. In the first chapter, I describe different forms of evolution—how we came to know them, and how we can go beyond them in a logical evolutionary way.

In the second chapter, I establish the logical evolutionary theory by going beyond Gödel's incompleteness theory. I describe in a formal way *how* a logically consistent system can evolve into a system that is interrelationally more advanced than its antecedent. Logical evolution is not something new; only the formulation of the logical evolutionary theory is. Logical evolution is implicitly learned in grade school in the advancement of number systems in mathematics. This theory establishes that evolution is not only biological, it is also logical. I show how logical evolution existed and was operative from the time of the big bang, long before biological evolution developed.

In the seventeenth century, Isaac Newton developed the theory of fluxions or calculus and used it as a new tool, which opened a whole new way of looking at our gravitational world. In a similar way, I will show how the logical evolutionary theory provides a new philosophic tool which yields an advanced evolutionary way of looking at our world. I will show that logical evolution and Darwinian evolution are parallel in their processes—Darwinian evolution in the kinetic, physical order, and logical evolution in the potential, metaphysical order. I will show that the evolutionary tree of biological life and the evolutionary tree of ideas are parallel and analogical, arising from the kinetic and potential sides of energy, respectively.

Philosophers and medical researchers have long used analogies to study the things in our world. But whence comes the truth found in analogies? The logical evolutionary theory establishes why and how partial truth can be found in analogical comparisons of the things on different branches of the evolutionary Tree of Life. The linking of logical evolution and physical evolution via the two sides of energy provides a means for linking the intellectual worlds of philosophy, thought, and theology with the physical worlds of the material and psychological sciences. Because of energy's distinctive potential and kinetic aspects, energy establishes a common ground for relating the spiritual Almighty before the big bang with the physical activities found in our material world after the big bang.

But there is a major problem. The strongest argument against the existence of God is the presence of evil in our world. If the God of our world is all-good, how can evil exist in the world God created? How could even the possibility of evil be found in a world that an all-good God created? The presence of evil in our world is one of the biggest questions in the lives of many people. Their struggles with the apparent contradictions of evil are the greatest reasons many people reject the existence of God. I spend an entire chapter discussing and resolving this paradox. I deal with the problem of evil in several ways: philosophically, scripturally, and personally. In this chapter, I present a whole new way of looking at evil, founded on the way that energy operates interrelationally in our world. With a new interrelational view of evil in our world, the foundational section of this book is complete.

The second major section of this book begins with a recognition that every evolutionary process must begin somewhere. I propose that all evolution begins with the singular Primal Absolute Potential Energy (PAPE) which was antecedent to the singular kinetic big bang. Energy is interrelational. I will show that even in the metaphysical order, the interrelational process of energy is progressive in four formally distinct stages: receiving, holding, orienting, and releasing. Consequently, the chapter on the Almighty is progressively divided into four sections. Since potential energy is forward leaning and has four distinctive, progressive interrelational phases, the parameter of time must exist in the Almighty or PAPE. I further show that the parameter of time must have a minimal value and therefore be quantized. Because potential energy is always forward leaning, I explain why the parameter of time must be unidirectional. The first section also describes the initial state of Primal Absolute Potential Energy as *unformed* and therefore not conceivable as a substantial idea. This provides an explanation why some advanced spiritualities recognize God is primally ineffable.

The second section progressively describes the Almighty as an intrarelational, integrated reality, giving foundation to the belief that the Almighty is formally *One*. This view is also held by several dominant religions. The third section describes the Almighty as energetically interrelational, capable of forming an enduring, reciprocal, logically consistent, absolute system. This provides the philosophic foundation that the absolute One is also interiorly interrelational with four interdependent operations and three interrelational elements. This Trinitarian view of the Almighty is held in Hinduism and Christianity. Finally, because the systemic phase of the Almighty is logically consistent, it is capable of logical evolution through an inverse progenitor. This evolutionary progenitor would be inversely not absolute but finite in its operations. I show by the logical evolutionary process that this progenitor is formally distinct from the Almighty. Rather, this first finite receiver of energy and its progeny must be *panentheistically* "in" and "of" the Almighty.

Then, I show how this energized, finite progenitor can generate the first finite system. Because systemically all subsequently generated finite, partial terms must be enduringly distinct from each other, there must a new formal parameter: "distance." Then I show that the parameter of distance must be quantized. Also, within this logically evolved, first, finite, logically consistent, energetic, reciprocal, interrelational system, the parameter of distance is bidirectional. Furthermore, because of the quantum requirements of this proto-spatial field, I show that this interrelational proto-spatial system is 3-dimensional. In this way, the first evolved finite system is a quantized, 3-dimensional spatial "field." Because of the carryover character of the evolutionary process, all subsequently evolved systems will operate in this 3-dimensional field with four, formally distinct dimensions, $\{x, y, z, t\}$. By its systemic symmetries, I will show how this first evolved spatial field is quantized Euclidean.

Because there are minimum quantized values of both time and distance in this proto-spatial system, I show that communication of energy between neighboring space-points in this proto-spatial system has maximum speed, which is traditionally expressed as c. Then, by logical evolution, this maximum speed of energy communication in the proto-spatial field is carried over to all subsequently evolved interrelational systems. This explains why in the subsequently evolved gravitational system, the maximum speed of gravitons in space is c, and why in the logically evolved electromagnetic system, the maximum speed of photons is also c—the speed of light in empty space.

Experimental physicists have observed that there are four formally different forces that guide the operations of our universe. In a logically deterministic way, Albert Einstein proposed that there must be a unified field theory, which has a single, unchanging algorithm. He claimed that by logical deduction, the four fundamental forces of physics can be logically *derived* from this single algorithm. Some thinkers identify this algorithm with God. There have been several different attempts to establish the singular algorithm of the *unified field theory*; however, they have all been unsuccessful. In contrast, I propose an *evolutionary field theory*, in which the four foundational forces of physics are logically *evolved*—rather than logically *derived*—from the logistic of the proto-spatial system, with its formally distinct, 4-dimensional parameters. I will show in four successive chapters how the four forces of physics can be logically evolved from the 4-dimensional logistic of the proto-spatial system. Since the proto-spatial system is logically evolved from the logically consistent Primal Absolute Interrelational System (PAIS), this links in a logical evolutionary way the fourfold logistic of the Almighty to the four forces of physics via the proto-spatial system.

That is, I show how linear acceleration in the 1st dimension in proto-space results in the logical evolution of gravitation, the parameter of mass, and the formation of the universal gravitation formula. Next, I show how the breaking of the planar, 2-dimensional, quantized symmetry in the proto-spatial system results in the logical evolution of quantized spin, the electromagnetic force, and the parameter of electric charge. The breaking of the 3-dimensional symmetries of the proto-spatial results in the *strong force* and the new parameter of color charge. Finally, because of the unidirectional character of time, the breaking of mirror symmetry results in the evolution of the fourth force, namely, the *weak force*. Thus, in four chapters, I show how the four foundation forces of physics—gravitational force, electromagnetic force, the strong force, and the weak force—logically evolved from the breaking of the symmetries found in the four dimensions of the proto-spatial system. I close the fourth chapter with a general, comparative description of the relationship of these four distinct physical forces to other fourfold energetic phenomena in our world.

The third major section of this book is less technical and will be more familiar to most readers. Its chapters describe how our world has evolved from the Almighty since the big bang. Because humans are the most interrelationally evolved and intelligent of all species, special attention is given to the origin and operations of human intelligence. Emmanuel Kant raised a significant argument, claiming that all human knowledge is subjectively biased and cannot be objectively true to exterior reality. This line of thinking has played a major role in the self-centeredness of modern philosophy and popular relativistic thinking. IPhil overcomes this major conundrum using logical evolution. I show how human knowledge is not only subjective, but how it *can* know objective truths about our physical world.

Most philosophers and psychologists today say there is no difference between the human brain and mind. While potential energy and kinetic energy are very similar, they are significantly different. Similarly, by means of logical evolution I show that the physiological operations of the human brain and the conceptual operations of the human mind are similar, but they are formally different. To say that the brain and the mind are the same is to say that physical kinetic energy and metaphysical potential energy are the same. No physicist would agree with that. Just as potential energy pushes toward expressions of kinetic energy, so too the mind's metaphysical operations push the brain toward advancements in its physical structure and operations. I will argue that the evolved structure of the physical brain originates from the evolving operations of the metaphysical mind, rather than the reverse. And what of the human soul? I will argue that the human soul is not in the human body, but inversely the human body is in its *logistic-soul*. IPhil has its own understanding of one's soul as one's enduring logistic or substantial form. I also will explain how after the death of the material, physical body, the associated metaphysical logistic-souls of individuals endure.

The category of "person" is very important in human society. The traditional essentialistic definition of a person is "a rational individual." Contrariwise, IPhil replaces this standard essentialistic definition of *person* with an interrelational one. There are different levels of human communications, and most of them are impersonal. Nonetheless, some communications are personal. What is the difference between impersonal and personal communications and interactions? This leads to the question: Is the Almighty's energetic interrelationship with our world only materialistically impersonal, or can it be spiritually personal? Again, logical evolution provides an energetic way of answering this question.

Humans are the most intelligent species. However, other animal species display high levels of intelligence, problem-solving, and communications. So, what makes humans truly unique from other animals? While animals display different levels of intelligence, they do not display religiosity. Throughout history, humans have displayed their religiosity in a variety of ways. The establishment of different religions is a historic fact; their establishment is a part of how our world evolved from God. Modern studies in comparative religion expose many common spiritual characteristics. IPhil provides a philosophic explanation for the diverse, progressive ways humans have sought to know and become united first with natural spirits, then conceptual gods, and then with a singular Almighty.

The fourth major section of this book pertains to the end of our physical world. Currently, entropic scenarios describe how our expanding world will end in an icy, isolated state. In this scenario, communications between all material particles will wane into lower and lower energy states, until there are no interrelational communications between all physical things, leaving all things in our cold universe in total darkness. However, this scenario assumes that space is eternal, and energy is infinitesimally divisible. In contrast, IPhil recognizes that just as the spatial system of our universe world had a quantized beginning, so too the spatial system of our universe will have a quantized end. Like the skin of an expanding balloon, space itself has an interrelational coherence only up to a certain point. After that point, the fabric of space will suddenly fragment in a big collapse. IPhil explains why and how this will happen. Nonetheless, just as the end of kinetic energy does not mean the end of potential energy, so too, the physical, material end of our world will not be the metaphysical, spiritual end of reality. Energy is conservative and therefore eternal. Can anything more be said of energized existence after the big collapse?

Belief in the possibility of an afterlife has been a part of the evolving history of humanity from the time of the Neanderthals. But from IPhil's energetic perspective, how could an afterlife happen? Since the logistics of finite individuals are enduring and logically consistent, IPhil shows how an evolutionary afterlife beyond our current state is logistically possible for individuals and for our world. IPhil shows how the logically consistent nature of energy and the unending metaphysical endurance of the logistics of individuals and communal systems allow for the possibility of the logical evolution of our current world into an advanced afterlife. IPhil then logistically projects some of the transcendental characteristics of a logistically evolved afterlife, freely re-energized by the Almighty.

Part I
Foundations

Chapter 1

Evolution of the Notion of Evolution

Domestic Evolution. Natural Evolution. Genetic Evolution.
Conceptual Evolution. Logical Evolution. Summary.

In human history, understanding of evolution has advanced in stages as humans gained a deeper understanding of the inner structures of things.

Domestic Evolution

Humans became aware of evolutionary changes in an informal way through their domestication and breeding of plants and animals by means of artificial selection. The first animal to be domesticated was the dog in Europe around 15,000 years ago, and in central Asia around 12,500 years ago. Sheep, pigs, goats, cows, cats, chickens, Guinea pigs, donkeys, and other animals were domesticated in the Middle East between 10,000–5,000 BCE (Before the Common Era).[1] Domestication of plants, fruits, and nut trees may have started earlier. However, since seeds are smaller and are more prone to decomposition than skeletons, it is difficult to document early horticultural experimentation and the domestic development of fields and orchards. The selective evolution of maize in Mesoamerica from small grains into large ears began around 2,500 BCE and is well documented.[2]

Increasing size and desired characteristics through selective breeding usually involved a simple four-step process: (1) From collected animals and plants, (2) humans selected offspring and seeds that had more desirable characteristics. (3) Then by mating and growing of the progeny of the selected desirable individuals, (4) humans were usually able to obtain crops and animals having more of the desired characteristics.

Domestic evolution focused upon external, beneficial attributes—such as larger kernels, more kernels, more meat, more wool, stronger oxen, faster horses, etc. Subsequently, people applied this knowledge analogously to human generation, where "good breeding" and "family traits" within the aristocratic members of society became important factors in choosing a "proper" mate for one's children. Similarly, groups began to recognize and augment racial and cultural distinctions, separating those individuals who

---

1. Clutton-Brock, *Domesticated Animals from Early Times*, 49–52.
2. Roney, "Beginning of Maize Agriculture."

conformed to the physical and behavioral traits of one's advanced, "civilized" population from others, who were disparagingly called "inferiors," "outsiders," "backwoods hicks," or "barbarians." In other words, the practice of artificial or human-caused breeding was not reserved to domestic plants and animals, but it was applied to humans as well, thereby creating castes and elite classes within societies. Being a "blue blood," having "good genes," or being of "good family stock" are still active concerns in many cultures today. In Japan, for instance, documented and valued health histories of families over many generations play a significant role in choosing one's mate.

In summary, our oldest knowledge of evolution pertains to artificial selection by humans who strove to produce *external* traits that were materially more valued and socially more significant. However, humans did not have a detailed knowledge of how these desired traits were actually produced through artificial breeding.

Humans learned from experience that "correct breeding" did not always produce better offspring. Sometimes selective breeding produced deficient progeny. So, from early times, humans knew that artificial breeding for better traits was not a certain thing. They attempted to give reasons for this, and in the absence of easily observable reasons, their most common explanations were: "This is God's will," "This is Karma," or "This is Fate."

Humans discovered that there are limits to artificial breeding. Sometimes crossbreeding of different species of plants or animals produced a hybrid with traits of both parents, but these progenies were sometimes sterile and unable to produce another generation. For example, the mating of a female horse and a male ass results in a strong mule, but all mules are sterile and unable to breed their own kind. Selective interbreeding worked only when the interbreeding took place between similar strains. That is, it was recognized that new species with advanced traits occurred only in small steps. As it is said, "The apple does not fall far from the tree."

From such experiences, early pre-Socratic Greek philosophers, like Anaximander and Empedocles, proposed that one type of organism could descend from another type. Such proposals survived into Roman times. However, Plato maintained that there were ideal, perfect, fixed forms of all things in the heavens. What we observe are imperfect expressions of those ideal forms, which do not change or evolve. Aristotle similarly proposed that natural things had enduring individual substantial forms and enduring communal natures. As part of his teleological understanding of nature, he maintained that all things have an intended ideal place in our world, and each tended toward their ideal place. However, he did not demand that physical types always correspond one-to-one with exact metaphysical forms, indicating how imperfect things could become more perfect. Unfortunately, the Neoplatonic notion that all things tend toward their fixed, ideal form became the standard understanding of species into the Middle and Renaissance Ages. Only with difficulty was the Greek notion of static, ideal, essentialistic forms in nature gradually replaced in modern times with a developmental model of nature.

Natural Evolution

Science in the eighteenth century developed an increasingly mechanical view of our world. However, a few natural philosophers, such as Pierre-Louis Maupertuis in 1745 and Erasmus Darwin in 1796, suggested proto-evolutionary ideas. The methodical, scientific study of evolution really began in the mid-nineteenth century.[3] Comparative studies of fossil records and the diversity of living organisms in various, newly discovered parts of the world, convinced many exploring scientists that species did indeed evolve naturally. In the early nineteenth century, Jean-Baptiste Lamarck's theory of the transmutation of species was the first fully formed scientific theory of evolution. While it had many worthy features, his description of how traits were communicated from one generation to the next by the physical development of a parent, was found wanting.

The development of enduring evolutionary ideas was slow in coming. Georges Cuvier's catastrophism associated changes of animal forms with catastrophic environmental changes, such as the biblical flood. Adam Sedgwick recognized that worldwide annihilations and the subsequent formation of new species tended to happen in response to the changes in environments that *followed* such cataclysms. James Hutton's uniformitarianism maintained that the earth was shaped by the same gradual processes of climactic change that we observe today. Charles Lyell adapted these ideas in *Principles of Geology*. He was a close and influential friend of Charles Darwin.

As evidence mounted, some natural scientists began to describe patterns in nature not individually but statistically. They described changes in terms of distributions and probabilities in populations rather than in individuals. Thomas Robert Malthus applied this statistical approach to human populations in *An Essay on the Principle of Population*. This book strongly influenced Charles Darwin's analysis of his extensive observations.

During his journey on the *Beagle*, Darwin's studied the external differences he found in finches and other animals on the separated Galapagos Islands. His reflections led him to formulate the seminal idea of "natural selection" in 1838. He was still developing his theory in 1858 when Alfred Russel Wallace sent him a manuscript proposing a similar theory. They each presented their ideas in separate papers at the same meeting of the Linnean Society of London.[4] Then at the end of 1859, Darwin published *On the Origin of Species*, which explained natural selection in detail. It presented much evidence supporting his model of biological evolution, which he maintained took place over long periods of time through miniscule changes that promoted increased longevity and dominance of particular traits within a species.

---

3. Mason, *History of the Sciences*, 43–44.
4. Beddall, "Wallace, Darwin."

Thomas Robert Malthus's study of human populations focused on the relationship of group survival to available food supplies.[5] Darwin applied some of these ideas to other species, explaining how, with limited and changed food supplies, certain individuals had an increased, statistical likelihood of surviving and generating. These individual members were said to be more "fit" than other members of their species and more likely to reproduce. Though the principle of "survival of the fittest" is an element in Darwinian evolutionary theory, it is not one he emphasized. Rather, he emphasized that evolutionary changes are usually small and generally go unnoticed from one generation to the next. Darwin also recognized that there were other mechanisms that influenced natural selection besides a male's prowess to gather many females under his dominance. A hen shows favored sexual selection in response to a cock's looks, his performance in mating rituals, his ability to make a pleasing and secure nest, etc.

Others picked up Darwin's ideas on evolution and applied them beyond the field of biology to human behavior. Soon people applied the principle of "survival of the fittest" to economics, social competition, leadership competitions, and the rise and fall of entire peoples and nations. Such extensions were called "Social Darwinism." In this worldview, certain humans and groups are more "fit" to survive and advance in the competitions and the ruthless jungles of daily human life. Those humans who are less fit do not survive or propagate well. So, it was said that by the evolutionary laws of nature, the "less fit" tend to die off more quickly, leaving more "fit" humans to lead humanity to a more evolved, better way of life.

Darwin's book and ideas were well received internationally. Nonetheless, he was not able to explain the physical process by which variant, superior individuals developed within a given population. He and others maintained that variations of *secondary* traits developed very slowly over centuries. Then, through many changes of "secondary attributes," evolution progressively resulted in the formation of different species. This model discarded the classical metaphysical distinction between the singular essentialistic form of a species and its myriad accidental forms. In this way, these evolutionary scientists recognized only progressive changes within species, and they abandoned the classical philosophic category of unchanging, ideal essences.

Genetic Evolution

Explaining how those structural variations developed in newborn progeny required the discovery of the microscope and other instruments by which scientists were able to study the foundation of biological life on a molecular level.

After years of study, the Augustinian friar Gregor Mendel established in 1865 that peas transmitted variant characteristics to succeeding generations in a predictable manner. He laid out the principles of hereditary mutations. His ideas, however, were ignored by the scientific world in his lifetime. Hugo Marie de Vries, unaware

---

5. Malthus and Flew, *Essay on the Principle of Population*, ch. 7.

of Mendel's work, rediscovered the laws of heredity in the 1890s. He suggested the concept of genes as part of his mutation theory of evolution. Initially, the notions of genetic heritage and mutations appeared to contradict Darwin's theory of evolution by natural selection. However, by the 1920s and 1930s those apparent contradictions were resolved by the work of evolutionary biologists such as J. B. S. Haldane, Sewall Wright, and particularly Ronald Fischer, who set the foundations for the establishment of the field of population genetics.[6] In the 1940s, Oswald Avery and colleagues identified DNA as the foundational genetic material, and in 1953 James Watson and Francis Crick described the helix form of the DNA molecule and demonstrated it to be the physical basis of inheritance.[7] Since then, genetic biology has become the core of evolutionary microbiology. In the 1960s, scientists, such as W. D. Hamilton and George C. Williams, pioneered a gene-centered view of evolution, using concepts such as kin selection.[8]

"Genetic evolution" differs from classical "natural evolution" in the scale and mechanism of its changes. Natural evolution classically focused upon the easily observable phenotypic traits of plants and animals, and it used these to explain changes in terms of natural selection and the survival of the fittest within populations of the same species. In contrast, genetic evolution focuses upon mutations and variations on a molecular level, where differences can be observed only microscopically.

Conceptual Evolution

From the beginning of human history, ideas have affected the way people have thought about the world, about advanced education, and about developing innovative systems in our world. In 1975, E. O. Wilson's book on *Sociobiology* established a significant place for an evolutionary theory in psychology, giving rise to the field of "evolutionary psychology." Critics argue that the hypotheses of evolutionary psychology are difficult or impossible to test, but psychologists say that these evolutionary hypotheses can be both corroborated and contradicted by data. The fields of psychology and neuropsychology have become increasingly materialistic, and there has been increased interest in the physical evolution of the brain and its associated sensations, feelings, images, and thoughts.

Other scholars have focused on the application of evolutionary thinking to the realm of ideas themselves. Studies of the historical meaning of words in the Bible and in biblical commentaries written through the centuries clearly show that words and concepts do not have singular, unchanging, universal meanings. The same word and concept can have different meanings in different historic contexts. In fact, words and concepts can even change and develop during a single individual's life—as the

---

6. Fisher, *Genetical Theory of Natural Selection*, 131.
7. Watson and Crick, "Structure for Deoxyribose Nucleic Acid."
8. Newman and Harris, "Scientific Contributions of Paul D. MacLean."

words *gay*, *cool*, and *bad* have in our own lifetime. Essentialistic philosophy grounds itself on the eternal truth of words and concepts. However, that idealistic view of words and concepts has been greatly shaken as our historic knowledge of the evolution of ideas has increased.

As semantic studies of biblical texts became more analytical, and as the writings of other authors began to accumulate and be ordered, scholars began to study words in an orderly way within their changing historic contexts. The historical-critical method seeks to discover the original contextual meaning of texts, especially significant, revered religious texts. The phrase "history of ideas" was coined by the historian Arthur O. Lovejoy, who initiated its systematic study in the early twentieth century. The first chapter of Lovejoy's 1938 book, *The Great Chain of Being*, laid out a general overview of what he intended to be the program and scope of the study of the "History of Ideas." Lovejoy's premise takes as its basic unit of analysis the unit-idea or the individual-concept. These unit-ideas he recognized to be building blocks of the history of ideas. Even when spoken words remain relatively unchanged over time, the meaning of these unit-ideas can change significantly, especially as they recombine in new patterns and express themselves differently in changing historical contexts. It has become clear to many individuals that like biological species, intellectual ideas also evolve over time.

The British evolutionary biologist Richard Dawkins coined the word "meme" in *The Selfish Gene* (1976) to explain the spread of ideas and cultural phenomena in an evolutionary way. The concept of meme comes from an analogy: as *genes* transmit biological information, *memes* transmit ideas and beliefs in a similar way. A meme is a culturally established idea, symbol, or practice, which can be transmitted from one mind to another through writing, speech, gestures, rituals, or other forms of communication. As examples of memes, Dawkins mentions melodies, catch phrases, fashion, and the technology of building arches, etc. Advocates of the *meme* idea say that memes may evolve by natural selection in a manner analogous to that of physical evolution. Memes that propagate less prolifically than others may become extinct, while other memes may survive, grow, spread, and mutate—for better or for worse. Memes that replicate most effectively enjoy more success. The evolution of memes takes place in the domain of ideas, while the evolution of genes takes place in the physical order. Though their development and evolution are in many ways similar, no one has yet proposed a direct link between the evolution of metaphysical ideas and the evolution of physical entities.

Logical Evolution

This book goes beyond the evolution of individual words and concepts to the evolution of logical systems. That is, logical evolution describes how evolution is not only biological but can also be logical. Furthermore, I will show that because energy exists in both kinetic and potential states and because potential energy is metaphysical and knowable, it is capable of logical evolution. Consequently, it is the formal distinction and connection of the kinetic and the potential sides of energy that provides the linkage between the evolutionary kinetic, physical order of material things and the evolutionary potential, metaphysical order of ideas and spiritual interrelationships in our energized world.

Summary

History shows that the notion of evolution has advanced through five major stages. In each successive stage, our understanding of the notion of evolution has become increasingly interior.

1. Initially, *domestic evolution* developed close to the family. Selective breeding and selective planting were guided by people's desire for greater external attributes—especially for more and better food. Greater size and strength were also significant factors that influenced their selections, as were breeding faster or stronger horses, making warmer or more elegant clothes, having better dogs for hunting different types of animals, etc.

2. Explorers identified *natural evolution* by looking at differences in skeletons and secondary operational features, such as the shape of a bird's beak. These *structural* changes gave these animals physical advantages in their struggle to survive in the wild. "Natural selection" favored more "fit" members of a species in a changing, competitive environment. The cataloguing of progressive variations of the physical structure of species enabled paleontologists to organize in a progressive way a species' physical evolutionary history and correlate it with the earth's geological history.

3. Scientist identified *genetic evolution* by looking at the structure of molecules at the microscopic level. They focused on mutations of DNA molecules, whose genes specify the historical development of its biological holder. The structure of the DNA has a four-component genetic code, and each biological individual has its own DNA code. By promoting chemical and genetic mutations, different combinations *can* generate in an evolutionary manner, individuals who are healthier, more beautiful, and resilient to diseases.

4. *Conceptual evolution* is grounded in the study of the *transformation of words* through the centuries. The history of ideas shows how the development of ideas and the splitting of trains of thought have resulted as humans deal with changing historical, political, cultural, and scientific advancements.

5. *Logical evolution* is grounded in the study of the interrelational relationships found in *energetic, logically consistent systems*. It shows how the interrelational

reciprocity found within a logically consistent system can evolve into a more advanced system. It also shows that the push for the logical evolution of systems into more advanced physical systems arises from potential energy through actions of the symmetry-breaking of the logical boundary of a given system by an advancing progenitor of the previous system. This leads to an awareness that the origin of logical evolution is found in the potential energy antecedent to the kinetic energy of the big bang. I will spend much time explaining how the primal potential energy found before the big bang is the enduring root cause of the logical evolutionary of the physical things and metaphysical ideas in our world.

In summary, the above list highlights how through the centuries, human understanding of evolution has advanced from external physical traits, to interior physical traits, to molecular physical traits, to verbal metaphysical concepts, and now to metaphysical systems.

Note that the presentation in this book does not follow the above historical order of our progressively deeper and deeper understanding of evolution. Rather, this book inversely charts how our physical world and biological evolution energetically evolved from the nature of potential energy from a time before the big bang.

Chapter 2

Logical Evolutionary Theory

Gödel's First Incompleteness Theorem. General Logical Evolutionary Theory. School Examples of Logical Evolution. Inverse Logical Evolutionary Theory.

The seminal idea for the logical evolutionary theory came from an extension of Gödel's incompleteness theorem.[1] Like Gödel's incompleteness theorem, logical evolution first requires the existence of a logically consistent system. A logically consistent system has the following:

1. Logical terms. Logical terms can be things, concepts, or statements. These can be concrete things, like, apples, electrical charges, humans, atoms, etc., or logical statements, such as, "It is raining outside," "The sun is shining," "The dog has a bone," etc.

2. Operators. Logical operators, such as "and," "or," or "not," connect the above terms and statements according to specific rules. For example: "It is raining outside, *and* the sun is not shining."

3. Axioms. Systemic axioms are fundamental statements, inductively recognized to be true because they are consistent with ordinary experiences. Axioms establishing a system cannot be self-contradictory or logically inconsistent within that system.

4. Conclusions. Logical conclusions are statements that can be deduced from axiomatic premises according to a set of logical rules pertaining to the operations of logical operators. These logical conclusions draw their truth-value from the truth-value of the system's founding axiomatic premises.

Gödel's First Incompleteness Theorem

Kurt Gödel is best known for his two "incompleteness theorems," published in 1931.[2] These theorems ended a half-century of attempts—beginning with the work of Friedrich

---

1. Goldstein, *Incompleteness*.
2. Gödel, *On Formally Undecidable Propositions*.

Frege and culminating in Alfred North Whitehead and Bertrand Russell's *Principia Mathematica* and David Hilbert's formalism—to find a set of axioms sufficient for all mathematics. The first and more famous incompleteness theorem states that for any logically consistent system powerful enough to describe the arithmetic of the natural numbers, {1, 2, 3, 4 . . .}, there *can* be true propositions about natural numbers that *cannot be proved* from the founding axioms of that system.

By extension, this theorem implies that if an axiomatic system is logically consistent, it may not be complete—because there always *can* be a statement true to a given logically consistent system that cannot be logically derived from the foundational axioms of that system. This proof contradicts the idea that, in a logically consistent system, all statements which are true to that system are provable from the axioms of that system. The realization that every logically consistent system is incomplete and that there *can* be statements that are true to that system that cannot be logically derived from that system, shook the mathematical and logical worlds.

General Logical Evolutionary Theory

Gödel's incompleteness theorem provided for me the initial idea for logical evolution. Consider the following: If there is a logically consistent system, S_1, there *can* be a Gödel statement, G_1, which is true to the axioms of S_1 but which cannot be logically derived from those axioms. Since G_1 is non-derivable from the axioms of S_1 but is logically true to those axioms, that statement can be added to the axioms of S_1 to produce an expanded logical system, S_2. Since S_1 is logically consistent in itself, and the non-derivable G_1 is logically consistent to S_1, the combination of those two would produce a new, expanded system, S_2, which is also logically consistent. But that advanced system is formally distinct from the antecedent system, just as that Gödel statement, G_1, is logically not a part of the antecedent logical system.

Repeating the process with S_2 as the foundational logically consistent system, there *can* be another Gödel statement, G_2, which is true to S_2 but which cannot logically be derived from that system. Folding the new statement G_2 into the axioms of S_2 produces an expanded, new logical system, S_3, which is logically true to the antecedent systems, S_1 and S_2, but which cannot be logically derived from those two antecedent systems. Continuing this process would produce an unending, advancing, and expanding sequence of systems, all of which are logically consistent. Each new, advanced system would contain the premises of all antecedent systems, but its new statements cannot be logically deduced from the antecedent systems. This sequence of progressively more complex systems would be "logically evolutionary."

In the development of such a logical evolutionary sequence, however, there is a problem. Since each new Gödel statement, G_n, cannot be logically derived from the antecedent system, S_n, how can a rational person discover even one non-derivable Gödel statement? That is, how can a single, non-derivable Gödel statement, G_n, be discovered if it cannot be logically derived from the antecedent system? One way might be to mix and combine at random all the terms and operators of a given system and then test each statement to determine whether it is a statement that is true to the given system but not derivable from its premises. Such randomly formed statements would have to not contradict the system that provided the terms and operators in the proposed statement. Still, it would have to be logically consistent to the premises of foundational system *and* not logically derivable from the premises of that system. Unfortunately, the number of such combinations of terms and statements could be infinite, and verification would be an impossible task. So how can a person produce even *one* Gödel statement from a set of premises of any logically consistent system? Intuitively, I knew there had to be a simple way. Surprisingly, I discovered a way to generate non-derivable Gödel statements in the elementary school next to my office.

School Examples of Logical Evolution

Consider the following example: When Johnny enters the first grade, he knows how to count. He has a rudimentary understanding of addition, and the teacher expands and hones his skill in addition using whole integers. He learns how to add sets of apples, sets of oranges, sets of cars, etc., using bigger and bigger integers, by using the "constructive operator" of addition. Then, in the second grade the teacher introduces addition's "inverse operator," namely, subtraction. "If you have eight apples and you give away three, how many apples would you have?" That question is easy for Johnny: five.

In that setting, however, someone could propose, "If you have three apples and you take away four apples, how many apples do you have?" Johnny's response is simply, "It doesn't go." Or to put it in more sophisticated language: When you are dealing with positive integers, or natural numbers, the subtrahend cannot exceed the minuend. For pedagogical reasons, the teacher lets this question and Johnny's judgment go. But in third grade, when addition is applied to examples of distance, Johnny comes to realize that the operation of addition can be applied to movement toward the right on a line, and the inverse operation of subtraction would then mean movement to the left on that line. In this setting, it is easy for Johnny to see how it is possible to subtract a larger number from a smaller one, because that would simply mean movement to the left of the "0" reference point. By breaking past the 0 mark to the left, Johnny's mathematical world is suddenly doubled. Each positive integer now has a corresponding negative integer on the opposite side the 0-reference point. By means of the inverse operation of subtraction (-), negative numbers (-N) can be produced. All negative numbers are outside the domain of natural numbers, which are always positive. In this way, the number system of both positive and negative integers shows itself to be an *evolved* set of numbers in which natural numbers remain logically consistent parts of the expanded real number system, *and* in which inverse negative numbers now are also a part of a new numeric system, even though they are not part of the natural number system. This evolutionary advancement of the natural number system into the signed number system takes place when the inverse operator (-) *breaks the logical boundary* of the antecedent natural number system, producing new negative numbers that now bear the sign of that negative operator, namely (-N), in relation to the natural numbers (N).

The above illustration is a very common experience in elementary schools, and making this logical evolutionary expansion is so easy that it appears very natural. However, something very profound has happened here. This example holds the key to the logical evolutionary theory. In the above process, the following crucial point is usually passed over and ignored. In the generation of the whole integer system through the constructive operator of addition, the operation of subtraction *is defined* as the inverse operator of the constructive addition operator. The operation of that inverse operator (subtraction) is initially *symmetrically* bound by the operation of the constructive operator (addition), but when *that reciprocal symmetry is broken*, there is a *logical evolution* of the positive number system into the signed number system.

In second grade, when Johnny was challenged to take four apples away from three apples, he said, "It doesn't go," and he was right! Operating in the "natural number" system, there can be negative operations, but there cannot be negative numbers. However, it is possible to break the inversely defined, symmetrical boundary condition of the defined inverse subtraction operator. When the symmetrical, reciprocal boundary condition of the logical inverse of the constructive operator of addition is broken, new, expanded, negative integers are possible.

The new, evolved, real (positive and negative) number system is logically consistent with the antecedent natural (whole integer) number system, and it contains that antecedent system, even though the antecedent system of whole integers system does not contain the negative integers of the real number system. The premises of the expanded system cannot be *logically deduced* from the premises of the antecedent system, but they can be *logically evolved* from the premises of the antecedent system through the breaking of the reciprocal boundary condition imposed by the definition of the inverse operator in the antecedent natural number system.

Let's get back to Johnny. By the time he is in the fourth grade, Johnny learns the multiple addition process called multiplication. After he learns his multiplication tables and becomes competent in multiplication, the teacher introduces the operation of division, which is defined as the inverse operator of the constructive multiplication operator. As he is learning division, he is initially given numbers where division by the divisor into the dividend yields an integer. Then the teacher quietly breaks the original definition of division as the inverse operation of multiplication and raises the possibility of a dividend being not "evenly divisible" by the divisor. This leaves a remainder after the division process, which initially the teacher sets aside in a detached way.

It is at this point that the teacher introduces the notion of fractions. This leads to the establishment of the rational (or fractional) number system, which contains the antecedent integer system as well as all new integer-divided fractions. The antecedent natural number system and the signed integer system, however, do not contain the rational number system. Nonetheless, in an inverse way, the rational number system contains the antecedent natural number system and the signed integer system. Consequently, the axioms of the subsequent rational number system cannot be logically deduced from the axioms of the former real number system, but the axioms of the natural number system can be deduced from the axioms of the rational number system. While the logical evolution of the signed whole number system doubled the population of the natural number system, the logical evolution of the rational number system establishes an infinite set of fractional numbers between each successive natural and signed integer. This evolutionary advancement happened when the symmetrical boundary condition of the definition of division as the inverse operator of multiplication is broken.

Going further, a square-operator, N^2, is a constructive operator equal to N x N. When the square root operator, $\sqrt{N}$, is symmetrically defined as the inverse operator of the constructive square-operator, there is a boundary condition on the operation of that inverse operator within the antecedent natural number system. However, when the symmetric boundary condition of that definition is broken, a new, expanded, advanced system of transcendental numbers is logically evolved. This new system draws its logical consistency from the consistency of the antecedent natural number system, where the product of two real positive or negative numbers is always a positive real number.

In the real, signed number system, the product of two negative number is always positive. When that boundary condition is logically broken, however, there is generated another new, advanced, expanded system, namely, the imaginary number system, where ($i = \sqrt{-1}$), and the product of two imaginary numbers is a negative number. Since the axiomatic foundation of this logically evolved system depends upon the incorporation of the antecedent real number system, the newly evolved system will have complex numbers in the form of (A + Bi). This new system contains all antecedent number systems. Also, all conclusions of the antecedent mathematical systems can be logically derived from the imaginary system. However, many conclusions of the imaginary system cannot be *logically deduced* but only *logically evolved* from the premises of the antecedent natural, signed, rational, and transcendental number systems.

In this way, in elementary and middle school, Johnny repeatedly experiences the logical evolution of progressively more advanced number systems. Teachers do not present this process as "logical evolution," but that is what it is. In a clear way, the repeated use of this process displays the basic process of logical evolution in the domain of arithmetic. Here the breaking of reciprocal, symmetrical boundary conditions inherent in the definitions of inverse operations provides the inverse catalyst for the formation of more new, distinct, logically consistent, advanced number systems. However, is logical evolution limited to arithmetic systems?

PART I: FOUNDATIONS

Inverse Logical Evolutionary Theory

The logical evolutionary process found in arithmetic can be applied generally to the domain of symbolic logic, which can be described in terms of Boolean algebra.[3] This system of symbolic logic involves statements using terms describing well-defined specific things, specific actions, and/or specific systems.

Consider a consistent logical system (#1) which has the statements A and B and a constructive operator (*) such that for all A and B, there is a statement C where

$$A * B = C. \qquad (1)$$

In defining the inverse operator ($*^{-1}$) of (*), there is a logical boundary condition upon the operations of that inverse operator to keep all the inverse operations within the original logical system. The operation of the inverse operator ($*^{-1}$) is within the original system *if and only if* (iff) there is the antecedent operation of the constructive operator (*) in that system for each operation of the inverse operator. For each constructive operator (*) in statement (1), all operations of the inverse operation ($*^{-1}$) are true within the system developed by the constructive operator *if and only if* (iff) a certain condition is met, namely,

$$C\,(*^{-1})\,B = A \quad \textit{iff} \quad A * B = C. \qquad (2)$$

An illustration: In the natural number system, the equation $4 - 3 = 1$ is true *if and only if* $1 + 3 = 4$. In the natural number system, $3 - 4 = \emptyset$ or "nothing," because in the natural number system, "nothing" $+ 4 = 3$. The range of the inverse operator, defined in terms of a constructive operator, is *bound* by the antecedent operations of the constructive operator. That is, the "iff" (if and only if) condition defines a *symmetrical boundary condition* for the values of C and B in the statement

$$C\,^{(*^{-1})}\,B = A. \qquad (3)$$

In the first elementary arithmetic example, the constructive operator is addition (* = +), and the inverse operator is subtraction ($^{(*^{-1})}$ = -). In the positive integer system, the "if and only if" boundary condition requires that C be always greater than B. *Nonetheless*, one can logically evolve the negative integer system from the integer system using that inverse operator by breaking the symmetry of the boundary condition of the definition of the inverse operator in (2) and allowing C to be less than B. That is done by transforming the definitional "if and only if" (iff) condition of the inverse into a simple "if" condition and "letting go" of the limiting "only if" condition.

---

3. Boole, *Investigation of the Laws of Thought*.

$$C^{(*-1)}B = A \quad \textit{if} \quad A*B = C. \qquad (4)$$

The "letting go" of some internal controls within a system is a very important concept, because that "sacrifice" enables the practitioner in each system to advance to a more advanced, interrelational system.

The "if and only if" condition establishes a *symmetry* between the operation of the constructive operator and the inverse operator in a logical system. Then, when the "*only if*" condition is "let go," the system's reciprocal operational *symmetry* is "broken," allowing for the logical evolution of the antecedent system into a subsequent, advanced, asymmetrical, but logically consistent system, some of whose terms are formally distinct from the antecedent system. That is, when logic takes the symbolic form of Boolean algebra, it is clearly seen that the "symmetry breaking" of an inverse operator of a constructive operator of a reciprocal, logically consistent system transforms that inverse operator into a *catalyst* capable of logically evolving the antecedent system into a more advanced, logically consistent, logically evolved system with some logically distinct terms.

Geometric "symmetry-breaking" will have an important role to play in subsequent chapters, especially in the logical evolution of the four fundamental forces of physics. As this book advances, I will show how this "letting go" of the "*only if*" boundary conditions will produce many evolutionary advances, not only in mathematical and in physical systems but also in everyday human situations.

In summary,

Logical evolution requires the existence of an antecedent, logically consistent system,

> which has at least one constructive operator,
>> from which an inverse operator can be defined,
>>> which definition has an "if and only if," symmetrical boundary condition.

By "letting go" of the symmetry of the "if and only if" boundary condition for an "if" condition,

> new, less restrictive, more interrelational statements involving the inverse operator can be generated which are true to the antecedent system but cannot be derived from it.
>> When combined with the axioms of the antecedent system, there can be produced a new, advanced, logically consistent, logically evolved system,
>>> which in a logical evolutionary way, formally contains and is formally more interrelational than the antecedent system.

Furthermore, by logical evolution every energetic system that is arithmetic in its foundation will be evolutionary and oriented by its potential energy to form a new, more interrelational consistent system, capable of further logical and physical evolution. Going further, logical evolution demands that there be at least one term that breaks the boundary of the logically consistent antecedent system. In a progressively expressed system, the first extraneous term is important, like the number -1, for it triggers the generation of the entire logically evolved system, and it is properly called the advanced system's "progenitor."

The evolution of a new, logically evolved system follows a four-phase process, which has modern evolutionary terms:

1. The logically inverse progenitor can and does interrelate with members of the antecedent system such that their logically produced progeny endures, and they potentially can interrelate, producing a proto-population having the advanced characteristic of the progenitor. This process may be called *hybridization*.

2. As this proto-population generates, members of that population will more probably increasingly reject less talented members of the antecedent system, and they are more able to increasingly interrelate with only individuals which are logistically more advanced. As dissociations of evolved members from non-advanced individuals increase, the percentage of members with the evolutionarily advanced logistic grows, until only evolved members are interactive in the proto-population. This process may be called *speciation*.

3. As the logistical consequences of the inverse, evolutionary characteristic(s) of the evolved species increase, new "constructive operators" can be defined. This increases the number of new systemic interrelationships, and it expands the logistic of that evolved system in new ways not found in the antecedent system. Here is systemic *diversification*.

4. The logically consistent but quantitatively differentiated system, having new "constructive operators," will have corresponding "inverse operators" which are not found in the antecedent system, establishing a *universal logical evolutionary process*, which can be extended indefinitely.

Chapter 3

Initial Consequences of Logical Evolutionary Theory

Evolutionary Trees of Life and Logic. Comparison of Biological and Logical Evolution. Origin of the Principle of Analogy.

As described in the previous chapter, the logical evolution of the non-signed, natural numbers system *can* happen through breaking of the symmetry of the defined, reciprocal boundary of the inverse operator of subtraction (-), leading to new numbers acquiring the new parameter of the negative sign (-N). Likewise, the logical evolution of multiplication of whole numbers introduces the possible extension of the inverse operator of division and the need for the new parameter of fractions (/). The logical evolution of square numbers introduces the inverse operator of square root ($\sqrt{}$), which introduces the imaginary parameter ($i = \sqrt{-1}$), as in (A + Bi). These examples indicate that each step in the logical evolutionary process generates a new *parameter*, characteristic of the logical inverse character of the progenitor that broke the symmetry of the inverse operator of the antecedent system. Later, I will show how in the logical evolution of each physical system there will generate a new parameter, like: distance, mass, electrical charge, color charge, consciousness, rational idea, etc.

Since the novelty of a new evolved system compared to an old system is through an inverse operator, the new system will not only be interrelationally more complex, it will also be inversely complementary to the antecedent system. This gives a preliminary explanation of why, in human history, subsequent generations of humans and of truly innovative products will tend to be inversely contrary to the behavior of antecedent members of the parental generation.

Also, the logical evolutionary process demands a *carryover* of the foundational principles and interrelational characteristics of the antecedent system into the new system. For example, all the premises of the signed integer system are carried over into the rational (fractional) system. This insures the new, evolved system's logical consistency and compatibility with all its antecedent systems. The *carryover* characteristics of the antecedent system are important in the evolved system to the advanced operations of the progenitor and its generated "progeny"—but in a subordinated way. In this subordinating process, the axioms of the antecedent logical systems become "nested," one inside the other, in a subordinated way to the axioms of their subsequently evolved systems. This explains why total rejection of the ways of a previous system and starting from scratch rarely results in an enduring advanced system, but regularly repeats the deficiencies of the former system or regime. For example, the success of the expanding Roman Empire and the rebuilding of Germany after World War II relied heavily on the carryover of past bureaucrats from their antecedent, defeated regimes. If they were willing to work under the rules of their new ruler, their ability interrelationally to hold together their antecedent political system can become the stabilizing base for the new, advanced social order.

In a series of logically evolved systems, not all new terms in the logically evolved system can fully operate as members of or within the antecedent systems. For example, the number ½ cannot be a member of the antecedent natural number system, but every natural number, like +2, is a member of the subsequent, evolved rational number system.

Nonetheless, advanced terms from an advanced system *can* interrelate with the terms within an antecedent system *provided* that all advanced differences are "covered" or "let go" by members in the advanced system. This is possible if these differences are considered "interrelationally inconsequential." This is what the teacher of division did when a division problem yielded a whole number with a remainder, which was put to the side for later consideration. Similarly, in daily life, a human can learn to understand the various verbal and nonverbal communications of their pets and other animals. However, humans must "cover" or "let go" of any expectation that their pets will communicate using human words. That is, according to the rules of logical evolution, a self-limiting human can learn how to communicate with a dog in doglike ways, but a dog is limited in the ways it can communicate with a human in a humanlike way. That is, terms in a less evolved system can communicate fully in the more evolved system within the terms of the more advanced system, but the interrelational terms in the more evolved systems can only *partially* communicate in ways that are received and interrelationally valuable to members of a less evolved system.

In the realm of interrelational theology, the Primal Absolute Potential Energy of the Almighty has all potentialities and therefore all natures. For the Almighty to interact with humans, however, it is logically necessary for the absolute to "cover" attributes that are beyond the human logistic or nature and only interact with humans on the current level of human development. In this way, logical evolution provides the foundation of the classic Scholastic dictum: *Quidquid recipitur, in modo recipientis recipitur.* "Whatever is received, is received in the mode of the receiver"—not in the full mode of an advanced giver, but in the mode of the lesser receiver. Consequently, every human reception of revelation from a transcendent Almighty will be partial because of the finite limitations of the current historical situation of the non-absolute, partial human receiver. Thus, every revelation received by a human from the Almighty is not the total truth but only that partial amount of the truth that the limited human receiver is then capable of receiving intelligently. As humans evolve intellectually, humans will be able to understand revelations more fully from the Almighty through the absolute communication Spirit in our shared potential energy.

Also, given a logically consistent system, logical evolution *can* happen, but it does not necessarily *have* to happen. If the range of a given system already includes all its possible terms, the inverse of the operators in that system cannot yield any new inverse values—because there are none available. This is seen in a cyclic system, such as a clock, where going backwards around a clock yields the same set of values as going forward around it. Here, the field of variables is the same for statement (2) as for statement (3), such that the evolutionary process through that particular inverse would go nowhere new, and the logical evolutionary process through that inverse would end.

Evolutionary Trees of Life and Logic

The logical evolutionary theory indicates that in a logically consistent system, logical evolution *can* go into as many directions as there are constructive operators. Each of these constructive operators has a corresponding inverse operator, each having its own symmetrical boundary condition. Then, by breaking the symmetry associated with that inverse operator, each logical inverse *can* be a catalyst into the logical evolution of a new, formally distinct, logically consistent system. The more advanced the system, the more constructive operators it will have, the more evolutionary inverses there could be, and the more distinct logically consistent systems it could logically evolve. Thus, the logical evolutionary theory would produce a branching logical evolutionary tree, similar to the one that Darwin sketched for physically evolving biological species in mid-July 1837 in his "B" notebook.[1]

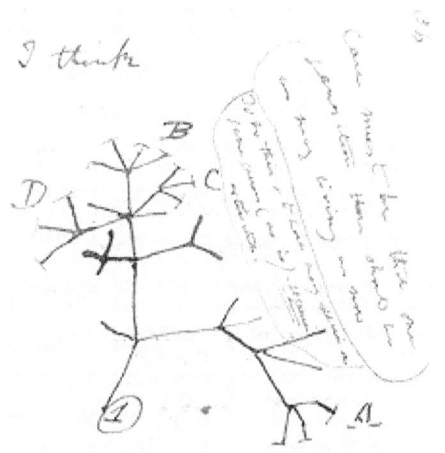
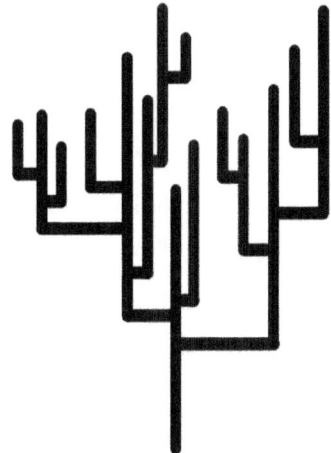

Darwin's Evolutionary Tree of Biological Life.

IPhil's Evolutionary Tree of Logical Systems.

1. Darwin and De Beer, *Darwin's Notebooks on Transmutation of Species.*

Darwin recognized that evolution does not begin accidentally with an individual's evolution of a special qualities and survival skills. That is not enough for systemic evolution. Any new, preferred trait must be passed on to the individual's progeny, producing a line of descendants that inherits those same special qualities and abilities. Furthermore, he realized this would produce a branching evolutionary tree of biological life, in which the lines of descent with those special qualities would branch out from the antecedent line of inheritance. Likewise, in the Evolutionary Tree of Logical Systems, each new branch *can* come from but does not necessarily have to come from a logical inverse which breaks away from the reciprocal symmetry of the antecedent, logically consistent system. In this way, the renegade inverse operator *can* produce an advancing constructive progenitor operator for a new branch in the Tree of Logical Evolution. However, the existence of that advanced trait in just that *individual* does not mean an advanced *system* or species would result. According to the logical evolutionary theory, the progenitor must not only receive a new inverse characteristic; its inverse operations must also be embedded in the logistic of some of the members of the antecedent system in order to generate an advanced logically interrelated population that is logically consistent, will endure, and will grow into an evolved system.

Natural evolution and logical evolution are parallel because a thing's kinetic energy in its physical order is a partial actualization of a thing's antecedent potential energy. This characteristic of logical evolution links metaphysical, logical evolution of the knowable interrelationship in the domain of potential energy with the physical, material evolution in the domain of kinetic energy. This is why there is a correspondence between the intellectually known equations of physics with the observed physical behavior of material objects in our world.

Comparison of Biological and Logical Evolution

Thomas B. Fowler and Daniel Kuebler in *The Evolution Controversy: A Survey of Competing Theories* summarized the fundamental principles of Darwin's evolutionary theory this way:[2]

1. There first must be a population that sustains itself as a species, providing a common descent for the subsequent different strains of that species.

2. Through random variations in birth—not by divine intervention—an individual is born that has one or more physical traits which are different from the general population and which gives that individual some type of greater ability to survive and dominate.

3. This mutated individual cannot simply exist alone but must breed with the other non-mutated members of its species such that the advantageous trait(s) or

---

2. Fowler and Kuebler, *Evolution Controversy*.

mutation(s) *can* be carried over into subsequent offspring. When this happens the generated, mutated progeny will become a new, advanced subspecies, having generally the advanced trait and logistic of its progenitor.

4. Mutated offspring then are able to interact in a more advanced way to establish collectively, through small incremental steps, a species which in some way(s) is uniquely different from the antecedent species and better adapted to the given environment.

Within the above scenario, the word "evolution" can be understood in two different ways: (a) Narrowly, the term *evolution* can refer to the initial mutational change(s) that emerge(s) in the progenitor individual within an existing species, as described in step #2. (b) Broadly, the term *evolution* can refer to the development of a new, advanced species as an enduring society, as found in step #4.

In speaking about physical evidence in the field, the broad use of "evolution" applies more appropriately. The chance of physically finding the first evolutionary mutation is extremely slim. Nonetheless, physical evidence usually points toward the conclusion that there was one special progenitor and probably only one. Two distinct evolutionary progenitors would tend to produce two distinct lines of development. Nonetheless, physical evidence for the same spontaneous mutation in different parts in the world at the same time is occasionally found, but it is very rare.

When people talk about evolution, they usually have large species in mind, and evolutionary changes are thought to be dramatic. In contrast, the historical survey in chapter 1 demonstrates that our advancing knowledge and understanding of the evolutionary process has gotten smaller and smaller in scale. On a macroscopic scale, changes usually take long periods of time. On the chemical level, mutations can take place in short periods of time. On a microscopic scale, molecular mutations and evolution *can* occur at a rapid pace, once all the conditions for physical and logical evolution have been met. In the metaphysical and mental domain, logically innovative ideas *can* advance with lightning speed, as sudden, innovative, human inspirations attest.

Origin of the Principle of Analogy

When two things are both similar and different, they are said to be analogous. For example, a human is similar to or analogous to a chimpanzee. They have the same basic anatomy; they feed themselves in similar ways; they raise families, they squabble in ways that are similar, etc. They are also different insofar as humans and chimps cannot breed with each other; chimpanzees are more agile than humans, and humans are able to think and put things together in ways that chimpanzees cannot.

The major stages of humans (birth, adolescence, adulthood, and old age) have long been recognized to be analogous to the seasons of the year (winter, spring, summer, and fall). In nomadic societies, babies were usually conceived in the spring when youths were able to leave the protections of their winter homes. The interrelational life of an infant then began with winter, when people protectively stayed close to home. Spring is an adolescent period, when the physical beauty and fertility of nature are at their peak. During summer, plants grow slowly. They produce better fruit when adult humans patiently groom them to be more productive. In autumn, the leaves fall from trees back to earth, and humans gather the harvest after long toil. In the autumn of one's life, humans become weaker, fall, die, and are buried in the earth.

Although there is no apparent connection between the seasons of the year and the stages of a human's life, people have long recognized much truth in this analogy. Why is much truth found in analogies?

For centuries, people have used analogies to express their ideas more clearly and to guide their thinking and explorations in parallel ways. Fables, parables, and stories were used to teach wisdom to youth and the general population. However, teachers seldom tried to explain why many partial truths are found in such analogies. Focused logically on true versus false statements, essentialistic, rationalistic thinkers have not been able to securely ground the general truthfulness found in analogies. In contrast, IPhil can pinpoint the origin of analogies through the characteristics of logical evolution and why there is partial truth in analogies.

In logical evolution, subsequently evolved systems incorporate the fundamental premises of their antecedent systems. When a logically consistent system has two different constructive operators or traits, logical evolution is possible from the logical inverse of each trait. The result is the possibility of logical evolution of two differing systems, which have a common, *similar* characteristics carried over from the same antecedent system, *and* these two evolved systems are *different in*sofar as they have different progenitors and systemic development. That is why they are analogously partially the same and partially different. Parallel comparisons of the logistics of two systems show their common truths, because they come from the same logical evolutionary stock through the carryover process of logical evolution. That is, while there may be many logistically different traits between two systems with different progenitors, because they have the same logical evolutionary axiomatic foundation, there also will be many similar, shared, interrelational characteristics in the two systems. This is what makes them analogous in many truthful ways.

The closer two species or systems are to each other on the logical and biological evolutionary Trees of Life, the more metaphysical and material traits they will have in common, and the more their analogies will be truthful. The principle of analogy is regularly applied in medical research. The more an animal species has organs and chemistry similar to or analogous to the organs and chemistry in a human, the more likely applying that analogous medical research to humans will be successful. That is what the logical evolutionary theory predicts, and that is what human experience shows.

INITIAL CONSEQUENCES OF LOGICAL EVOLUTIONARY THEORY

Chapter 4

Interrelational Logistic

Examples of Ongoing Physically Consistent Systems. Prototype of Interrelational Process of Energy Exchange. General Interrelational Process. The Interrelational Logistic. Energy Communicators. Ownership. Wave-Particle Dualism. Human Fourfold Analogate. Interrelational Illustration.

Logical evolution as described in the previous two chapters exists in the metaphysical order. Yet the first chapter indicates that evolution was first discovered in the physical order, and knowledge of it is grounded in physical experience. Etymologically, the word *metaphysics* comes from *after physics* and refers to what is abstractly known after discovering something physical. Metaphysics looks for an antecedent reality or characteristic from which subsequent physical events or characteristics can develop. Consequently, potential energy is a metaphysical reality, antecedent to subsequent physical manifestations of kinetic energy emerging from it in a formally distinctive way. Our knowledge of metaphysical potentiality is drawn from what is observed physically. Inversely, physical, kinetic displays are subsequent to their metaphysical potentialities. Thus, physical, kinetic activities and metaphysical, potential energy are intimately bound together; yet, they are formally distinct from each other.

The content of our knowledge, philosophy, and science comes from experience. Everything I know experientially has some physical relationship to me. If an event or system does not have a relationship to me, I cannot experientially know it or physically interrelate with it. In addition, I am not satisfied with simply knowing about a thing. The interrelational potential energy received in my physical body pushes me to want to interact with others. In other words, my world is not only relational to me, it is also reciprocally interrelational with me. Interactions tend to go beyond an intrarelational "me" toward a systemic "we."

The modern general theory of relativity focuses upon observation of others either moving or static with respect to oneself. Relativists hold as true only what is physically known by an observer. However, atomic physics points out that the very act of observation impacts what is observed. Thus, the findings of nuclear physics point beyond an observer. Systems founded upon potential energy have metaphysical interrelational characteristics beyond individualistic, self-centered interests.

IPhil recognizes the importance of other individuals as they know and cooperate within the interrelational systems of which they are vital, interactive parts. This advancement moves the thinker from an individualistic, diametric either/or point of view to a systemic, dialogical both/and point of view.

Observations show that when things interact with us, we know them and ourselves as participants in a system. Atoms are able to produce electromagnetic waves not as a collection of parts but only as an interactive system. Horses are able to run not as individual biological cells but as a living, holistic animal. The behaviors of corporate wholes are greater than the sum of the behaviors of their reductionistic parts. In other words, the world I observe and know, things are not only *relational* with respect to themselves, they are also *interrelational* with respect to others.

The senses of my body not only receive sensations; they also have complementary nerves that are reactive to these sensations. I know not only my received sensations, but also my body's reaction to them. Similarly, my conscious mind outwardly reacts to sensations and memories which are antecedently unconscious. It is wrong to consider as real only what we are conscious of. Electrically sensitive monitors demonstrate that our senses' unconscious responses precede many of our deliberate responses. Still, some outward, conscious, free self-initiatives in myself and in radioactive decay occur independently from inward stimulations. Systems endure when their relationships and exchanges are reciprocal to some extent. "You scratch my back, and I'll scratch yours." That saying is true not only in enduring friendships and business deals; it is also true in the exchanges of all known physical and mathematical systems.

However, experience shows that these exchanges do not have to be equal in reciprocity, especially when a system is in the process of growing or decaying. For example, the reciprocity of a child and a parent is asymmetrical, with resources going primarily from parent to child in the early years. Then in later years, the movement of resources and care go from adult children to the ailing parents. When symmetry or equality *does* occur, that relationship continues to endure and remains strong. However, when asymmetry occurs, the disparity of output to input can have a significant, debilitating impact upon the giver, receiver, and the endurance of their interrelationship.

Observations show that physical systems are not perfect systems. They often do not manifest precise regularity and order because of the freedom of the self-initiatives arising from their potentialities. Nonetheless, because of the progressive interrelational character of most systems, they often have internal correctives that enable them to develop and endure within adversarial environments.

Most conscious human experiences rarely take place at the extreme beginning or ending of a life or interrelationship. Most of our conscious experiences occur in the middle of interactions and interrelationships. Since this philosophic study is grounded in physical experience, it is better that the induction of its principles begins with intermediate systems with which we are more familiar.

PART I: FOUNDATIONS

Examples of Ongoing, Physically Consistent Systems

Once I demonstrate a good grasp of the interrelational process in familiar settings, I will propose generalized principles. Then, the induced principles need to be logically pushed to their extremes, where they must continue to be logically consistent within a given system to be posited as true. Consequently, I begin the discussion of the principles of interrelational philosophy (IPhil) with everyday experiences of enduring physical systems. Often, these systems are not physically perfect, but their consistency of form does point toward a potential metaphysical ideal.

- In the formation of a corporation, communications are exchanged between the founders, leaders, and workers. Not only words but money and goods are exchanged between the owners, suppliers, and customers. Information and work are exchanged between laborers and management. If there are no exchanges of ideas and resources, a business cannot begin or endure. Without effective communication of information and exchange of goods, a corporation will quickly collapse. Then, gathered resources and information will usually disperse in many directions.

- A telephone conversation requires information exchanges between two (or more) parties. That information, however, is not communicated face-to-face but via electromagnetic signals connecting the parties of that conversation. If the conversation is mutual, it endures. The more reciprocal it is, the livelier and more animated it is. When one party does most of the talking, the other party soon tires, and the communication soon terminates, and each party goes their own way.

- In the operation of our physical bodies, communications and exchanges are necessary for the operations of all its organs. If that reciprocity is hindered or halted, parts of the body will atrophy, and the life of the whole system will become limited, the body may soon die, and its once interactive elements will disperse.

- Richard Feynman, the noted physicist, described the electromagnetic force between two passing atoms as a series of photon "shells" fired between electrically charged battleships (atoms) passing in the night. If there is no wave or particle exchange of energy between the two atoms (or battleships), they will simply go apart in their own way.

- On the nuclear level, scientists describe the formation of protons and neutron into atoms through the interrelationship of quarks connected by interactive gluons. If nuclear particles lose their internal cohesion, their energy will disperse into space.

- In walking, an individual first becomes balanced on two legs. Then the individual extends a foot, throwing the body off-balance. Third, the individual plants the extended foot firmly on the ground at a comfortable forward distance. Fourth, the back foot is brought forward, either up to the location of the extended foot such that the individual regains a state of balance, or by extending the back foot past the planted foot. Then, the individual *can* begin the new, repeatable cycle of walking.

All the above systems display a common pattern in which distinct terms communicate some form of energy/matter between them. It is the reciprocity of these exchanges that allows these physical systems to interrelate in an enduring way. If the reciprocity is imperfect, the available energy of those senders is reduced. Such systems soon collapse, and their energy/matter disperses in all directions.

Prototype of Interrelational Process of Energy Exchange

The systems described above are quite complicated. As I looked around my daily world, I found a clear and simple example of the interrelational process. It was the game of catch on the elementary school playground next to my office. The game of catch contains a very simple, reciprocal, symmetrically balanced, four step interaction between two people. Student #1 is the initial thrower; student #2 is the initial receiver. In the process of the ball exchanges in the game of catch, those roles are reversed.

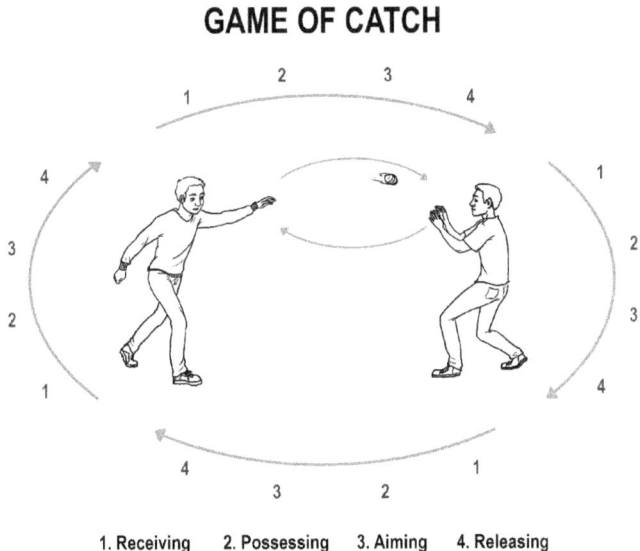

GAME OF CATCH

1. Receiving   2. Possessing   3. Aiming   4. Releasing

The following description of this interrelational process is made from the standpoint of the receiver #2 because in real life, we usually start not as a giver #1 but as receiver #2. How giver #1 has the ball initially is not an inherent part of this interrelational exchange, but it is a question that needs to be discussed later.

1. Student #2 waits expectantly to receive the ball.
2. Student #2 catches the ball and takes control of it.

3. Next, student #2 looks around and figures out how to throw it back.
4. Then student #2 throws the ball, letting go of it.
5. Meanwhile, student #1 waits expectantly to receive the ball back.
6. Player #1 catches the returned ball and takes control of it.
7. Next, student #1 looks around and figures out how to throw it back.
8. Student #1 then throws the ball again toward student #2, letting go of the ball. Then the cycle repeats itself in the game of catch.

In the game of catch, each student goes through four distinct stages of operation: (1) receiving, (2) possessing, (3) aiming, and (4) releasing. The reciprocal action of the second individual produces a symmetrical feedback system, which allows the catch exchange to be ongoing. These basic phases can be identified analogously in all the examples given above, ranging from operations of subatomic systems, molecular systems, business ventures, galactic systems, and ordinary daily activities. Thus, on all levels of physical reality, this fourfold, reciprocal pattern of energy/matter exchanges appears to be fundamental to the interactive systems in our energy driven world.

Nonetheless, that game can be interrupted at each stage for various reasons. The student may not be "open" to receiving the ball; the student may miss the ball; the student may choose to throw it to a third person, who then holds on to the ball. The student may not "let go" of the ball but may instead hold on to it and then go home with it. Many different pieces of that process must be in place in order that a game of catch or any interrelational system can endure. However, if all the steps of the exchange are in place, the game of catch goes on in an enduring, systematic, and enjoyable way.

Notice that the game of catch cannot be predicated on either player. The game of catch is a systemic reality that is bigger than its participants. The game of catch is an example of a "whole" being greater than the sum of its "parts."

Let's apply the principle of logical evolution here. First, it is possible for an individual initially to play a game of catch with *oneself* by throwing the ball up in the air and then catching it as it falls. Notice that in the game of catch between two individuals, the notion of "another" is the logical inverse of the notion of "oneself" as receiver. That is, an "another" is the logical inverse of "not-oneself." Consequently, the game of "catch with another not-oneself" is a logically evolved form of the game of "catch-with-oneself." Interrelationally more complex and evolved than playing catch-with-oneself, the game of catch-with-another requires a greatly expanded set of interrelational skills. We experience greater happiness in successfully playing catch-with-another beyond playing catch-with-oneself because the advanced game is interrelationally more demanding, leading to greater satisfaction and happiness when it continues—and greater disappointment when it does not.

A crucial part in the above game or interrelational process is regularly taken for granted and ignored. In the 4th phase of that process, the thrower must "let go" of the ball in the throw. We humans have a very strong tendency to hold on to everything and resist releasing things from our control. By "letting go," however, the player *breaks the boundary* of self-possession. Only by a self-initiated "letting go" can an evolved interchange take place. In other words, the dynamic game of catch is an example of a logical advancement from a state of static self-possession to a state of dynamic exchange and interaction. The surrender of one's possession of the ball for the benefit of another is at the heart of the game. By refusing to "let go," the holder has some joy in increasing one's possessions, but the individual who retains possession of the ball is not actually a player but a holder. Only by "letting go" of the ball in a transfer to another does the individual increase the joy of playing with each other in the game of catch.

The game of catch is normally between two people. However, when a third player comes forward, the game of catch can undergo another logical evolution, for the throwers *can* now throw the ball to another who had *not* thrown the ball to the current receiver. Notice how the form of a two-person game of catch is broken by a logical inverse insofar as a receiver in a three-person game of catch *can* be with a receiver who had *not* just been a thrower.

When more individuals arrive, the game of catch between three individuals can be logically broken. In this way, playing catch with oneself can advance to playing catch with another, which can advance to playing catch with a third person, and then to playing with a whole team, and the game of catch becomes some form of a ball game.

Consequently, the game of catch is a good illustration of the fourfold pattern of active interrelationships: (receiving, holding, aiming, and releasing). Also, the game of catch displays itself as a reciprocal, enduring, and logically consistent system, capable of logical evolution.

General Interrelational Process of Energy Exchange

In describing the interrelational process, I use hyphens, as in the terms: "receiver-giver," "giver-receiver," and "giver-receiver-giver." By doing so, I emphasize the temporal, progressive character of the term in an energetic interrelational system. The term after the hyphen indicates a subsequent state of an individual in that interrelational process. An interrelational exchange begins with a "giver" and ends with a "giver-receiver." This hyphenated terminology makes clear that the "giver-receiver" and the "receiver-giver" refer to two very different phases in the interrelational process.

In addition, because energy can exist in a potential state and in a kineteic state, energy's interrelational prototype can be viewed from two different perspectives. There are the four *exterior*, dynamic, physical stages of the transfer of energy *between* terms, and there are four *interior*, transformed, metaphysical stages *within* the terms in that interrelationship. The following is more detailed exploration of this interactive, systemic process,

1. The interrelational process begins with an energized holder oriented toward sending some energy/matter toward a "virtual receiver," which is not physically distinct outside the giver, but is logically distinct *within* the giver. The virtual receiver is potentially, metaphysically open to receiving that energy/matter. In the 1st phase of this process, the interactive terms of sender and receiver are only potential characteristics. At this point, the virtual receiver is not energetically actual, but exists projectively, metaphysically, potentially. Nonetheless, the virtual receiver is known to be logically/formally distinct from and within the initial potential reality of the potential-giver, whose potential energy is pushing toward freely becoming an actual-giver.

2. When the energy/matter is actually "let go," the potential sender becomes an actual sender, and the virtual receiver becomes an actual receiver outside of the sender. The actualization of a virtual receiver can be an already existing open individual who has the potential to become a receiver, or the sent potential energy can have the potential to become a receiving reality at a distance from the sender. In this 2nd phase, the receiver intrarelationally claims the sent energy/matter as its own, either consciously or unconsciously. By intrarelating the received energy/matter within itself, the received energy/matter becomes part of the receiver, who then owns and integrates it into the receiver's energized individuality. If the sent energy/matter is not actually received, there is no actual receiver, and there can be no reciprocal, systemic exchange.

3. The energized receiver now undergoes a logistical transformation from being an energized receiver into interiorly becoming a receiver-potential-giver. Here, the energized-receiver now can actually—either randomly or deliberately—initiate a transfer of that received energy/matter back toward the original giver. Meanwhile, being emptied of some of its interrelational energy/matter, the initial giver logistically becomes an open giver-potential-receiver.

4. Next, the energized receiver-actual-giver needs to "let go" of some of its received energy/matter, which is then physically conveyed toward the intended giver-potential-receiver.

5. The original giver is open to the possibility of receiving energy/matter from the receiver-giver. This begins the response phase of the symmetrical fourfold interactive process when the original giver in the mind of the receiver-giver interrelates with the original giver as its virtual giver-potential-receiver.

6. Then the giver-potential-receiver takes possession of the sent energy/matter from the receiver-giver and intrarelationally integrates the received energy/matter into itself. This reception of the sent energy advances the giver-receiver into a new state of integrated wholeness as an empowered individual once again, capable of advancing the interrelationship of the two.

7. This giver-receiver will not only have immediate, experiential knowledge of the energy/matter received, but it has also remembered knowledge of its former giving. Because the original giver is again re-energized as a giver-receiver, the giver-receiver can compare the subsequent receiving with the original giving in that exchange.

8. If the energy/matter given matches the matter/energy received, something logistically new has been accomplished and established, namely, a reciprocal, interrelational, energetic *system*. That is, this exchange is now more than two opposite throws of the ball; it is now a game of catch. The actual reciprocity of this system establishes that this exchange is logically consistent, and as such, this interrelational system is capable of repetition and even of logical evolution.

The Interrelational Logistic

The comprehensive interrelational process described above is what I call a "logistic." The term *logistics* is a military term describing the movement of supplies and information from a home base to a distant destination as well as the return of transformed materials and information back to the original base. I use the term *logistic* rather than the Aristotelian philosophic term, *form*, because in English the term "form" conjures up a static, perfect mental image of an individual reality, while the term "logistic" implies a dynamic, reciprocal process within an interrelational system of interactive persons or terms.

When the interrelational process described above pertains to the physical parts within or associated with an individual, that interrelational pattern of behavior I call the "individual logistic." It has a definite, dialogical structure that involves the four stages of receiving, possessing, ordering, and sending as experienced within each *individual*. Furthermore, I use the term "communal logistic" to refer to the four stages of receiving, possessing, ordering, and sending to the interaction of dialogical individuals or terms within the enduring *system*. In the game of catch, the singular "communal logistic" of the game contains the two or more "individual logistics" of its participants. They individually operate in distinctive ways within the range of potential possibilities of the individual logistics of the energized participants, who can operate in different ways within the broader range of possibilities within the communal logistic of the interrelational system. Consider this example. The equation $y = 3x + 4$ is the communal logistic for all values in a system; the individual logistical values of one of the interrelational terms in that system can be $x = \{5, 8, 11\}$. That is, the equation and the variable values are formally different—systemically and individually. Yet, they are closely interrelated, as in the communal and individual logistics of every interrelational system.

Every logical argument is preceded by statements, which are preceded by observations, which are preceded by physical interactions, which are preceded by some type of energetic/material existence. Observed energetic interactions have four formally distinct phases: receiving, holding, projecting, and releasing. Consequently, in logically consistent, reciprocal, energetic systems, there are four formally distinct types of energetic statements: (1) Statements of existence arising from the presence of potential energy. [There is a ball.] (2) Statements of interior, intrarelational qualities. [The ball is red.] (3) Statements of exterior, interrelational associations. [The red ball is in my hand.] (4) Statements of future energetic possibilities. [The red ball in my hand can drop to the ground.] Each type of statement describes energy/matter in its own formally distinct, progressive way. This type of experiential analysis can be generalized into the foundations of classical logic and Boolean Algebra.

Energy Communicators

In the game of catch, there are not only the receiving and giving players, there is also the ball that is thrown between them. Looking at many different types of systems, there are different types of *energy-communicators*. Energy-communicators can be a ball, a photon, a graviton, a sound wave, etc. Energy-communicators do not have to be a solid object, for energy sometimes can be communicated between givers and receivers via an energy-wave as well as an energized particle. These energy-communicators also go through their own fourfold interrelational process. Refer to the catch illustration above. Here is a description of the four phases of an energy-communicator.

1. The intermediate energy-communicator receives energy/matter from the sender after the sender "lets go" of some or all of its energy. If the giver is a ball-thrower, the intermediate energy-communicator is the ball. If the giver is a speaker, the intermediate energy-communicator is a sound wave, etc. Each energy-communicator is energized according to the logistic of the sender. That is, in a logically consistent system, the energized giver, the energized-communicator, and the energized receiver must all have the same communal logistic. Only together in an interdependent way can they fully actualize the logistic of the logically consistent interrelational system.

2. The intermediate energy-communicator after receiving energy/matter from the giver *intrarelationally* takes internal possession or ownership of the transferred energy. Here, the energy-communicator can be an integral solid object or a wave embodying the received energy in an integral, individualized way.

3. In traveling from its giver to a future receiver, the energy/matter of the energy-communicator is directed in a well-defined way according to the received energized logistic of the giver.

4. When the energy of an interrelational communicator reaches a receiver, it communicates its energy to the receiver by "letting go" of its energy. A sound wave "lets go" of its transverse wave energy when it impacts a person's eardrum or any other part of one's physical body. The ball itself possesses energy during the flight, and when it lands in the receiver's mitt, the ball "lets go" of its kinetic energy as the receiver takes possession of the ball's energy/matter.

In other words, in the full interrelational process, there are not only two dialogical terms which go through the four stages of the interrelational process (receiving, possessing, aiming, and giving), but there also are intermediate energy-communicators (e.g., a ball or a sound wave) which also go through the same four-phase interrelational process. Recognizing that the communicator-to and the communicator-from are distinct from the two distinct interrelational terms, the total number of interrelational elements in a logically consistent energetic interrelationship is four. Since each participant goes through four phases in his or her interrelational process, the total number of logically distinct phases in an energetic interrelationship is 4 x 4 = 16 progressive, interactional phases.

Ownership

Consider this example: On the day of a student's graduation from college, the parents take the youth outside their home. Pointing to a new automobile parked out front of their home, the parents say, "This is our graduation present to you. Congratulations! It is now yours." In that action, the car does not move, but ownership is shifted completely from the parents to the youth. Change of ownership is not a physical change of location; rather, it is a change of the *intrarelational* holdings of a receiver.

In the interrelational logistic, the receiver first receives the energy mass communicator. In the 2nd phase, the receiver intrarelationally takes control and possession of what is received in an integrating, intrarelational way. The 1st phase of receiving of the ball is receptive and *external*. The 2nd phase of holding and temporarily taking ownership of the ball is *internally* metaphysical and integral.

In the 2nd interrelational phase, the incoming interrelational energy/matter is intrarelated in the receiver, and a change of ownership by the receiver takes place. When an individual is being physically formed, that intrarelational integration of the incoming miscellaneous parts must involve some type of knowledge of the compatibility of the parts around the logistical "seed" of its individual logistic. In this way, the "seed" becomes the knowledge-center—the psychic center—for all the energetic-material parts of the intrarelating individual. It can be said that an intrarelating individual at least unconsciously knows the possibility of future "ownership" of all the elements received in its forming body. As a person advances to its 2nd phase of ownership, the individual becomes conscious of the ownership of one's body and the things received by its own body. This owner then recognizes the interrelational potential of what was owned in the 3rd phase. Finally, in the 4th phase of the interrelational logistic, the individual can "let go" of that ownership. This can be deliberately directed toward another receiver, or can be unconsciously released, as at death. Other aspects of ownership will be discussed in a subsequent chapter.

Wave-Particle Dualism

The 2nd phase of the interrelational process of the energy-communicator is particularly significant. As the energy-communicator goes through the four phases of interrelational logistic (receiving, integrating, ordering, and expanding), the energy-communicator expresses itself progressively in a wavelike manner. Nonetheless, in the 2nd phase, the energy-communicator integrates and owns that communicating energy intrarelationally and individualistically. That is, in the 2nd phase of the interrelational process, the energy of the communicator takes on a substantial, integrated form, and in doing that, the communicated energy takes on the character of an "energy-particle."

Also, in a complementary fashion, each dialogical term has its own four-phase logistic. The overall wavelike character of the logistic of each interrelational term gives each substantial term a wavelike character.

Thus, in interrelational energetic systems, the interrelational logistic of everything energetic gives particle-styled terms an overall wavelike aspect, and the fourfold interrelational logistic of energy-communicators has in its 2nd phase an intrarelated particle character. In this way, from the nature of interrelational energy in systems, IPhil explains, the origin of particle-wave complementary in all energized substantial terms and energized wave-communicators in all energetic interrelationships. Recall the maxim *Quidquid recipitur, in modo recipientis recipitur.* "Whatever is received, is received in the mode of the receiver." Recognition of an energy-communicator's wave-like quality requires the "particle receiver" to view the incoming energy-communicator according to the receiver's wave-like aspect, and recognizing an energy-communicator's intrarelational particle character must be according to the receiver's particle character. Is it a wave or a particle? IPhil explains why it is formally both, according to the phases of energy exchanges of the terms and communicators.

Human Fourfold Analogate

In the previous chapter, I explained how progressively evolved, logically consistent systems are analogous. Because we are most familiar with human physical development, I will frequently describe the progressive stages of the energetic development of individuals, groups, and things using the following analogy from the four stages of human development.

Level 1. *Child*. A child's organs and senses are open to receiving matter and information from many directions. The child initially receives things in a random way, after which the child's body interrelationally strives to sort things out in an integral, orderly fashion. Initially, these experiences are nebulous and confusing. On this level, the open, receiving, naïve child is passively reactive to incoming matter and information, rather than consciously proactive. Because a child's self-identity is not yet well-developed, choices are not rational or deliberate but reactive and random. Because a "thesis" is a proposition which is only preliminary and needs interrelated evidence to slowly demonstrate its truthfulness, I call this 1st stage of development the *thetic* stage.

Level 2. *Adolescent*. As children advance in years, they try to make sense out of what they experience. Adolescents especially are concerned about "getting their heads together." To a certain extent, they succeed in putting into an orderly way much of what they have experienced and know thus far in life. However, because their experiences and their learned information are limited, the intrarelational formation of their ideas is usually inexperienced, partial, and naive. Still, eager to push forward, adolescents easily and quickly jump to conclusions that are narrow-minded, idealistic, and ideological. Because of their search for greater personal security and identity, when extraneous information does not fit into their recently formed idealistic categories, adolescents regularly ignore, reject, or even oppose extraneous ideas. Consequently, adolescents regularly believe they know everything about a topic, and they strongly oppose parents, teachers, ideas, and ideals that are contrary to what they have thus far put together in their minds. Striving to get things right, their judgments are regularly absolute, closed, idealistic, black-and-white, and dualistic. Thus, the adolescent, second level of development is strongly *antithetical*.

Many people have a negative, pejorative understanding of the word "adolescent," because during this phase, the forming individual often makes a lot of dumb mistakes, which may be beneficial only in the short run, but which can gravely hurt themselves or others in the long run. Looking at this matter positively, during this maturation period, it is fitting to call 2nd level activists "adolescent," because they are "adult-ascendant." They are still maturing from an adolescent, self-centered, intrarelational, personal perspective into an adult, other-centered, interrelational, communal worldview. Cultures recognize the transitional character of immature adolescents, and they have special laws and courts that deal with adolescents

Similarly, IPhil recognizes that adolescents' dissociative mistakes should not be held against them into adulthood, hoping that their early mistakes will provide the maturing adolescent with lessons for greater understanding of the complexity of our communal world. Consequently, in this book, the term "adolescent" is understood in a positive way as an intermediate, intrarelational, maturing phase of life. Unfortunately, some middle-aged individuals do not "grow up"; and they fail to mature out of their adolescence. These individuals may become bodily more mature, but they remain psychological and intellectually immaturity, self-centered, and 2nd level. This can happen to politicians, philosophers, institutions, and cultures. Physical adults can remain narrow-mindedly, idealistically, and stubbornly adolescent, because they fail or resist developing into the adult 3rd level phase of their human logistic.

Throughout this book I will be critical of essentialistic thinkers who, in strong self-centered, individual-centered, intrarelational ways, believe that there is only one right way for things to interrelate—their way. Drawing only from their own limited, intellectually-closed experiences or the narrow concerns of their associates, they staunchly hold that all true ideas must be "this way and only this way." This makes their logically consistent thinking closed-minded. All other ideas are judged wrong. Sometimes, religiously minded individuals will consider their conception of perfect interrelations to be decreed by God and absolutely true. Like Socrates, Plato, and other idealistic, essentialist figures, they are constantly looking for the *one and only one* true understanding of different philosophic concepts. They are *antithetical* to every other position, which they judge false.

Level 3. *Adult*. The words "intrarelational" and "interrelational" look very similar in writing, but these two terms point to two very different things. *Intrarelational* is inwardly, individually oriented, and characteristic of 2nd level adolescent development. *Interrelational* is outwardly oriented toward 3rd level building of mutually interactive organizations. Adult, 3rd level individuals are concerned with the state of one's society, striving to correct whatever is dissociative and hurtful, turning problems into opportunities for advancing in wisdom and community development. The most vocal people in society tend to think and operate in a very proud manner on the 2nd level, demanding in loud cacophonous ways that everything be done *their* way. In contrast, adults on the 3rd level are more other-center, humbly, and quietly willing to support the slow process of interrelating with others in mutually supportive, societal ways.

Life is developmental. In contrast to 2nd level, adolescents who want things to become static and perfect, adults are other-centered, 3rd level, and interrelational, striving to include all information and all groups, especially those that have been previously excluded. Synthetically, adults try to bring different individuals and groups together into public, communal realities in orderly, reciprocal, and respectful ways. They are tolerant and respectful with individuals who are different from them, believing and hoping that somehow all things can eventually interrelate harmoniously in mutually beneficial ways. The *synthetic* pursuits and understandings of adults push toward a 3rd level holistically systemic world.

Level 4. *Elder*. Individuals and systems in the 4th stage of development generally experience a deterioration of their physical bodies. They increasingly "let go" of physical possessions and strenuous activities. Yet, thoughtful individuals recognize the possibility of an advanced, metaphysical, spiritual life. Thus, while they are "letting go" of material, physical concerns, maturing elderly are increasingly embracing the spiritual, metaphysical aspects of life. Innovative thinkers who are 4th level, push to advance beyond previous concepts and interrelationships toward a highly potential future that is socially more interrelational and of one loving spirit. Their thinking displays a "paradigm shift" in which their focus changes from what is material, physical, and local and to what is spiritual, universal, and eternally enduring in a higher transcendental way. In doing this, the distressing dissociation of what is material in one's life is seen as an opportunity for these people, who intuit that there is a future, better place or state. For these reasons, I call this the *transthetic* phase of material deterioration and of transcendent, spiritual advancement.

Interrelational Illustration

To close this chapter, look at the box below and tell me what you see.

1. The first thing an individual *actually* sees—not consciously but unconsciously—is *not* the letter "G" or the printed box surrounding it, but the piece of white paper of this book. We tend to ignore the physical backdrop which uniformly stimulates the whole field of our vision. Most people take the physical background of life for granted in an unconscious way because of the tranquility of our senses in settings where there are no contrasts. We recognize when there is perfume in the air, but we fail to recognize the air in which the perfume is an addition. We notice when a large stone falls by gravitation on the top of our foot, but we are oblivious to the gravity at play in this event. We recognize kinetic changes in our world, but not the potential energy background from which all expressions of kinetic energy come.

2. Against a given background, an individual quickly becomes conscious of differences. The observer becomes consciously aware of the black letter "G" in the middle of the printed box. Antithetically, an individual becomes conscious of it because of the contrast of the black ink on the white paper. The closer the color of ink is to the color white, the less the individual will be consciously aware of the contrasting letter.

3. Next, individuals experience the mental interplay between the black box and the black letter in the box. At this stage of awareness, there is no single point or part of that rectangular box and letter to which the individual can point in her awareness that the letter "G" is in the middle of the box. Does "in" have a *physical* existence? No. Rather, the interrelationship of the observed elements metaphysically goes beyond the physical sensations received by one's eye from

this page to a metaphysical awareness of the spatial interrelationship of the different elements of this picture.

4. Going beyond this, the thoughtful individual realizes that what has been printed on the page is metaphysically more than what is physically there. There is a lesson here. There is *meaning* here. This illustration is a doorway to a generalized, transcendental philosophical insight, which applies not only to this illustration but transcendentally to all of reality.

The many visual and verbal illustrations in this chapter indicate that the fourfold interrelational logistic is *everywhere*, and a universal philosophy can be built around the principles of the interrelational logistic of energy and its logical evolution.

Chapter 5

Interrelational Philosophy (IPhil)

Ur-Stuff of Reality in Ancient Times. The Metaphysical and Physical Aspects of Energy. Establishment of Energetic Individuals. Establishing Individuals with a Communal Logistic. Self-Initiative & Freedom. Foundations of Interrelational Philosophy (IPhil). Summary Comparisons.

Besides the interrelational logistic and the logical evolutionary theory, several other foundational elements are needed for the establishment of a comprehensive philosophical system. Since interrelational philosophy (IPhil) presents a paradigm shift from most traditional philosophies, it is important to induce its foundational premises from common human experiences in a large variety of contexts. These premises must be formally distinct yet interrelated in a logically consistent way. This chapter examines each major premise in depth, and at the end of the chapter these major premises are laid out in a well-ordered, progressive way, which echoes the fourfold logistic of all energetic interrelationships.

Ur-Stuff of Reality in Ancient Times

What is the ur-stuff that is the foundation of the structures and interrelationships we observe in our world? That question was a major catalyst in the foundation of Western philosophy in eastern Grecian colonies in the sixth century BCE. This question did not deny that the gods above played a significant role in the *extraordinary* events in our world. Rather than highlighting dramatic extraordinary events, as contemporary mythologies did, these Greek thinkers inversely had the intuition that there were natural explanations for the *ordinary* events and things in our physical world. To the best of our knowledge, Thales of Miletus gave the earliest answer to the question above. He claimed that our world's ur-stuff, or fundamental element, is water. Cold water becomes solid, and hot water becomes gaseous. Water comes from the sky above and from springs below. Water is essential for all growing things.

PART I: FOUNDATIONS

Still, other early philosophers proposed other elements.[1] Empedocles (ca. 450 BCE) classically grouped together the four elements of earth, water, air, and fire. In modern times we recognize these as analogous to the four states of matter: solid, liquid, gas, and plasma. Democritus concluded that our world must be made of indivisible atoms of matter, which modern science has shown to be basically true. With time we have come to realize that many of the insights of these early philosophers are partially true, but not totally true.

As Greek philosophy matured beyond its materialistic 1st phase, it became increasingly abstract, conceptual, and idealistic in its 2nd phase. Two major objections which Greek philosophers had with the Olympian gods were that they were self-centered and fickle. The autocratic, 2nd level acts of the gods of the Greek poets were frequently just proud exercises of power for their own benefit. These gods acted like the tyrants of that time, where the welfare of the people was secondary to the joys and glory of aristocratic rulers. Ordinary people and nations were often the victims of the selfish pleasures and aspirations of the gods and human leaders. Inversely going against the revered poets, Greek philosophers went to the opposite extreme. Rejecting the disordered fickleness of many gods, Socrates and Plato placed the source of reality in the 2nd level, perfect, unchangeable, ideal ultimate principle of Good.[2]

Unfortunately, this reality was so perfect that philosophers had a hard time connecting the idealistic Good with the imperfect goodness found in our changing, physical world. Plato proposed in the *Timaeus* dialogue that the causes of change in our imperfect, transient world from the absolute good are Demiurges. In Greek society, *Demiourgoi* were the skilled, middle-class artisans between the aristocratic *Eupatridae* and the *Gormori* farmers. However, while Plato posited their intermediate existence and activity, he was not able to explain how these intermediate Demiurge artisans came to exist and operate between the unchanging, perfect ideals in the Good and the changing imperfect things in our world. Augustine of Hippo and other Christian thinkers rejected the Neoplatonic notion of intermediate Demiurges and called them daemons or demons. These spiritual mediators were blamed for all that was not perfect in the world created by the all-good God.

Aristotle claimed that the notion of *Being* provided the enduring foundation in all existing things. He also proposed the existence of a Primal Being that is most perfect, never changes, and therefore is eternal. He considered Primal Being as existing in the unmoving sphere which encompassed the moving spheres of the planets and stars in the heavens. Being was the perfect Unmoved Mover which caused material things in our world to move, not through physical, efficient agency, but through mental, final causality originating in Being.[3] In addition, Aristotle proposed another foundational reality, Prime Matter. While Being is the one, perfect, and eternal Unmoved Mover,

---

1. Burnet, *Early Greek Philosophy*.
2. Gaiser, "Plato's Enigmatic Lecture."
3. Aristotle, *Metaphysics*, V, 1013b, XII, 1072a.

Prime Matter is the eternal reality that is moved and is transformed into the many imperfect, temporal material things in our physical world. Since Prime Matter changes form as different material things grow and die, it is the bearer of all potentialities. This is in contrast to Being, which provided the enduring, spiritual existence of being of all finite things, regardless their various material forms. However, if Prime Matter is pure potency, how does it have being and exist?

The division of reality into perfect, spiritual, designing causes and imperfect, material effects, matched the sociological structure of ancient society. In those days, the imperial ruling class was fearfully respected as "always right." They issued decrees from their elevated thrones and palaces. Imperial orders were verbally made by kings, who stood or sat on elevated thrones and lived in castles on the tops of hills. Their orders were physically executed by the lowly, sweaty, imperfect working class, who lived in hovels below and around the royal palaces. The division of rulers from servants in ancient cultures was their way of dividing management and labor in early cultures.

This also explains why people conceived their world as dualistically divided—with righteous heavenly gods above and profane, earthly things below. Then, philosophers and religious leaders extended this dualistic social order into the way they described in their writings how everything in nature operated.

Christian theologians easily embraced Neoplatonic ideas because they affirmed that Being as Good is the highest, eternal, nonmaterial, spiritual, heavenly principle, while all evil was aligned with Satan, the prince of darkness, who lived in the underworld and hell. This dualistic cosmological view of our world was antithetical, and it is still with us today.

In the early years of Christianity, the Neoplatonic and Aristotelian dualistic descriptions of Being versus Prime Matter stood in strong contrast to the monotheism of their Judeo-Christian tradition. The opening sentence of Genesis, however, states that matter already existed before the creation of our world. "In the beginning, when God created the heavens and the earth, the earth was a formless wasteland, and darkness covered the abyss, while a mighty wind (spirit) swept over the waters."[4] Thus, the opening sentence of the Bible quite clearly states that from the beginning there were two different primal principles: God's Spirit above, and the primal watery depths below. In response to this dualistic view of creation, philosophical Christian theologians in the third century responded by saying there had to be a "first creation" before the "second creation" as described in Genesis 1. Early Christian theologians explained that Prime Matter and the Primal Waters originated from an initial decree of God before the decrees of Genesis 1. Because God's words or breathed Spirit was not material, the "first creation" was a spiritual *creatio ex nihilo* (creation from nothing).[5] The making of the *something* of earth from *no-thing* was achieved by the spoken command by the Lord God—just as the other things were produced by the decrees of the Lord

4. Gen 1:1.
5. May, *Creatio Ex Nihilo*.

in the six days of creation in the rest of Genesis 1. Here, God is pictured as an imperial LORD who simply sits upon his throne in the heavens and gives commands to the things below. It is God's word of command, expressed in his Spirit, *ruah*, or "breath," that caused all things to be made. However, from a philosophical point of view, how is it impossible for something to be created/made from nothing?

In the contemporary, historical contexts of ancient Greek and Hebrew cultures, no one questioned whether God as King could by a thought or a word make all material and physical things that existed or happened. Why? Because in ancient times, slaves and servants were nobodies; they didn't count. Everything was attributed to the ruler. Slave laborers did the dirty work, had no legal status in society, and were ignored. In a similar way, the *Encyclopedia Britannica* even today says that the pharaoh Cheops (or Khufu) built the Great Pyramid of Giza. Yet the pharaoh probably never touched a stone. He only gave the command, and everyone else carried out that command. In this worldview, there is no efficient connection between a lofty, spiritual, managerial cause and the lowly, material, physical effect of events in our world. Because of the way their societies operated, the ancients saw no philosophical disconnect here. This was the cultural foundation of Aristotle's explanation that it was through final causality, triggered by a thought or word (*logos*) in the mind of Being as the imperial Unmoved Mover, that all physical motion happened in our finite world. This became the standard explanation of Christian philosophers for centuries. However, today, in our materialistic, scientific age, the ancient 2nd phase emphasis on nonmaterial, spiritual final causality has been replaced by a modern emphasis on 3rd phase formal, physical interrelationships and 4th phase efficient causality.

Even while God's imperial relationship with our world made sense in ancient times when the rich and powerful controlled everything, there were major philosophic problems with recognizing God as primal absolute being of everything material in our world. If God is absolute being and only absolute being, God is total actuality, and God has no imperfect potentialities. So how can God create anything that has potential and can be imperfect? There is a classic philosophic principle that states: "You cannot give what you do not have." If there is only actuality and no potency in a perfect Being, how can this perfect Being produce imperfect things that have the potency to change? Because God is perfect and unchanging, it was held that God knows each thing in history as it is "now." Since a perfect Being experiences no change, perfect Being cannot *experientially* know time as it is advancing in our physical world. Consequently, God as absolute being could not fully, *experientially* know an individual's repentance or change of heart. Not knowing the difference between past, present, and future, how could God know and judge whether a person ever had a change of heart in relationship to the good or bad deeds performed in history? Our singing would be heard by God not in melodious verses but only as a single cacophonous sound.

Furthermore, if changes happened through a thought in God's mind, there would have to be a *change* in the mind of God to establish by final causality a subsequent, final state that would be different from an antecedent, initial state. If God is simply Being, "subsequent" and "antecedent" could not really be known by God. If this were true, how could it be said that God as absolute being is the God of our changing world? In summary, philosophically claiming that the ur-stuff of our changing physical world is absolute being has several logical inconsistencies.

Potential Energy as Ur-Stuff of Reality

Modern historical and scientific studies recognize process as an essential characteristic of our world. Consequently, we need to advance from a one-sided view of God as Being to recognizing God as providing the fourfold dynamism of before, now, immediate future, and long-term future in our changing world. However, if *Being* is not the ur-stuff of physical reality, what is? Modern science has discovered that the foundation and propellant of all physical interactions and interrelationships in our world is *not being* but *energy*. This is not immediately obvious. It took science a long time to come to this conclusion.

The word *energy* is derived from the Greek *energeia*. In a philosophic context, the notion of energy appears philosophically for the first time in the work of Aristotle in the fourth century BCE. However, to Greek aristocrats and thinkers, work was relegated to slaves and servants. The god of work, Hephaestus, made the swords and armor of the other gods. He was pictured as a dirty, sweaty, crippled blacksmith, and he was forced to live apart from the other gods on Mount Olympus. This negative view of work and energy began to break down only in the Medieval Ages when guilds of artisans began to picture God as a craftsman and artisan, producing the beautiful and good things of Genesis.

Then, with the rise of the Age of Reason and the Industrial Revolution, energy became increasingly important, especially the energy produced by water wheels, coal, and other nonhuman, physical resources. Scientifically, the concept of energy emerged out of the idea of *vis viva* (living force), which Gottfried Leibniz at the turn of the eighteenth century defined as the product of the mass of an object and its velocity squared. He believed that total *vis viva* was conserved.

The first person to use the term "energy" in its modern sense, instead of *vis viva*, was possibly Thomas Young in 1807. Following him in 1829, Gustave-Gaspard Coriolis described "kinetic energy" in its modern sense. It was not until 1853 that William Rankine coined the term "potential energy." However, philosophers argued for some years about whether potential energy was a substance (the caloric) or merely a physical quantity, such as "momentum." In summary, the scientific declaration that *energy* exists in two forms—potential and kinetic—is only about 150 years old.

What is the relationship of potential energy to kinetic energy? A 2nd level essentialistic description is idealistically, antithetically "all or nothing at all." Here potential energy and kinetic energy are opposites. When a quantity of potential energy is transformed into kinetic energy, the potential energy is reduced by that amount. In contrast, a 3rd level IPhil description is progressive. As in the example of a ball kinetically thrown up in the air, that kinetic energy is reduced as the ball gains in potential energy. Then, the potential energy is transformed into downward kinetic energy, until it is stopped and has only potential energy. The gravitational potential remains potentially present during the kinetic action of the ball. From IPhil's progressive, evolutionary perspective, a thing's continued existence comes from potential energy, even as it takes on different kinetic values. The forward-leaning self-initiative of potential energy breaks from its passive boundary when a thing takes in an active kinetic state. Throughout any change, the emergent kinetic energy retains that antecedent potential energy in a metaphysically subordinate way, and when the kinetic energy of a thing is "let go," the continuing, subordinated potential energy underneath again shows itself.

As in the logical evolutionary process, the existential foundation of potential energy is carried over into each new kinetic state as its existential foundation. Then, when the four-stage logistic of the kinetic energy is completed, it "lets go" of its self-initiative, while its passive potential foundation endures and shows itself again. In the 2nd level antithetical view of energy, the sum of the energy remains the same because the dissociative sum of these two opposing states remains the same. In a 3rd level synthetic view of energy, kinetic energy emerges from an enduring potential base, and the dominant form of the potential energy in the physical order is reduced such that in the physical order the associative sum of current potential energy and kinetic energy is always the same. In the metaphysical order, it is the potentiality of the system what gives it existence and it subordinately remains the same, even as some of the potential energy is transformed into kinetic energy, introducing something new into the system. IPhil maintains that the transformation of potential energy into kinetic energy is a logical evolutionary process, where the existential potential energy is carried over in a subordinate way into the new, logically evolved, kinetic energy state.

Consider an oriental illustration in which 2nd level kinetic energy is like an island, which emerges from the seabed of 1st level potential energy. The kinetic energy pops out of the surface of the sea into the physical world for a time, and then eventually returns to the original ocean bed of potential energy. This is similar to the way that atomic scientists observe nuclear particles emerging from potent space for a time and then returning back into potent space. In life and death, we are regularly reminded that our bodies have come from the earth, and they will return to the earth. In a similar way, IPhil recognizes that as our kinetic world emerged in the big bang from antecedent potential energy before the big bang, and IPhil will argue that our kinetic world will universally return to a final potential state in the end.

In summary, rather than seeing potential energy and kinetic energy as diametric opposites, IPhil recognizes kinetic energy to be an advancing logical evolutionary expression of potential energy. Similarly, in daily life, IPhil recognizes the free initiations of physical, bodily deeds as advancing logical evolutionary expressions of metaphysical, mental ideas. Physical deeds are our creative little big bangs. In this way we are co-creators with the Almighty.

In 1924 Edwin Hubble discovered that the observed changes of the frequency of bands of red light coming from celestial bodies indicate that our universe is expanding. In 1927, Georges Lemaitre, a Belgian physicist and Roman Catholic priest, theoretically proposed that the inferred recession of celestial nebulae was due to the expansion of the universe, as indicated in Einstein's general theory of relativity. In 1931 he went further and suggested that if the observed expansion of the universe is projected backward in time, the further into the past one goes, the smaller the universe was.[6] Logically then, at some finite time in the past, all the mass of the universe would be concentrated in a single point. He described this point as a "primeval atom," from which the fabric of time and space came into existence. English astronomer Fred Hoyle derogatorily coined the term "big bang" in 1949 on BBC radio, and the name stuck.

Einstein's work on relativity described the relationship between mass and energy.[7] The atomic bomb and other atomic laboratory experiments demonstrated how matter and energy are in fact interchangeable. The discovery of the equivalence of substantial matter with dynamic energy brought together the classical philosophic opposition of absolute being and prime matter. Einstein's equation showed that invisible, metaphysical potential energy and visible, physical, enduring matter are not antithetical but are complementary forms of the same thing: energy.

Furthermore, experimentation established that not only can potential energy be transformed into kinetic energy and vice versa, the sum of potential energy and kinetic energy in an experiment in the physical order is constant. That is, energy is conserved through time, leading to the principle conservation of energy which is traditionally described as: potential energy + kinetic energy = constant. IPhil, moreover, recognizes that in the metaphysical order, starting on the atomic level, kinetic energy is a logically evolved form of potential energy, with antecedent potential energy carried over into the kinetic state. That is why the sum of kinetic energy and potential energy is constant also in IPhil, albeit for different reasons. Being conservative, energy is eternal—not only in the physical order since the big bang, but in the metaphysical order of energetic potentiality before the big bang and after the big collapse.

Some people may argue that time began only with the kinetic big bang. But where does that assumption come from? They may argue that time can only be observed from physical, kinetic actions after the big bang. However, the forward leaning of potential energy, which advanced from a potential state to a kinetic state in the big bang, had to have the parameter of time antecedently to that transformation. That is, the parameter of time had to be metaphysically inherent in potential energy before it was transformed into kinetic energy starting in the big bang. If the parameter of time did not exist in the world's potential energy before the big bang, then transformation of

---

6. Lemaître, "Beginning of the World," 706.
7. Stachel, *Einstein from "B" to "Z."*

that potential energy into the kinetic energy of the big bang could not have happened. If the principle of logical evolution is operative in the big bang, the parameter of time found in antecedent potential energy would be carried over via logical evolution into the subsequent kinetic energy of the big bang and into the temporal activities of all finite individuals and systems logically evolved after that event.

In summary, many modern experiments and many theoretical syntheses have established firmly that our expanding, evolving world began from a singular, kinetic, energetic big bang event approximately 13.8 billion years ago. Since energy exists in either a potential or kinetic state, the ur-stuff of everything prior to the kinetic big bang had to be in the form of potential energy. This line of reasoning reinforces the idea that the ur-stuff of all physical reality in our material world is the singular, metaphysical potential energy antecedent to the singular kinetic big bang.

The Metaphysical and Physical Aspects of Energy

The word *metaphysical* comes from the Greek *meta* + *physika* and literally means "after physics." It is said that when Andronicus of Rhodes (ca. 60 BCE) published his famous canon of Aristotle's works, he physically placed the treatise on this subject after Aristotle's treatise on Physics.[8] Consequently, this treatise came to be called "After-Physics" or "Metaphysics." Regardless of the truth of this story, the word "metaphysics" is appropriate, for these insights and inferences about reality come not from direct physical observations but from intellectual inferences drawn *from* and *after* what has been physically observed.

Unfortunately, different philosophers have applied the term *metaphysics* to various fundamental aspects of reality. To many people, the word *metaphysics* is applied to what we as yet cannot rationally or fully explain. So, the content of "metaphysics" in many people's mind is only about what we really don't know much about yet, and to many its inferred knowledge has less integrity and value compared to current verifiable physical sciences. Furthermore, historically when some topic in the domain of metaphysics is well understood, that subject is dropped from the scope of metaphysics and is embodied in its own discipline.[9] Professional metaphysicians currently attempt to answer two basic questions in the broadest terms: What really *is* there? And what is it *like*?

In this book, however, I will use the term *metaphysics* in its root sense, namely, to mean what can be inferred after and from physical observations or from other inferred mental ideas. From this point of view, abstract mathematics is a metaphysical discipline. In fact, every discipline has its metaphysical side, as professionals push beyond observed data into possible, theoretical projections that currently lie beyond current, experienced data.

8. Smith, *Dictionary of Greek and Roman Antiquities*.
9. Zetta, "Metaphysics."

In this book, I will often compare the physical order with the metaphysical order, contrasting what is physically observable versus what is metaphysically inferred. It is because potential energy and kinetic energy are closely interrelated through the principle of conservation of energy that metaphysics and physics are intimately tied together. The term "theoretical" is applied to projections that are well established by experimentation. What is "theoretical" is really not known via "physical" observations, and so without physical proof, theoretical projections are, strictly speaking, "metaphysical."

Establishment of Energetic Individuals

The term *intrarelational* is different from *interrelational*. IPhil recognizes that energetic individuals are formed in integral ways from their material and energetic parts. Where does the distinction between intrarelational versus interrelational phenomena come from? This formal distinction arises from the logically distinct, progressive stages and aspects of potential and kinetic energetic interactions in enduring, logically consistent systems.

Observations show that the material, energetic parts of animals, plants, rocks, celestial systems, atoms, and nuclei intrarelate in different integrating ways to form holistic individuals of different classes.

- All the organs in the human body are both distinctive and intrarelated in a specific, orderly way according to the genetic logistic of one's DNA. Deoxyribonucleic acid (DNA) has four different nitrogen-containing nucleobases: citrine (C), guanine (G), adenine (A), and thymine (T). Consequently, an individual's physical DNA can be described metaphysically as a code with four symbols. Each organ in the body has its own role to play intrarelationally, and if it doesn't play a meaningful role, that organ will soon atrophy. The health and operations of the human body holistically depend upon the harmonious operations of its parts. Together they intrarelationally form a body, which then is capable of interrelating systemically with other bodies. This is an example of an intrarelated whole reality being greater than the sum of its parts.

- Each organ in the human body requires a certain set of cells which intrarelate in a harmonious way for that organ to be enduring and a systemically meaningful part of the human body. Also, material parts can wear out and need to be periodically replaced by other similar parts. Nonetheless, even with material replacement of parts, the systemic operations of an organ remain the same. This indicates that while material parts come and go, there is a metaphysical, intrarelational *individual logistic* which remains during the life of that individual, ordered to operate holistically as one.

- One of the significant stages in history of biological evolution was the formation of single-celled organisms. They can take in nutrients and expel waste through breaks in their cellular skins. Their internal structures are intrarelationally well-defined logistically by their RNA and DNA. Observations show that the generation of an integrated body usually arises from the progressive development of a body around the "seed" of the individual. That seed intrarelates with surrounding elements such that the individual logistic of the DNA of the "seed" of the new organism specifies the parts' operations in the organic whole.

- On the atomic level, scientists have learned the logistics of different individual nuclear particles. When they are combined, the integral combinations of these particles *can* produce an atom, and its individual logistic is greater than the logistics of their pieces. We know this because the Pauli exclusion principle for atoms is not found in any of the atomic constituents but is clearly present in the atom as a systemic whole.[10]

- For me, the best evidence that potential energy has an inherent intrarelational tendency is the evidence that the kinetic energy of the big bang came from a single spacepoint. This indicates that the potential energy PAPE antecedent to the big bang also had to be one intrarelationally.

Some materialistic scientists and philosophers have a reductionistic view of our world and say that all things are just aggregations of parts. They argue there are no such things as integrated individuals whose logistics are formally more advanced than the logistics of their parts. However, randomly mixing together all the physical elements and molecules that go into a living animal does not automatically produce a living animal. The parts must be combined in a way that maintains the advanced intrarelational structure of the animal.

Consider an automobile that can travel sixty miles per hour. Then take all the car's pieces apart. None of those pieces can go sixty miles per hour by itself. When reassembled in a random way, the assembly probably won't run at all. Only when all the parts intrarelate according to the individual logistic of that car can it again travel at that speed. What has happened here? All the parts that go into the car have their own logistic. They have their separate, limited range of activities. Yet each part *can* become a part of a more advanced system through breaking the limited, closed boundaries of the distinctive parts by combining them with other parts such that they can become parts of an advanced, logically evolved vehicle, whose communal logistic contains and formally goes beyond the individual logistics of its parts.

A major problem with reductionistic thinkers is that they only view everything in our world in the material, mechanical aspects of the parts. They affirm strongly that things *only* have a specific material, physical nature. They do not recognize as real the metaphysical structure of material things. They impose an "if and only if"

---

10. Massimi, *Pauli's Exclusion Principle*.

(iff) condition on the physical aspects of the materials that go into a closed system, and they reject recognizing the metaphysical possibility that a unified material system can be open to advanced behavior. However, each part has an interior structure that is consistent in operations and therefore has a metaphysical logistic. Gödel's incompleteness theory recognizes that the logically consistent logistics of each part is not closed. Rather, by logical evolution the supposed closed, materialistic logistics of the parts *can* be broken to combine with others to form integrated things that are logistically more advanced, thereby producing a whole that is logistically greater than the logistics of all their parts.

In addition, in a complex organism or organization, its enduring parts have their own logically consistent logistic, which can logically evolve. When an inner part evolves in itself, the whole also will experience an *internal* evolutionary change in the subsystems of the whole system. In the evolution of humans from chimpanzees, there are both evolutionary changes of its systemic whole in the logical evolution of its intelligence, *and* there are evolutionary changes in its parts, like the length of its arms, the hair on its body, etc. Thus, logical and physical evolution can happen interiorly with respect to its parts as well as exteriorly with respect to its interrelational whole.

Darwin recognized that *individuals* could evolve a trait that inversely is more "fit" within a population, thereby giving the "fit" individual a greater success rate of reproduction within a given environment. This gives the more "fit" individual a greater probability of developing a society of progeny with that advanced trait. These interior evolutionary changes normally are very small. Larger accidental changes would be difficult to sustain by drawing upon resources that interrelated with other members of the antecedent species or group. The smallness of those changes increases the likelihood of interbreeding with the members of the antecedent system and the formation of a new class or group with that inverse change. Consequently, small evolutionary changes with respect to the *individual logistic* of a particular part of individuals would be more favored.

Establishing Individuals with a Communal Logistic

Nonetheless, IPhil also recognizes that logical evolution *can* take place with respect to the *communal logistic* of the members of a group or species. Because of the magnitude of these communal changes, and because such changes would be more strongly opposed by leaders in the antecedent logistic, this mode of evolution would be more difficult to achieve. Nonetheless, IPhil recognizes that logical and physical evolution through the boundary-breaking of one's communal logistic is logically possible, and occasionally it should be physically observed. This is IPhil's explanation for what is called *Punctuated Equilibria*. It was Ernst Mayr who first proposed the possibility of sudden, major evolutionary advances in 1954.[11] Backed by field data, Niles Eldredge and Stephen Jay

11. Mayr, *Evolution in Process*.

Gould more thoroughly presented this proposal in a paper in 1972. While classical Darwinians claim that the only way evolution can take place is through very small changes in individuals through mutations of small DNA elements of their individual's logistic, Eldredge and Gould showed that evidence indicates that occasionally evolution can make sudden, large changes in a punctuated way.[12] The physical track of evolution for some species shows an occasional major change, followed by an extended period of little or no change. In this way, IPhil recognizes that small evolutionary changes in the individual logistic of a thing will be more frequent, *and* large evolution of the communal logistic of a thing can happen but will be less frequent.

Both small and large evolutionary changes are more easily recognized in the domain of ideas and inventions. Most intellectual advancements take place in small changes of existing ideas, scientific studies, and physical products. These small logical evolutionary advancements are generated as variations of traditional ideas and standard products. Yet occasionally, there is a significantly new idea or invention that advances intellectual knowledge and technology in a very large way. Examples of macroscopic logical evolutionary advances include: the Copernican model of our solar system, Newton's development of fluxions/calculus and his postulation of the inverse square character of gravitation and celestial mechanics, Einstein's special and general theories of gravitation, the invention of the radio, the invention of transistors and computers, etc. These are not small changes in parts but large innovations of systems, which can produce significant paradigmatic shifts in our worldview and the ways we interact in daily life.

A *holon* is something that is simultaneously a whole and a part. An example of a holon is the hydrogen atom in a water molecule. The hydrogen atom intrarelationally remains an integral whole at its core, logistically surrounded by one outward-reaching electron. In the water molecule the hydrogen nucleus "lets go" of exclusive control of its electron such that control of the electrons is surrendered to the logistic of the advanced interrelational system of the water molecule. Then, the water molecule can become a holon within an organic cell—where it simultaneously is both an integral water molecular and also is a crucial interactive part of an organic cell, which is a holon in the whole organic body. Philosophically, a holon demonstrates the simultaneous distinctness and interdependence of the individual logistic and the communal logistics of individuals with interrelational energy.

Self-initiative and Freedom

"Free will" is a classical expression. In common parlance, freedom is normally associated with consciously making rational choices and then putting them into action. Consciously and deliberately exercising one's free will is considered the height of human and evolutionary development. By contrast, some great thinkers have

---

12. Gould and Eldredge, "Punctuated Equilibria," 115–51.

argued that there is no real freedom in our world. In the seventeenth century, Baruch Spinoza was a leading modern rationalist. To him, everything followed deductively from founding principles.[13] To him, God embodied the fundamental principle of our world. Consequently, he argued that everything in our world proceeds rationally and deductively from God, who ordered all things in our world in the most perfect, reasonable way possible. He claimed that our choices are not by self-initiatives but in response to numerous environmental influences. Following this line of thinking, Einstein said that if it could be said that he believed in God, it would be the God of Spinoza.[14] As a determinist, Einstein proposed the existence of a unified field theory of the four forces of physics from which all physical reality came into existence in a deductive, totally determined way. If that is the case, from a philosophic point of view, when all finite activities happen in a rational way from God as first principle, we all have the same logistical nature as God. Because of this, some philosophers have said that Spinoza and similar rationalists are pantheists, although Spinoza never said this. For this reason, Einstein and other modern rationalists hedge greatly in calling their proposed primal algorithm: "God."

Academic study of material freedom and the laws of probability arose from the study of games of chance. Gerolamo Cardano in the sixteenth century wrote an elementary work on probability. Pierre de Fermat, Blaise Pascal, and Christiaan Huygens developed a more thorough doctrine of probabilities in the mid-seventeenth century. This was followed in 1663 by the first presentation of statistics in a study of human populations as recorded in *Natural and Political Observations upon the Bills of Mortality* by John Graunt. Thomas Malthus used principles of statistics in his 1798 treatise on human societies in *An Essay on the Principle of Population*, which inspired Charles Darwin to apply similar principles to nonhuman populations. This led to his theory of natural selection, the foundation stone of his 1859 *Origin of Species*. Thus, Darwin's classical work on evolution depended upon insights from probability theory, which assumes universal random freedom.

Many macroscopic systems display random statistical distributions of behavior. These studies indicate statistically the presence of spontaneous, random freedom rather than deliberative freedom. Double slit interference experiments indicate that light displays random freedom in the behavior of its photons. Geiger counters testify that the spontaneous emissions of alpha particles from radioactive substances occur in a random way. Random statistical distributions discovered in many different scientific experiments, seriously challenge the strictly rational determinism of Spinoza, Einstein, and others. While it is possible to predict the observed random distributions in systems, it is impossible to predict the behavior of individual elements in those random distributions. The random distributions found on the atomic and nuclear levels—in radioactive emissions, split light experiments, and nuclear interactions which

---

13. Kisner, *Baruch Spinoza*, 48.
14. Einstein, "On Science and Religion," 605.

are correctly predicted by assuming random interactions—seriously challenged belief in the total non-free, deductive character of the physical order.

This leaves a question: If there is only random freedom in the earliest and lowest orders of material reality, how can it be said that deliberate freedom exists in higher evolved orders, like humans? Atomic-level systems are well-ordered and logically consistent. Therefore, by the logical evolutionary theory, they can logically evolve through a not-free logical inverse. If random self-initiatives are logistically characteristic of elementary atomic-level systems, then by logical evolution there *can* be an advance operation and a logically evolved system where its self-initiatives are not-random but deliberate. The exact character of logically evolved systems, whose self-initiatives are not only random but also deliberate, will be described in detail later.

But this leave a more elementary question: Where does the ability to self-initiate changes from potential energy into kinetic energy come from? The primal example of such a self-initiative is found in the big bang. Because of the conservation of energy, the singular kinetic energy event of the big bang came from an antecedent singular potential reality. The self-initiation of the transformation of that singular potential energy reality into a singular kinetic big bang could not come from any other physical reality because there were none. Therefore, the faculty to self-initiate the transformation of that antecedent potential reality into the kinetic energy of the big bang had to be in the Primal Absolute Potential Energy (PAPE) antecedent to the big bang. Since there are no other energetic realities to influence that self-initiative, that new self-initiative must be truly free. That is, the transformation of Primal Absolute Potential Energy into the kinetic energy of the big bang establishes that PAPE is truly free. Then, it can be argued that finite energetic systems evolved from PAPE would by the carryover characteristics of logical evolution also have true freedom. There are some complexities in stating this, and this matter will be discussed in greater detail in a future chapter.

Thinkers tend to push ideas to extremes. Just as rationalism led to a deterministic view of our world in which there is no real freedom, so too some scientists claim that the random freedom found in statistical observations of atomic level reality is the *only* kind of freedom in all phenomena and systems. Claiming that there is random freedom and *only* random freedom in all finite individuals locks these thinkers into a bounded, closed, random view of everything in our world. The logical evolutionary theory, however, breaks such closed, symmetrical boundaries. Logical evolution projects the logical possibility that inversely there *can* be a freedom that is *not random* but deliberate.

Just because something *can* happen does not mean that it *will* happen. In the throw of a six-sided die, there are six possibilities in a random throw. That does not mean that in a given throw all six sides will top the die; rather, only *one* will. In a double slit experiment, a spectrum of light-hits is possible on the wall target long-term. However, when sensors are put on that wall, the full spectrum is not sensed at a given time. Rather, only one photon at a time actually hits the wall in a distinct part of that possible spectrum. In radioactive decay, particles are emitted at different random intervals. Over time these emissions display a spectrum of possible intervals. However, at a given time, the emissions occur not according to that full spectrum but only at randomly different intervals. That an event *can* happen is of the metaphysical order in a spectrum of possibilities that is physically observed only over a time. Actual physical events, within that spectrum of possibilities, usually occur only one event at a time. Why that is will be discussed later.

Physicists speak about a cloud of possibilities around each particle. Unfortunately, many regularly speak of that cloud as physically real, rather than metaphysically real, because the word "metaphysical" is not in their active vocabulary. IPhil recognizes that when an event is only potentially energetic, it is not yet kinetically energetic. Similarly, cosmologists write about multiverses, but those multiverses are in the realm of theoretic possibility and not physical actuality. It is so easy for scientists, like eager teenagers and religious sages, to claim possibilities as actualities. Without physical evidence, such statements are possibly, metaphysically true but not experientially, physically true. In the throw of a die, there are six possibilities, but only one actuality. Around a nuclear particle there is a cloud of possibilities, but at a given moment of time there is only one actuality. Exactly which possibility is actual can only be determined by some type of interactive observation, which unfortunately disturbs that possibility. That is why we cannot directly physically observe potential energy but only secondarily infer it. This is how IPhil establishes that self-initiatives and freedom are characteristic of inferred potential energy, which can never be directly physically observed. That matches our experience.

Foundations of Interrelational Philosophy (IPhil)

Let's put all the above together in a concise orderly presentation. The examination of physical experiences of energy shows that energy in intrarelational individuals and in interrelational systems has a fourfold logistic, both internally and externally. Thus, it is expected that the founding principles of interrelational philosophy (IPhil) are fourfold, like the formally distinct, advancing stages of the game of catch. Progress is not totally dissociative; each successive step is not totally independent in itself and totally distinct from the antecedent step. Rather, progress consists of stages where each successive step builds upon and goes beyond the former in a developmental way, where the antecedent stage is secondary to and remains with the subsequent stage or aspect.

1. *Potential Energy*. IPhil maintains that the ur-stuff of reality is the potential energy that existed before the big bang. By the principle of conservation of energy, Primal Absolute Potential Energy (PAPE) must exist before the kinetic big bang. Also, that energy must potentially be able to self-initiate the transformation of that existing absolute potential energy into the kinetic energy of the big bang.

That potential energy inherently exists, but that existence is not perfect, complete, or static. Rather, potential energy is forward leaning and therefore has both *energetic being* and *potential becoming*. Potential energy is oriented toward subsequent actual states. So, PAPE is the origin of existence, change, and time. By logical and physical evolution, these characteristics are carried over into the logistical characteristics of all the subsequently evolved finite physical things and into their metaphysical aspects in our universe.

2. *Intrarelated Individuals.* In the process of Primal Absolute Potential Energy (PAPE) pushing toward forming interrelationships, the energetic PAPE first pulls internally toward intrarelational formation of integral individuals. In an integrating way, an intrarelating individual becomes increasingly self-knowing. Establishing and maintaining its integral identity from received potential energy/matter, the individual intrarelationally integrates that received energy in an interdependent way around an evolved progenitor "germ," thereby giving all the received parts the same intrarelational logistic as the individual logistic of the corporate germ. The progressive, intrarelational linking of received potential energy produces a formed, integrated whole that is intrarelationally greater than the sum of its energetic interrelating parts.

3. *Interrelational Systems.* Energy is not only intrarelational toward itself. By logical evolution, it is also potentially oriented toward becoming interrelational with other individuals, thereby forming enduring, reciprocal, interrelational systems. The progressive formation of enduring systems with other individuals is a metaphysical, interrelational advancement beyond the formation of integral individuals. Moreover, the establishment of an enduring system indicates the metaphysical presences of an enduring, organized communal logistic. In the interactions of distinct individuals by means of common energy communicators, reciprocal, systemic interrelationships are fourfold and logically consistent, such that these systems *can* logically evolve.

4. *Self-Initiatives.* Like the Primal Absolute Potential Energy before the big bang, by logical evolution, an energized individual *can* by its own self-initiative progressively transform its received potential state into a subsequent, formally different, dynamic, kinetic state. This new kinetic state occurs through the energized breaking of the interrelational boundary of the potential logistic of the antecedent state. In this way, it retains the potential energy of the antecedent state *and* expresses that potential energy in a new, interrelationally advanced, kinetic way. The difference between potential energy and kinetic energy is not the *quantity* of energy but the logistical range of the *quality* of its interrelationships. Here is the origin of the difference between existential *quantity* and interrelational *quality*.

Because potential energy pushes to be more interrelational, it has the power to self-initiate physical changes. This require some level of knowledge of the interrelational, connecting components of a system. In the initial 1st level receiving stage of receiving diffused potential energy, the forming interrelational realities are yet not oriented toward 3rd level interactions, so early knowledge in elementary individuals is not interrelationally conscious but intrarelationally unconscious. Then, progressively pushed by evolutionary potential energy, knowledge and self-initiatives can advance from diffused unconscious randomness, to self-conscious self-centeredness, to community-conscious other-centeredness, and then to transcendental aspirations and efforts.

Thus, IPhil recognizes one's received potential energy has four different giving aspects: existence, knowledge, logistic, and self-initiative. Furthermore, the progressive energy stages has correspondingly more advanced forms of knowledge.

1. Intuitive knowledge of one's existence in time.
2. Sense knowledge of one's integral, intrarelational individuality.
3. Intellectual knowledge of one's interrelationships with others in space.
4. Projective knowledge of one's self-initiatives into some logical-evolutionary advancement.

One's received freedom allows for inversions within one's logistic, and these inversions are often analogously associated with different body parts.

1. Being closed-minded rather than open-minded with respect to ideas.
2. Being stiff-necked and closed rather than open and universally respectful.
3. Being hard-hearted rather than tender-hearted with respect to others.
4. Being stubborn rather than progressive with respect to innovations.

Summary Comparisons

The following comparisons briefly indicate how different concepts in interrelational philosophy (IPhil) are related.

State of energy:	potential	kinetic
Types of reality:	possible	actual
Orders of reality:	metaphysical	physical
State of existence:	being	becoming
Qualities of existence:	spiritual	material
Domains of operation:	interiorly intrarelational	exteriorly interrelational

Perspectives:	subjective	objective
Knowability:	by energized intuitions	by energized sensations
Known:	by induction & evolution	by observation & deduction
Primary disciplines:	philosophy	science

Finally, some scientists who work with equations have identified "God" with a primal algorithm behind the big bang. From this perspective, Steven Hawking asked: "Where is the fire that gives those equations life?" IPhil locates that fire as the Primal Absolute Potential Energy behind the big bang. Potential energy provides the oomph which animates bodies to act physically according to the metaphysical equations of physicists.

In many religious traditions, God is considered the Almighty, which is the primal metaphysical, spiritual cause of our physical, material world. However, before I can seriously consider whether God really is the primal absolute energetic reality behind the big bang, I need to discuss another matter.

Many individuals reject the notion of God as creator and sustainer of our world because of the problem of evil in our world. Before I can talk about God the Almighty as the One from whom came the big bang and every evolved aspect of physical reality, I need to resolve the classical "problem of evil." What is the relationship between interrelational potential energy and dissociative physical evil?

Chapter 6

Resolving the Problem of Evil

I. The Classical Philosophical Problem of Evil. Induction of Evil in IPhil. Temporary Evil vs. Enduring Evil. Right & Wrong vs. Good & Evil. Means and Ends. God of Good & Evil. II. The Biblical Problem of Evil. Genesis 1. Genesis 2-3. Tree of Life. The Testing of Abraham. The Testing of Job. III. Dealing with Personal & Material Evil. How to Forgive Oneself and Others. IV. Humor and Evil. Thank God. Trial and Error. Death.

This long chapter is divided into four main sections. First, it describes the classical philosophic problem of evil, followed by IPhil's solution to that problem. Second, many people turn to the Bible for help in understanding and dealing with the problem of evil. Several biblical accounts are given, followed by IPhil's response to them. Third, many people struggle in a personal way with the problem of evil, especially the physical and psychological pains their loved ones are forced to endure. IPhil presents paths for dealing with personal experiences of evils in one's life. Finally, this heavy discussion on evil in our world ends on an uplifting note by showing the relationship of evil and humor.

I. The Classical Philosophical Problem of Evil

One of the strongest arguments against the existence of God is the philosophic problem of evil. Formulations of this problem are all basically the same. David Hume's formulation consists of three propositions.[12]

- God is almighty.
- God is all-good.
- Evil exists in our world.

It is easily argued that these three propositions are logically inconsistent; they cannot simultaneously all be true. If God is almighty and all-good, God could not have

---

1. Hume, *Essays and Treatises on Several Subjects*.
2. Pyle, *Hume's Dialogues concerning Natural Religion*.

created a world in which there is *any* evil. If there is evil in this world and the God of this world is all-good, then God must not be powerful enough to eliminate that evil. If God is almighty and there is evil in the world that God created, then the Almighty cannot be all-good in the expression of that might. Therefore, if there is a God, one of the above propositions cannot be true. But which one?

Augustine as a Neoplatonist maintained that the third proposition (evil exists in our world) is not true, for he maintained that evil is not a reality in our world but a privation.[3] In his essentialistic world, everything that God ordered in this world has a certain set of perfections, and a thing is evil insofar as it *lacks* one or more of those perfections. A person with a broken arm has a bone that lacks the perfection of the form it was meant to have. Stealing is evil because in the exchange of property there is give and take, where taking more than is one's right is a lack of equity in that exchange. However, when a person scalds an arm with boiling oil, the pain experienced from the lightest touch is not a privation but a real physical pain, which is felt by the person and can be measured by an electric monitoring device. The labeling of all evils as privations appears to be a clever rationalization. It is not an adequate description of the real painful human experiences judged to be evil.

Some people identify the cause of evil in our world to be human free will. However, our free will's potentiality to self-initiate evil comes from God. So, if our free will has the potentiality to be not all-good, the God who gives us our free will also must be not all-good. Besides, geological evidence shows that earth has experienced several mass extinctions before the evolution of humans, so the evils of those extinctions in the history of our world could not have originated from human free will.

Once people start talking about evil, they tend to highlight and describe the worst evils recorded in human experience. This is a 2nd level, all-or-nothing, idealization of evil. When a tsunami kills hundreds of thousands of people, or when a building with thousands of people inside collapses, or when a communicable disease infects and kills thousands of people, the question immediately arises: "How could God let this happen? If God really cared about us humans, God would have prevented this tragedy from happening." On a smaller but more personal level, when a loved one becomes very sick, a person frequently prays to God for help. When the condition gets worse and the person dies—as was the case for Charles Darwin's beloved daughter, Ann—one's faith in the existence of a caring God can be deeply shaken. When God does not respond to heartfelt prayer and expectant hope, a person regularly becomes very angry at God. During the Holocaust, a Jewish prisoner in a Nazi concentration camp scratched into his bunk, "If there is a God, he is going to have to *beg* me for my forgiveness."

Because of the magnitude of the impact of evil upon us in our most trying times, people tend to look for the origin of evil on a grand scale. However, despite the magnitude of the devastation of the atom bomb blasts at Hiroshima and Nagasaki, the actual

---

3. Hick, *Evil and the God of Love*, 137.

physical cause of that blast happened on the atomic level. The Black Death plague killed one-third of the population of Europe, but its cause was a microscopic organism transmitted by fleas riding in the fur of rats. Consequently, it is probable that the cause of evil in the world is not something big but something quite small.

Induction of Evil in IPhil

Consider some of the ordinary evil things of life. By cutting a person with a knife, a surgeon physically disrupts the normal operations of that patient's body. The cutting separates bodily cells, which function best when they are interconnected rather than divided by the cut. The surgeon's invasive procedure, however, is temporary, removing the infected parts to correct a malfunctioning part as quickly as possible. If the cutting and removal lead to an improvement of the health of the patient, the surgery is considered a success. In contrast, a thug's villainous stabbing is considered a serious evil—even while the same cutting action is considered right when a doctor does it to remove an infected appendix. Lying is considered a serious wrong when it occurs in court. However, when a person lies to get a friend to a location for a surprise birthday party, people find that acceptable. Driving through a red light is considered wrong and can result in a ticket and a bad accident. However, when the vehicle is a fire truck on the way to a fire, it is acceptable and indeed encouraged.

We are painfully aware of the presence of psychological pain in a debate, and of the pain of losing a football game. Some hurts are found acceptable and others are not, depending upon the age and size of the combatants. Experiences of pain and the judgment of what is evil are very closely connected in most people's minds. From these and many such examples, a preliminary observation can be made: Our experience of "evil" arises from some type of *dissociation*, experienced within an interrelational activity. Contrariwise, our experience of "good" arises from *associative* experiences within an interrelational activity. The degree of dissociation is commensurate with the intensity of one's sadness, pain, and horror—which are labeled bad or evil. Greater associative combinations produce greater levels of pleasure, happiness, and joy. Thus, it can generally be said that material evil and moral evil arise from the dissociative aspects of energetic interactions.

Some people object to the universal way that I apply the term "evil"; they would prefer that I apply the moderate term "bad" to lesser evils. However, the dividing line between what is bad and what is evil is different for different people and is thus ambiguous. Consequently, I uniformly label all dissociations "evil," and I ask the reader to stop demonizing that term, and to moderate their understanding of that term.

Consciousness of an evil is not essential for objective, dissociative evil. Our physical world existed long before humans did, and scientific research has shown that random material dissociations in the material world and mass extinctions took place long before dissociations took place in the world of conscious humans. This indicates that material evil arose before moral evil. Consequently, it is fitting to discuss the origin of material evil first.

If our painful experiences of evil come from dissociations, where do dissociations come from? The following four things can be said from an energetic point of view.

1. IPhil recognizes that the ur-stuff of the reality of our world is the potential energy before the big bang. Potential energy is oriented toward a subsequent state that is somewhat dissociated from the antecedent state. From IPhil's perspective, if dissociation is the most primitive form of evil, then the first disassociation can be found in the metaphysical dissociations inherent in the interrelational logistic of primal potential energy before there was any kinetic action. Activations deplete the potential energy of a giver, making one's potential energy less potent than it was before. Note that because conscious knowledge arises only in intrarelated, integrated individuals in the 2nd stage of their development, there would be no conscious awareness of any psychological pain in 1st stage dissociations, only inner distress. Similarly, IPhil recognizes that the first type of evil is not physical but metaphysical. Some may object, saying this is not really evil because there is no sense of physical pain here. However, when a stabbed patient is in shock, the patient feels no pain, but everyone recognizes a stabbing as a dissociative evil. So even in the absence of conscious physical pain, logistical dissociations are metaphysically distressing and somewhat evil. In daily life, everyone experiences that changes are difficult. From IPhil's point of view, all changes from a potential state to a kinetic state are difficult. With respect to the antecedent state, each change is to some extent dissociatively evil. Anyone who struggles to get to the gym for exercise knows that pain of dissociative evil.

2. In the formation of an individual, material elements can be intrarelational in a somewhat dissociative way with respect to the DNA logistic of the seed of the individual. For example, some children are born with arms but not hands, and that is evil with respect to one's individual logistic. A child can be born with a congenital heart defect. Consequently, with respect to one's received metaphysical logistic, any physical dissociative malformation of a part of one's body or one's psyche with respect to one's metaphysical individual logistic is considered materially evil.

3. One's received energy is potentially directed toward some better association within a community as well as within oneself. Associative improvements result in better intrarelational integration and interrelational associations. Achievement of such associations feels satisfying and is judged good. However, when interactions with others are dissociative and disruptive to the development of a friendship, a corporation, or a society, such dissociation within those interactions are judged bad or evil.

4. Finally, for a system to advance in design, it is often necessary for an innovative individual to disassociate oneself partially from the orderly logistic of the recognized academic or social institution. By "letting go" of some of the association's communal logistic, the inversely dissociative individual *can* become the progenitor of a new, more advanced logistic. Innovations which display logical evolution are somewhat dissociative to the antecedent, commonly accepted system, and they are regularly judged by conservatives to be societally disruptive, bad, and evil.

Temporary Evil vs. Enduring Evil

After making a cut with a knife as part of an appendectomy, the surgeon closes the wound as soon as the infected appendix is removed, so that the patient's body will heal quickly. Such invasive medical procedures are physically evil. Nevertheless, these invasive dissociations are medically allowed because they normally have good results in the long run. However, if the wound gets infected with staph, the evil of that procedure is extended and magnified. If the infection leads to paralysis or death, it is judged an enduring evil or mortal evil. There is a significant difference between short-term evils, which are tolerated, versus enduring evils, which are considered the worst evils and condemned. Murder and stealing of significant amounts are judged mortal evils because their effects are enduring and have greater interrelational effects in people's lives.

The evil of a particular action is not the same in different interrelational contexts. Consider the following four examples of the act of tripping. Walking down a hallway, a person can accidentally trip on the carpet, but this is a minor dissociation, because the individual can quickly catch herself from falling and move on. However, if a person trips while trying to walk in physical therapy after an accident, the patient and therapist will be concerned about that failure, for it indicates a moderate glitch in the progress of the patient's therapy. In contrast, when a quarterback trips in front of the entire student body as the player is carrying a football for a touchdown, that tripping is most embarrassing and is considered to be a great societal evil, which is often remembered and regretted for years. Fourth, if a bride-to-be *deliberately* trips and severely twists her ankle on the aisle runner so she must be taken to the hospital rather than get married to the waiting groom, that deliberate fall is a major, interpersonal, spiritual evil, affecting many people for a lifetime and for generations after. In each of these examples, the dissociative physical action is the same, namely, tripping. However, the more interrelational the settings and consequences, the more that dissociative action is considered evil.

Right & Wrong vs. Good & Evil

When a kind person gives another person a piece of candy, it will taste *good* to the receiver. However, if the recipient is a diabetic, eating that candy could be *wrong* for the receiver. That is, for diabetics, eating candy, which physically tastes *good*, can be metaphysically and medically *wrong* because of the current logistics of their bodies. Rather than associatively taking the *good* candy, the diabetic should *rightly* dissociate himself from that candy, even if it is psychologically *painful or difficult* to refuse.

When a prankster golfer associatively pockets all the balls on a fairway, that person may feel *good* to have more balls in his pocket. However, when that action disrupts the golf game of the other players, it is *wrong*. It is against the rules of the game and the expectations of the other golfers. It could easily cause such a blow-up that the game is terminated. Even though a person may be tempted and would feel *good* about slipping his opponent's ball into his pocket, it is *wrong* to play the game that way. Then, while it may feel *bad* to admit one's dissociative fault and apologize, that admission and apology is the *right* thing to do, for it might placate the anger of the other players and get the game back on track.

In other words, just as energy exists in two different states—kinetic and potential—so too our energetic world and life operates on two levels: in the individual physical vs. the communal metaphysical orders. We experience dissociations and associations on the physical level as emotionally bad/evil and good, respectively, and we experience dissociations vs. associations on the metaphysical level as moral wrongs or sins vs. right or virtuous, respectively. In IPhil, the term "evil" properly applies to dissociations in the material, physical order, and the term "sin" properly applies to dissociations in the spiritual, metaphysical order. The type and degree of material evil or spiritual sin depend on the interrelational logistic of a given species or religion, respectively.

Physical evil or good may be right in one stage of an interrelational process and wrong in another. For example, in the 2nd phase of the game of catch, it feels both good and is judged right that the receiver catches and holds the ball momentarily. Then it is right for the receiver-potential-thrower who possesses the ball to throw the ball back to the antecedent thrower. There is a time for catching, a time for holding, a time for aiming, and a time for releasing the ball. No state or disposition in an energetic process is essentialistically always right or always wrong, always good or always bad. In the 4th phase of the game of catch, it may physically feel *good* and funny to continue to hold on to the ball, but it is metaphysically *wrong* with respect to the logistic of the game of catch. Holding and releasing at their proper times are essential for the game of catch to continue. The careful distinction of what is materially good and evil in the physical order from what is morally right and wrong in the metaphysical order is most helpful in discussions of both material and moral matters. In our progressive world, there is no inherently right or wrong physical action or mental decision. Rather, what is right or wrong, good or bad depends upon one's current state within the logistic of the activity in which one is engaged.

Here's another example: Despite the weakening of their bodies, many 4th level elders want to continue to act as 2nd level adolescents. It may be psychologically hurtful for waning elderly persons to realize they are no longer able to run a marathon, but it is probably good for their bodies if they don't. They often become frustrated at their bodies for not functioning the way they want them to. Yet, it is often metaphysically *right* and physically better for waning elderly persons to sit back and simply enjoy seeing young people run. "Letting go" of youthful aspirations by the elderly gives those who are young the opportunity to show off their skill to their grandparents, giving grandparents the opportunity to encourage and honor the efforts and care of the upcoming generations.

Similarly, learning advanced mathematics is *right* and *good* for a student who is learning to become a mathematics teacher, but it is pedagogically both *wrong* and *bad* for that teacher to try to teach advanced mathematical theory to middle school students. It is *wrong* because advanced theory is inappropriate for middle-school-aged minds, and *bad* because of the frustration the teacher causes in the students when the teacher's expectations do not match the current abilities of their students.

Most people learn what is appropriate or inappropriate to particular settings. In contrast, essentialistic thinkers tend to be inflexible in their concepts and in their rational judgments that certain behaviors are always morally wrong or morally right. IPhil recognizes that judging something to be morally wrong depends of the interrelational context of that action. The advancing interrelational logistic of IPhil provides reasons for why some behaviors are right at this moment and wrong at another. As the writer of the book of Ecclesiastes wrote long ago: "There is an appointed time for everything; a time for every affair under the heavens" (Eccl 3:1).

Means and Ends

It is commonly said: "The end does not justify the means." First, no one knows the origin or authority of that moral maxim. Second, that maxim has been questioned and debated by many philosophers over many centuries. Third, the lack of universality of that statement can easily be shown. Cutting a person with a knife is contrary to the integrity of an individual's body and therefore a material evil. Yet the common moral judgment is that it is morally right for a surgeon to slice open a person's skin to remove an inflamed appendix and prevent the person from dying. Who is to say that the end of the surgeon's effort to save the life of a patient does not justify cutting the patient open? Some essentialists would say that surgical cutting is always good but that one person stabbing another is always wrong. Such moral judgments are essentialistically grounded upon the application of pre-defined word-labels rather than upon the actions themselves.

The statement "Peter run fast" is not a good sentence because the singular subject, "Peter," is not in agreement with the plural predicate, "run fast." Similarly, "The end does not justify the means" is not a good maxim because the "means" refers to a 3rd level interrelational action, and the "end" usually refers to a 2nd level, self-centered goal. People use this maxim to teach adolescents that it is wrong to steal from another to get rich for oneself. Notice in that scenario that the stealing is a 3rd level interrelational activity, and getting rich is a 2nd level intrarelational consequence. In contrast, a valid maxim has a cause and an effect in the same order. The domain of the cause must match the domain of the effect. For stealing to be interrelationally justified, the result of that stealing must be reciprocal, such that the stealer and the victim are both well-off in the long run. If not, then that interrelational action is wrong because it has a less—rather than greater—interrelational effect. Furthermore, IPhil recognizes that the maxim, "The end *must* justify the means," is an application of the principle of conservation of energy to the topic of interrelational causality.

Consequently, the statement "Peter run fast" is an invalid sentence because there is disagreement between the number of the subject and the number of the predicate. So too, "The end does not justify the means" is an invalid maxim because there is disagreement between the interrelationality of the means and the end of that caused action. Adolescents typically do not consider the 3rd level social consequences of their 2nd level self-serving actions. So, this maxim is useful as an instruction of adolescents, for it jars them into reflecting on the dissociative, 3rd level social effects of their many 2nd level, self-centered, intrarelational actions. As with most 2nd level idealistic statements, maturity shows that this maxim is not absolutely true but only somewhat true.

IPhil's advanced form of that maxim is: The end justifies the means, unless the means is *consequentially* dissociative (or evil) to any element of the interrelationship or to the endurance of the interrelationship as a whole.

God of Good and Evil

At the beginning of this chapter, the classical problem of evil began with the proposition: "God is all-powerful" or Almighty. If God is almighty, then God is capable of making changes. IPhil recognizes that changes involve dissociations, which are always to some extent difficult, painful, and materially evil in the process of reaching a more associative, interrelational end. So, God is the primal source of material evil via the interrelational logistic of the Almighty's potential energy, as Primal Absolute Potential energy pushes toward more interrelational results.

Contrariwise, essentialistic, 2nd level individuals hold that God is all-good and *only* good. Where did they get that? Certainly not from the Bible. The Western philosophical tradition that God is all-good comes from the classical Greek philosophies of Socrates, Plato, and Aristotle, who believed that absolute good and being were unmoving and detached from our changing world. IPhil's model of God is realistically midway been the fickleness of the evil gods of Greek mythology and the static Good of classical Greek philosophy. IPhil's interrelational model of God is also closely consonant with the *righteous* God who performed both good and evil deeds as recorded in Hebrew Scriptures. IPhil's claim that the energetic Almighty is *both* associatively good and dissociatively evil demands that 2nd level idealists make a major paradigm shift in their view of God. In proposing that the Almighty is the Primal Absolute Potential Energy (PAPE) antecedent to the big bang, the Almighty must be both dissociatively evil and associatively good. If God is Almighty, God must cause some change, which demands some degree of dissociative materially evil to establish subsequently advanced associative good. Consequently, IPhil recognizes that the statement "God is almighty" demands that God be capable of making changes that are somewhat dissociative, and to some extent materially evil as an intermediate stage of producing a greater interrelational good. Therefore, from the perspective of the interrelational logistic of energy as found in our world, the second proposition, "God is all-good," is logically incompatible with the first proposition, "God is all-mighty." This position agrees with what is said of God in the Bible and in most other religious traditions.

II. The Biblical Problem of Evil

There are multiple stories in the Hebrew tradition and Bible of how God inflicted evil upon different peoples and nations, both Jewish and non-Jewish. The Jewish concept of God was also strongly associated with the image of an imperial king or emperor. This LORD was most rich and powerful, and he righteously inflicted both blessings and punishments on his subjects for the benefit of the nation and the people. Unlike the Greek poets, the Jewish prophets insisted that God's decisions to inflict punishment upon sinners were always fair, righteous, just, and wise. To the biblical sages, God was not always good, for he inflicted great, albeit deserved, punishment upon sinners, but God was always *righteous* in these judgments. Reflecting on the salvation history of the descendants of Abraham, Jewish prophets, psalmists, and sages proclaimed a dualistic morality in which their LORD God punished wrongdoers for their unfaithfulness and sins, but the LORD God forgave and blessed them when they repented and subsequently obeyed his will. That is, God did inflict material evil upon sinners, but God was always righteous and just in his infliction of evil. This is taken for granted in the Bible. However, can IPhil establish that God always acts righteously *philosophically*?

By the fifth century BCE, reflective Jews began to recognize that God did not always reward the obedient and punish the disobedient. Often the virtuous person experienced great sufferings, while the wicked often prospered. These observations turned many people away from God. If God does not always reward the good and punish the bad, how can it be said that God is always right and just in his actions? Let me approach this question gradually, starting with the early books in the Bible.

PART I: FOUNDATIONS

Genesis 1

From a non-reflective, literal reading of the creation story given in Genesis 1, many people conclude from Genesis 1 that everything God did was good. In their thinking, since everything that God *made* was good, then everything God *does* must likewise be good. These people look at the *ends* and ignore the *means* of those creative actions.

A fuller understanding of the creation story begins by appreciating the historic context and reason for its inclusion in the Bible. The cosmology of Genesis 1 and all the stories in Genesis 1–11 are Babylonian in background. This indicates that they were added to the books of Moses during the Babylonian Captivity, as Jewish sages made counterproposals to contemporary Babylonian religious beliefs. The combative mythology of *Enuma Elish* is set in a cosmological setting where the gods aligned themselves in opposing camps, which were regularly at war with each other.[4] Similarly, the dualism of Zoroaster, popular in that country, divided the gods into two camps: good gods and bad gods.[5] These myths said that humans and the earth were made from the sweat that fell from the brows of these gods in their celestial battles. Thus, this mythology provided an explanation of why humans and everything on earth are a combination of both good and bad.

When the Jewish leaders and artisans were taken from Jerusalem and made servants/slaves of the Babylonians, Jewish monotheism had to stand up against the dualism of the Babylonian religious beliefs. In response, in a logical inverse way, a Hebrew prophet was inspired to write what we know as Genesis 1. He used as its setting, the domed model of the cosmos as formulated by the Babylonian scientists of that time. The biblical story, however, was cast to support Jewish monotheism, opposing the popular Babylonian dualism in several logically inverse ways.

Affirming Jewish monotheism, there are repeated statements that creation was made by one God rather than by two opposing sides of gods as in the Babylonian creation mythology. The results of the creative actions of the Hebrew God were repeatedly described as good, rather than a mixture of good and evil as in the Babylonian warring myths. The orderly creation of the world by the God of the Hebrews stands in stark contrast to the chaotic creation of the things in the world within the Babylonian creation myth. That is why the dissociative (evil) elements in creation were downplayed in Genesis 1.

A closer study of Genesis 1, however, shows that hidden within the actions of God in Genesis 1, there were some dissociative aspects in its creation process. Looking at each of the days of creation in Genesis 1, there is a pattern in the way each successive order of nature came to be.

---

4. Dalley, *Legacy of Mesopotamia*, 213.
5. Al-Rawi and Black, *Journal of Cuneiform Studies*.

1. Before the creation of anything new on any given day, there already was some existing material reality. This was true even before the creative act of the first day in Genesis 1:3–4. "Now the earth was formless and empty, darkness was over the surface of the deep, and the Spirit of God was hovering over the waters" (Gen 1:2). Notice there are two main things, water and the Spirit, and notice the waters of the deep are passive and waiting, while the Spirit/wind above the waters is moving and active—like the passive and active sides of primal potential energy. While Babylonian mythology described the gods as combatively dialectic, Hebrew monotheism is constructively dialogical—as are the existential side and the dynamic side of potential energy.

2. Next, speaking as a 2nd level, all-powerful, oriental potentate, God directs that something new come into existence from the existing materials. For example, "Let the water under the sky be gathered into a single basin, so that dry land may appear" (Gen 1:9). These are dissociative commands. Then, the dissociative event happens, as God directed.

3. How those creative acts happen efficiently is not described in that story. In those days, the work of slaves and subjects was not considered significant. Only the commands and accomplishments of one's LORD and one's God were considered worth recording. Regardless how a transformation took place, the elements of the previous day are first dissociated and then associated into a new, more advanced reality.

4. In conclusion, God looks at the final product of each day and pronounces it to be good. The new level of reality was *right* with respect to God's plan, and it was *good* because of its harmonious association with the other things in creation. The pronouncements that the changes were "good" rather than "right" emphasized the harmony of the world created by the monotheistic Hebrew God, rather than the evil disharmony that came from the warring, dualistic Babylonian gods.

These four phases are repeated each day in Genesis 1, and these steps express very well the four phases in the interrelational logistic. The 1st phase affirms the existence of some energy/matter. The 2nd phase is dissociative in God's verbal, conceptual directive, which specifies that from what was already existing, something new is to be dissociatively formed. The 3rd phase executes the plan that had been specified, and finally in the 4th phase the completed reality is affirmed as good. Notice that only *the completed reality* is textually affirmed as good. The preliminary, existing materials are not described as good. The dissociative process of going from one state of perfection to a more advanced state of reality was taken for granted, as were the efficient cause(s) of the dissociative change. Thus, from the standpoint of IPhil's transformation process, there actually was a dissociative (evil) phase in the advancements of the antecedent matter that progressed to the final state of associative goodness, affirmed at the end of each new day in Genesis 1.

Rather than being chaotic in the sequence of the creation of the things in this world, as was characteristic of the Babylonian warring myths, the Jewish creation story describes creation of our world in a progressive, evolutionary sequence. Although not perfectly similar to today's findings, the sequence given in Genesis 1 amazingly matches closely what modern scientists have discovered 2,500 years later. In summary, Genesis 1 shows itself to be a logically inverse, polemical piece of literature, that gave the monotheistic Jewish people what they needed to hear to religiously stand up to the dualistic, polytheistic creation myths of the Babylonians, thereby enriching the religious tradition and faith of the Jewish people in the God of Abraham and Moses.

What are the primal waters and the land under the water before the first day of creation? In IPhil's model, the mythological deep water is analogous to the logistical potential of the potential energy of absolute God. The material lands beneath the waters are the existing, being, potential forms or logistics within God's potential energy. IPhil find the wind or spirit that moves over the waters to be analogous to the forward leaning, becoming of Primal Absolute Potential Energy. This energy as energy constantly pushes existing reality toward physical establishment of more interrelational realities in creation.

Genesis 1 continues to provide a lesson that is relevant today. Babylonian sages were polytheists, and they explained their world using the model of a dualistic battle between *two* camps of gods. Today, some prominent scientists have developed creation models in which there is *no* god. For them our world is the result of random chance. If that were the case, there would be no reasons why the algorithm of primal energy and evolution should be progressively well-ordered in the way it is. Furthermore, for them, there is no real basis for judging anything to be objectively good or evil, even though well-ordered, associative realities fill our physical world. Materialist scientists insist that right and wrong are only communal conventions, pragmatically established and changed for the sake of increased immediate, self-centered benefits. IPhil argues that right and wrong in our world are founded upon the evolution of the well-ordered, logically consistent logistic of the Primal Absolute Potential Energy antecedent to the big bang.

In summary, while the modern scientific view says there is no god, and the ancient Babylonian mythological view claimed there were two camps of gods, and other ancient myths claimed there were many gods, Genesis 1 continues to say that there is only one Almighty God from whom came a directive word, *logos*, or logistic that guided how our world developed and evolved in an orderly dissociative-associative process—through an interim dissociative, materially evil stage unto a greater, enduring good.

RESOLVING THE PROBLEM OF EVIL
Genesis 2–3

It is important to look at the story of the fall also from the contemporary, Babylonian context of its writing. In the ancient world, there were rulers and there were peasants. They lived by two very different codes of conduct. Because powerful leaders and rich princes had no one in authority over them, the rich and powerful acted with impunity, and they did good and evil as they wished. Their subjects and their slaves, however, were under the rule of the rich and powerful. If they did not follow the rules of their lords and masters, they would be severely punished.

Genesis 2–3 describes how, from the wasteland of the Middle East, the Lord, like a powerful Babylonian emperor, built for himself a wondrous, enclosed garden or paradise with rivers running through it. In his garden, he put all kinds of exotic plants and animals for his pleasure, much like the legendary Hanging Gardens of Babylon built by Nebuchadnezzar II. God then raised up a lowly Hebrew man, Adam, whose name is from the Hebrew word, *adama*, dirt. His name is an indication of the lowliness of this man under the feet of his elevated lord. The sign that Adam was still a child was his nakedness, for young children in that day innocently played naked. Prisoners of war were regularly forced to work naked for their conquerors. Clothes were a sign of rank, and in the ancient world, children and slaves had no legal standing. The Lord, however, would have been clothed according to his status as lord, and the nakedness of Adam and Eve was a sign of their being of no social account.

In Hebrew literature the verb, *know*, refers to experiential knowledge, and saying "a man knew a woman" was a polite way of saying he had sexual intercourse with her. These childlike humans of lowly earth did not *know* or have sexual relations with each other until after they had been expelled from the garden.[6] All these points indicate that the story of the Garden of Paradise begins with Adam and Eve in the childlike, 1st stage of their life. When they were expelled from the garden into the barren, difficult world, these children were forced to function as independent, 2nd level adolescents humbly clothed not in imperial silks but working skins from the animals they associated with. Here they struggled in an adversarial environment, and their sons became a middle-class farmer and a lower-class shepherd. In other words, the basic framework of the story of the garden and the fall may be considered a "rite of passage" story about advancing from childhood to adolescence.

Let's examine the story more closely. Adam was made to be a gardener of the Lord, responsible for the care and development of everything in the garden (Gen 2:15). Adam and Eve were allowed to eat of all trees in the garden but one. Their imperial Lord ordered them not to eat of the fruit of the Tree of the Knowledge of Good and Evil in the middle of the garden. It was reserved strictly for the Lord, whose imperial rank allowed him to *experientially know* or *do* both good and evil. Being slave-like children under

---

6. Gen 4:1.

the command of the lord, they were commanded to only *know or experience* what was good—which is the wish of every parent and ruler.

In this way, the Tree of the Knowledge of Good and Evil in the middle of the garden of Eden may be viewed as a "set-up." The Lord's command to not eat of that tree was a test, and such tests were common in those ancient days. Notice that if the Lord was strictly an all-good individual, the Lord would not have planted that dissociative tree in their middle of the garden as a test or temptation. In addition, in the garden there was a tempting serpent, which was "the most cunning of all the animals the Lord God had made."[7] Notice that it was the Lord who placed the tempting, dissociative serpent in the garden to confuse, tempt, and test these two naïve, innocent children. Also, Eve found the fruit of that tree good to look at. That means that even before their original sin, Eve had a concupiscent desire for that fruit before the fall.[8] Dissociative concupiscence already was placed in Eve by God as part of the test before Eve sinned.

From the standpoint of their contemporary caste system, which separated upper, rich, and powerful rulers from lower class working slaves, anyone who tried to live and act above their station would experience dire, life-threatening consequences. This was the way of life in those imperial days. This rite of passage story was canonized in Hebrew Scripture as a warning to all Hebrew children and servants to be obedient to their Lord's commands—or else!

By analyzing this story more objectively from our contemporary point of view, it quickly becomes clear that the individual who was more dissociative (or evil) in this story was not Adam or Eve but the Lord! But hey, that was all right, because in the ancient world, autocratic leaders could do evil with impunity, while the slightest dissociative disobedience by a child, a slave, or an inferior subject of their lord would be greatly punished by the lord. Clearly, this 2nd level story was directed to those who are childlike or adolescent and who are tempted to disobey their parents or their lord. This story warns them: If you are disobedient even in the smallest matter, you will be greatly punished. Although this prophetic warning is clothed in a Babylonian setting, this story analogously expresses that traditional 2nd level warning into our own day.

The fathers in the early Christian church wrote much about the garden of paradise and the fall. From it, Augustine of Hippo in the early fifth century formulated the doctrine of original sin, which became the standard Catholic interpretation of the fall. However, at the beginning of the second century, only seventy years after the foundation of Christianity, the renowned apologist, Irenaeus of Lyon, recognized that Adam and Eve in this garden story were inexperienced youth.[9] Their nakedness was a sign of their childlikeness, he said. Their experience of evil was for them an opportunity to mature. Similarly, IPhil recognizes that it was God who initially dissociative by making

---

7. Gen 3:1.
8. Gen 3:6–7.
9. Froom, *Prophetic Faith of Our Fathers*.

the garden the setting of the first test before the fall. Once they failed the test and were expelled from the garden, they passed out of the garden of childhood into the struggles of adolescent. They were thus forced to mature by facing multiple dissociative evil realities in the barren world of God's making outside the Lord's paradise.[10]

IPhil in a way does recognize the basic concept of *original sin*. Physically, doing something that is dissociative affects one's physical life and one's God-given metaphysical logistic, making the life of a sinner thereafter somewhat handicapped. This is much like a severe physical injury hampering the performance of the individual through their lives. Also, much like an acquired congenital flaw in one's DNA, an individual's material weaknesses can sometimes be carried over to subsequent generations. In this way, the effects of the wrongdoings of parents in the past can be logistically carried over to both one's remaining years and even to following generations—much like the Hatfield-McCoy feud. Nonetheless, such enduring material/spiritual defects can be corrected by some type of atonement (at-one-ment), restoring their ideal God-given individual and communal logistics.

In our modern world, life is materially more interrelational and pleasurable than the original paradise was. We want more. We are looking for better interrelationships with other humans, with the things in our world, and with the Almighty. Idealistic, 2nd level essentialistic thinkers continually look back and want to return to the idyllic past. Hebrew and Christian apocalyptic writings, however, describe a 4th level future kingdom that will be analogous to and transcendental to the storied paradise. So does IPhil.

Tree of Life

Before moving on, I want to discuss the relationship between sin and death as indicated in Genesis 2–3. In the center of the enclosed garden of paradise, the Lord also planted a Tree of Life. Not only Adam and Eve but also God ate of the fruit of this tree to extend their lives. In ancient Babylonian mythologies, the gods were born, and they also died. Carvings in stone, preserved even today, show Babylonian gods eating fruit from a Tree of Life so that their life would be extended, and they would not die.[11] The Tree of Life as described in Babylonian mythology is like the Tree of Life in Genesis 2–3. The fruit of this tree was so special to the gods, that they usually reserved it for themselves, as the Lord decreed in Genesis 3:23. The ancients realized that physical death is the natural conclusion of life—even for the gods. The presence of the Tree of Life in the Lord's paradise is a testimony that not only Adam and Eve, but also the Lord would naturally die without it.

After Adam and Eve sinned, the main reason they were exiled from the garden was to bar them from the Tree of Life. The Lord said, "See! The man has become like one of

10. Gen 2:5.
11. Giovino, *Assyrian Sacred Tree*, 120.

us, knowing what is good and what is evil! Therefore, he must not be allowed to put out his hand to take fruit from the Tree of Life, and thus eat of it and live forever."[12] Just as cherubim lined the entrance to the city of Babylon, God "stationed cherubim with their fiery, revolving swords, to guard the way to the Tree of Life."[13] After the fall, when Adam and Eve were denied access to the Tree of Life, their lives were not continually extended, and they eventually died a natural death. But when?

Earlier, the LORD threatened Adam that if he ate of the Tree of the Knowledge of Good and Evil, he would die immediately.[14] However, Adam ate of the forbidden tree, but he did not die immediately. In fact, it is recorded that Adam lived to the age of 930 years.[15] So, one cannot take the words of the LORD in scripture literally. Both statements are typical Jewish exhortatory exaggerations. If a Jew wants to make an important point, the Jew will exaggerate—as Matthew regularly did throughout his Gospel. Genesis does not say that sin caused death; rather, it *followed* Adam's sin and *shortened* his potential life span. Even today, our sins and misdeeds may not immediately cause our death, but they probably will lead to a shorter life than we were meant to have.

In ancient cultures where there were no jails. A typical moral warning was, "Do this and you will die." People say similar things like this even today. Not long ago I heard a teenage girl telling her younger sister, "You touch my new sweater, and you die!" Such warnings today, as in ancient days, are generally not to be taken literally. However, the magnitude of the threat does indicate there will be serious repercussions for doing something wrong. That is one of the lessons the paradisiac story of the fall: Be obedient to the LORD's commands or you will suffer much and probably die soon.

The Testing of Abraham

In his essay on friendship,[16] Cicero directed the reader to test one's friends to see whether they are fair-weather friends who will abandon you when times get tough. If they are true friends, they will remain loyal to you through thick and thin. Ancient rulers regularly challenged and tested the loyalty of those who wished to be "friends" of the king, close to the seat of power and authority. The Bible has many important stories of religious testing. These are like the trials in boot camp, which harden soldiers and bond them within their company. Successfully completing a test-program demonstrates to a commander that the soldiers under him are worthy of advancement. By overcoming difficult problems and physical evils, individuals show their superiors that they are ready and worthy to advance in rank and trust. Facing and conquering difficulty and evil are the stuff of which heroes and leaders are made.

12. Gen 3:22.
13. Gen 2:24.
14. Gen 2:17.
15. Gen 5:5.
16. Cicero and Edinger, *On Old Age; On Friendship*.

The greatness of Abraham's faith was manifested in his continued trust in the Lord, even when the Lord tested Abraham's loyalty by ordering him to sacrifice his only son on Mount Mariah. People today are shocked at the test God put Abraham through to show that his dedication to and love of the Lord was greater than his dedication to and love of his greatest hope—his son. Looking at all the actions of Abraham leading up to the point of actual sacrifice, if the angel of the Lord had not stopped him, out of faith and hope Abraham would have sacrificed his son. In the sacrifice of his dearly beloved son he would have shown that his obedience to the Lord was greater than anything else in his world. Such a test was common in the ancient world. There are archeological remains of children at the bases of the gates of ancient cities testifying to the practice of sacrificing children out of obedience to a god or to receive their god's special favor. Abraham believed he would be the bearer of the promise of the covenant between God and Abraham through his son. Abraham's actions displayed his obedient readiness to sacrifice his current son, trusting that God would somehow remain true to his covenant and somehow bring blessings to all nations through his progeny—even if that was not through Isaac.

"Letting go" of his paternal love of Isaac, Abraham had "hope against hope," where his hope for God's fulfillment of his covenant promise to him transcended his hope for that fulfillment to be through Isaac. That is why Abraham's faith in God is considered so great. His actions demonstrated his faith that God would somehow fulfill his side of the covenant between them, even though he did not know how. But God did provide a solution to that dilemma by sending an angel to stop Abraham just before he was about to sacrifice his son in total obedience to the command of the Lord. God provided a ram in his stead.

Another subsequent benefit came from that experience, for after that, all child sacrifices were henceforth banned among faithful Jewish followers of the Lord— unlike among the surrounding nations. The agonizing evil of this test was not permanent but transient and inconsequential in the long run. The dissociative test was for the purposes of a greater union between the Lord God and Abraham and all his descendants.

The Testing of Job

Another great tale of testing is found in the Book of Job. God allowed Satan to test Job, allowing him to be robbed of all his possessions, his children, and his health. All these physical dissociations were very painful. However, the greatest pain that Job experienced came from the breaking of his expectation of righteous judgments from God. As his Lord, the Almighty could do anything he wished. However, Job expected God always to be just and reward those who were obedient to the law, and likewise he expected God to punish sinners who were disobedient. Job could not understand how God could allow physical evils to come upon him when he had always done what was right. He held strongly to the principle of justice, which maintains that rewards and punishments should always be proportionate to one's obedience or disobedience to the LORD. In his dilemma, Job was acting like a reasonable, 3rd level individual.

After Job had expressed his torturous anger eloquently to friends and to the world, God spoke to Job out of a storm. He chastised and humiliated Job greatly by pointing out how much He had done in the world and how little Job had done.[17] God's imperial speech came from a time when "might made right." God did not explain to Job why God allowed him to be the victim of such physical and psychological evil. Because God was so powerful in the breadth and magnitude of His domain, God had to be right in doing whatever he wished, and he didn't have to explain himself to anyone. In the face of God's *tour de force* proclamation of his manifold greatness in the whole world, Job was totally humiliated. He admitted his ignorance, saying, "I know that you can do all things, and no purpose of yours can be hindered. I have dealt with great things that I do not understand. . . . Therefore, I disown what I have said and repent in dust and ashes."[18] In this way, Job "let go" of his judgment of God and his complaint about how God had treated him unfairly. He repented of everything he had said and covered himself with ashes, a sign of his lowly, earthly status before God who reigns supreme above.

When Job surrendered his deficient, human judgment to the superior wisdom and judgment of the LORD, he left himself open to whatever the LORD judged to be right. Then, voilà! The story suddenly changes. In a 4th level way, Job "let go" of his rational 3rd level argumentation, and he surrenders his puny judgment about himself to the unfathomable wisdom of God. With Job's 4th level surrender, the LORD's behavior freely advanced to a 5th level, which is an evolved stage in his relationship with Job. Now, Job received from God not what was equitable to his former station. He received greater blessings. He received from the LORD more property, more sons and daughters, and more flocks than he had before.

In summary, the story of Job is a classical "testing story." It contains in sequence the stages of the interrelational logistic. (1) The story began thetically with Job being a material receiver: a prosperous father with a large family and the owner of a big estate. (2) Then antithetically, Job physically lost most of the possessions and health he had. (3) In the third stage of his grief process, Job synthetically tried to make sense of this facture by rational arguments, expecting from God equitable and just treatment in the face of his past virtues. Job most eloquently argued with his friends and before God, attempting through reasonable arguments to resolve the cognitive dissonance between his virtuous behavior and his physical misfortune. (4) When that didn't work, he advanced to the transthetic fourth stage by "letting go" of his demands for justice and his unsuccessful attempts to argue his way to the solution he wanted. The LORD's reminder to Job of the greatness of His universe, jarred Job

17. Job 38–41.
18. Job 42:1–6.

to "let go" of his personal, rationalized expectations and to surrender his personal will to the will and greatness of the LORD. (5) By "letting go and letting God," Job opened himself to a fifth stage, in which the Almighty's potential energy overflowed and bestowed upon Job a greater fortune than before.

Some people reject the prose introduction and conclusion of the book of Job, claiming the original story of Job was only the central poetic, argumentative sections. However, in the discernment process of choosing which texts to include in the Hebrew canon, it was the current expanded version, which includes the prose introduction and conclusions, which was officially recognized as inspired by God and as the embodiment of religious truth. While the poetic section vividly describes Job's problem and angry response, it is the introduction that provides the origin of the problem, and the end describes Job's good fortune after he humbled himself and opened himself in a 4th level way to whatever God would send his way. God's 5th level *righteousness* is far beyond a human's 3rd level understanding of *justice*. Job was seeing things only from his personal perspective, whereas the LORD sees everything from the perspective of the universe. Whether it was *right* for God to allow Job to suffer so badly was not for Job or any finite individual to judge. The lesson of Job is clear: Because of the limited experience and knowledge of finite humans, we must ultimately surrender in hope and faith that in the end God will act wisely according the Almighty's universal knowledge of the past, present, and future.

When we are faced with bad times, we need to break the tendency to look only at our present moment from our perspective. Our problems cannot always be solved rationally and equitable. We need to strengthen our faith and confidence in God by recalling the good and great deeds God has done in the past—as God's speech pointed out to Job.

Recognizing God's greatness and goodness in the past gives the sufferer greater strength and confidence that these bad times are only temporary, and according to God's wisdom—not mine—everything will turn out better in the end. It is only in the end that the righteousness of God is experientially known. Unfortunately, Job remembered mainly his own past good deeds toward God; and he did not reflect on all God's good deeds toward him and everything in our world in the past. As a result, Job's faith in the *potential* goodness of the LORD was diminished and insufficient to support him during the temporary period of his struggle with his "undeserved" material evil. God's plan and power for advancement to a better, happy, associative life through temporary, painful, material dissociations are greater than those of finite humans.

That leaves the question: Is the Almighty all-righteous? The criterion for answering the 3rd level question whether the Almighty is all-good is based upon the interrelational logistic of energy, where the intermediate phase of change is dissociative and hence materially evil only temporarily. So, from physical and historical evidence, it cannot be said that in our energetic world the Almighty is *always* all-good. However, is God always *righteous* in imposing dissociative judgments? The question whether the Almighty is always all-righteous is 4th level, for it concerns the Almighty's free self-initiatives. The criterion of judgment, here, must be based on the becoming, metaphysical logistic of potential energy.

Since we do not know the future, it is impossible for humans to give a true answer to the question: Is the Almighty all-righteous? Humans don't know the future and cannot demonstrate conformity of the Almighty's free future actions with the ideal interrelational logistic of potential/physical energy. The best we can do is look at the past and current deeds of the Almighty and project those experiences into the future, leading us to a logistical knowledge of the way the Almighty will probably act. Like Job, many people only look narrowly at their recent events. However, interrelational righteousness is determined only in the long-run—from the whole interrelational logistic. The remonstration of God's message to Job was that Job was little aware of the full scope of God's activities in the world. Thus, we can at best say from a broad view of the history of our world and its scientific orderliness that the evolutionary advancement of our world indicates that the Almighty overall has been wise and all-righteous. This provides a concrete, reasonable foundation for saying *in faith and hope* that the Almighty will be all-righteous in the future.

Since the Lord is free, philosophically there is no logical certainty that God will always act righteously in the future. However, look at the many good and righteous deeds in the normal running of our universe, and recognize the *tendency* of the potential energy coming from God appears to be more interrelational in the long run. Thus, history shows there is high probability that the Almighty will always freely act in a well-ordered and righteous way—in the long run. Because of our limited experiences and knowledge, however, it is only by faith that a person can trust that the energizing Spirit of the Almighty will *always* push to act righteously. Job had a 3rd level understanding of justice in which right deeds were thought to result in commensurate just rewards. In the book of Job, however, God shows himself as not bound within our limited, human, rationally specified rules of 3rd level justice. Rather, the story tells of how the Lord calls a person to 4th level repentance, metanoia, change of heart. Then, the Lord's wisdom will freely give restorative mercy, making things better in a 5th level way, in which some material evils are allowed for a time as an inverse means of advancing to a more advanced material and spiritual life.

III. Dealing with Personal & Material Evil

Today most people are not concerned about the philosophic problem of evil or the biblical problem of evil. Rather, they complain most about the personal experiences of physical pain and evil in their own lives and in the lives of the people they care about. They ask: How can a loving God allow this to happen to me or to mine? That question quickly came to mind when hundreds of thousands of people died in the 2004 Indian Ocean Tsunami. Hurricanes and plagues regularly destroy the lives of many innocent people. When one's loved ones are very sick, one asks with concern, as Charles Darwin did about the sickness of his beloved daughter, Anna: Why doesn't God answer my prayers and cure my loved one? One of the worst tragedies in modern times was the Holocaust, in which millions of Jews were killed in Nazi Concentration Camps. Because of the horrors of the Holocaust, many survivors and their families have struggled to maintain their faith in God. Many Jews have become atheistic humanists.

Interrelational philosophy (IPhil) recognizes that there is no "silver bullet" that will get rid of all the pain and suffering in our energized world. Rather, IPhil provides a progressive path to emotional relief and spiritual resolution of the tremendous pains arising from one's dealing with the various evils in one's own life and in the lives of loved ones. The following process is similar to Elisabeth Kübler-Ross's description of the stages many patients go through in dealing with cancer in their lives.[19] These stages are also analogous to the phases of the interrelational logistic of all energetic events in our world. In a logical evolutionary way, the following stages for dealing with all the pain-filled dissociative changes in our energetic world describe the way many struggling individuals have processed their grief. These stages describe a way that many sufferers have used to turn their painful dissociations into *opportunities* to advance in life interrelationally and spiritually.

1. *Survival Response*. When a person is the victim of a great tragedy, there is no time to think. The body of the receiving person automatically and unconsciously reacts to the evil, often by physically fleeing from that evil. Overwhelmed by the intensity of the dissociative evil, an individual often goes into shock or denial, refusing to believe that this painful tragedy has happened. There is little time to think or ask: Why? One's focus is solely to get away from the danger or to rush to persons in need. Sometimes, when an immediate exit is not obvious or because the event traumatizes the individual physically, the individual simply freezes in place, as some animals do when faced with a great danger.

2. *Dissociative Anger*. Once the immediate crisis passes, and there is time to perceive the great evil that has happened, a person typically becomes angry. Very angry! "This is not the way it should be!" Because they perceive the Almighty or the Powers of the universe to be responsible for maintaining order in our world, people immediately protest, "Why did God let this happen?!" People want order, not disorder, and by their past good actions, they believe they do not deserve this evil, raising the protest question: "Why me?" If God is Almighty and can do everything, why didn't God foresee this great tragedy and prevent it from happening? Persons become very angry because their expectations are disappointed in the experience of this unexpected painful, evil dissociation. They say or think, "If God is all-good and all-powerful, the Almighty would not have allowed this great tragedy and this great pain to happen!" "I expect God to take care of me

---

19. Kübler-Ross and Byock, *On Death & Dying*.

and mine in the way that I expect him to! Because God has not lived up to my expectations of him in this evil, I am not only angry at him, I am tempted to push God out of my life completely."

In Job's black-and-white view of fairness and equity before the law, he expected that God would reward those who live as God told them, and God should punish or bring evil to those who acted in a way contrary to God's commandments. Job became very angry when he experienced much evil and pain despite having lived a righteous life. God's response to Job was: Your knowledge and thinking are too small and too short. Think bigger. There are reasons for this evil, greater than what you currently know and can know. Unfortunately, as you focus only on the problem, pain, and difficulty at hand, your vision becomes narrowed, rather than being broadened. The story of Job's struggle is not only about Job; it's about millions of people who over the centuries have been driven to reflect spiritually on the justice of their pains and struggles before God.

3. *Associative Assistance.* There are often family, friends, associates, and professionals willing to spend much time and effort searching for ways to resolve these problems. The horrors of the Holocaust shocked not only the Jewish people but the whole world. In response, the United Nations partitioned the promised land so that the Jewish people could return to their Abrahamic inheritance after two thousand years of exile. The Holocaust was against a people, and their subsequent return to their ancestral homeland was as a people. Blaming God and other bad people for the Holocaust and not crediting God and other good people for the establishment of the new state of Israel is not fair.

The problem of polio affected many individuals for a time, and it was the occasion for researchers to come up with a cure. The remedy for polio through the Sauk vaccine has subsequently affected all humans in the world for generations to come. So, think bigger—not only to you as an individual, but also on a national and historic scale. Hope intuitively arises from our interrelational potential energy, indicating that better times will probably come—if not for you, perhaps for the nation or in a future generation. Stop fixating on the dissociative evils in the past and exclusively upon those people you care about in your immediate orbit. Rather, look forward to better associated societal times in the future. With time, all dissociative evils *can* inversely, logically evolve into a better interrelational life—if not for you, then probably for others. If not in this life, then probably in the afterlife. Within the interrelational logistic, we are not to hold on to dissociations in unending anger but to "let go" of that anger while striving for new, possible, associative cures. If not now, we need to keep searching in hopes of a cure, for the interrelational logistic of our energetic world affirms that for every dissociation there is a subsequent associative solution that will bring about a more interrelational world.

4. *Hope-Filled Surrender.* Our greatest psychological and spiritual pains are caused by the narrowness of our understanding and expectations. When the solutions presented by self-examination, support from family and friends, and the best remedies presented by professionals fail, a person easily turns one's anger inward, and the individual goes into deep depression. When that happens, the individual must inversely oppose the lethargy that comes with depression and a sense of hopelessness. Rather, the individual must inversely counteract this lifelessness by keeping busy. Active engagement with others turns one's mind from oneself and one's troubles. Only with time, and after much arguing with God, did Job finally, humbly recognize that God's plan or the universal energetical logistic is greater than current human understanding. When Job went through a metanoia, a change of mind and heart, did he repent of his self-centeredness. Only when he put his painful situation totally into God's hands, did things become better for him in ways he never expected. In this way, an individual can advance from antithetical anger and alienation from God by recognizing that one's preoccupations with past dissociative evils have been making one physically and psychologically worse. Turning oneself and one's problems over to the full Almighty-given interrelational process allows for potential energy to come from another source to produce a significant change in the way one interrelates with our world.

5. *Going Beyond the Evil.* The depressed individual needs various supports and assistance as the individual waits helplessly for a change of fortune coming from outside one's current world. Then, from out of nowhere, relief and new inner strength can begin to arise or suddenly break through—not in a planned or expected way of stopping the problem, but in a new interrelational way of life that transcends that problem. Continued "holding on" to one's previous standards stops the process of our advancing potential energy going beyond past dissociations to greater, new interrelational associations. This often comes from an outside energetic source, from an inventor, a sage, a benevolent "angel," or even the Almighty. The hurting individual needs be open to receiving new potential energy from unknown sources and to trust in potential energy's invisible advancement beyond the current problem. Recovery through advancement will come as an inverse state that is logistically or evolutionarily other than and beyond one's current interrelational logistic. The potential energy of the world and the Almighty can go beyond the current situation in another direction to establish an unexpected, evolutionary resolution of the problem and a new way of life.

From the beginning of civilization, human intuitions, religious practices, lessons from history, and the logical evolution theory have maintained that a better time is truly possible. It may not be seen in one's individual immediate physical future, but it is part of the potentials that are inherent in the interrelational energy of our world. Here the individual puts trust in eternal potential energy and the logistical pattern of potential energy to go from a state of dissociation to a state of association and sometimes in a new, logically evolved, transthetic way. Potential energy is constantly pushing forward toward more interrelational states of life. That is why "hope springs eternal." As the German philosopher Friedrich Nietzsche said, "That which does not kill us, makes us stronger." Beyond this, IPhil urges sufferers to have faith that according to the interrelational energetic logistic of the Almighty, all dissociative things will get better—if not in this life, then in the next.

So, (1) when dissociative evils occur, (2) allow yourself to be angry because your idealistic expectations have been broken. Then, (3) use these dissociations as opportunities for putting together new associations, using all resources at hand. When you are not able to resolve and advance beyond these problems by your own current efforts or the efforts of others, (4) "let go" of debilitating options, and be open with transcendental confidence that (5) by the nature of energy and/or the Almighty, these problems *can* inversely evolve into a better, unitive, energized life in the future.

How to Forgive Oneself and Others

I have met many people who say they consider themselves forgiven by God for certain egregious wrongs in the past. They also have experienced being forgiven by others. However, they have a hard time forgiving themselves for past mistakes and misdeeds. In not forgiving themselves, they keep replaying the memory of these misdeeds in their heads. The repetition of the images of those misdeeds or omissions prolongs the disassociation of those misdeeds in one's mind and judgment. This prolonged dissociative mental habit turns a bad misdeed into a self-destructive evil.

A way out of this destructive feed-back process is for the individual to determine the logical inverse of that dissociative wrong deed. Then by repeatedly doing inversely interrelationally good deeds, the individual gains an increased number of positive memories of inversely good deeds This is what steps 8–12 of AA's recovery program advocate. Replacing bad habits with good habits gradually breaks an individual from a stubborn, debilitating bad habit and fills one's memory and imagination with associative rather than dissociative images. As the individual takes the time to reflect and counter the bad habit of self-unforgiveness with recognition and affirmation of inversely associative deeds, the individual builds up inner self-esteem and inner confidence in doing what is right. Unfortunately, occasionally in certain settings, memories of the misdeed may come back strongly, pushing the individual "off the wagon." At this point, the individual must again quickly "let go" of those negative memories and return to one's transforming, inverse, positive, progressive, associative program. This can be done by going back to support meetings and forcing oneself to help others in similar needs. Veterans who experienced traumatic injuries are able to function better and more happily by helping other veterans struggling with the same type of problem. In doing this, the negative past deed will slowly sink deeper and deeper in a peaceful, reconciled way into one's unconscious memory. In doing this, the pattern of judgmental unforgiveness will be distracted, dissipate, and weaken. The problem never totally disappears from one's past or from one's memory, but its negative effects within oneself can become manageable. Then, the proportionate impact of a past problem will become less and less, until it becomes effectively an "inconsequential evil" in one's life. While our inertial materiality is prone to make us turn backward repeatedly toward the memories of our past dissociative problems, our spiritual potential energy is forever forward leaning to a better interrelational future.

On a group level, when individuals and groups experience a traumatic event or a series of offensive events from another individual or group, vivid memories of these past events make reconciliation between these two parties very difficult. What are some of the ways to move from this 2nd level, self-focused, protective, antithetical state of mind, heart, and deliberate actions to a 3rd level, cooperative, synthetic way of thinking, feeling, and mutually beneficial interactions? (1) An individual on either side can take the initiative to break the closed boundary of this standoff and become the leader/progenitor of a new, corporative win/win situation, where both sides benefit from growing material sharing, respect, and increased appreciation. (2) Sports leagues provide a venue of competition for competing parties in an environment that is strict controlled by the rules of the game. (3) Restorative justice is the name given to the formal process of the bringing together opposing sides in a meeting room under the guidance of a professional group counselor to facility their storytelling, their sharing of past hurts, and their searching for mutually acceptable activities that deal with their needs and their hopes for a more peaceful life with less physical and verbal violence. (4) The parties can be brought together in a prayer service in which the parties in God-centered faith and hope can ask for support, guidance, and motivation from the communication Spirit of the grounding, potential Almighty, remembering that the Almighty supports the potential in all. Primal Absolute Potential Energy calls all to be interrelational unto a union that transcends history and the current situations of both struggling parties.

IV. Humor and Evil

The essence of humor is inconsequential evil.

I remember an incident that happened when I was a missionary pastor among the Lakota on the Pine Ridge Indian reservation in South Dakota. A group of full bloods was gathered for a monthly meeting of their religious society. As tradition dictated, they were seated in a circle around the room. It was the custom at the meeting that when an elder was so moved, that person would stand and share with the group an edifying and uplifting thought. Well, one day there was little business to be handled in the meeting, so everyone had turned to his or her neighbor to talk and share community news and gossip. Then an elderly man stood up to say something to the group. He started to speak on a religious topic, but no one paid any attention to him. They continued talking to their neighbors. The elder continued talking over the din for a while but recognizing that no one was paying any attention to him, the old man decided to sit down and be silent. I looked at his eyes. I could tell he was angry. But then suddenly his aspect changed. He stood up again and started to tell a dirty, filthy story. Immediately the conversations in the meeting stopped as everyone turned in shock toward the old man. The women complained, "This is a religious meeting. You shouldn't be talking like that!" The old man replied: "I tried to tell you something religious, but you were not interested. So, I decided to tell you something more to your interest."

In that humorous story, there is much evil. The gossiping group refused to recognize the religious traditions of that meeting. There was the telling of a sexy, dirty story at a religious meeting, and the subsequent negative criticism by the women against the man. Lastly, the man's response was a stinging reproach to everyone who had ignored his spiritual message. Nonetheless, the man's foul story had its intended effect: it returned everyone's focus to the religious traditions of the meeting.

Here, the man's story was a temporary evil, and its dissociative stimulus pushed the group to a more associative and spiritual state. Notice how this outlook is very different from an essentialist view of good and evil. Essentialists want things always to be good, rejecting everything that is in the least bit tainted by evil. The initial good preaching of the elder, however, had no positive effect. It was the man's dissociative, evil story that provided a long-remembered, positive religious lesson.

In IPhil, the important thing is not whether a physical event is associative or dissociative, good or evil, but whether the disassociation and association are metaphysically or spiritually *right* within one's current interrelational logistic. Moreover, the faster a given dissociation is resolved, the better a person feels. As comedians say about a good joke: It's all a matter of timing. Also, the stronger the dissociation is from the social norm, the stronger will be the laughter. A joke takes something ordinary and turns it awry in a clever way.

I live in the Twin Cities, and there is a long-standing rivalry between Catholic St. Paul and Lutheran Minneapolis. A friend of mine called Debbie likes to tweak people's categories. One day she came to work wearing a T-shirt that said: "Did you notice the Bible regularly mentions St. Paul but *never* Minneapolis?"

With my background in physics, I know that James Clerk Maxwell encapsulated all the characteristics of electromagnetism and light in four very concise but complicated mathematical equations. I smiled and laughed one day when my religious brother, who is an electronics engineer, came into the room wearing a splashy T-shirt on which was printed: "And God said, let there be . . . [followed by Maxwell's four complex equations]."

Why do we laugh? The essence of humor is inconsequential evil. A joke twists something, and we tense up and gasp. But the mind quickly realizes that the evil twist is inconsequential, so we relax and exhale. But then the mind thinks back on it, we tense up and gasp again. But then the mind again affirms it is meaningless, and we relax and exhale. This process spontaneously triggers the diaphragm to tighten and loosen in a belly laugh. The more dissociative a remark or a pratfall is from the social norm in an inconsequential way, the harder we gasp and let go, and the harder and longer we laugh.

At a party, a person with a plate of hors d'oeuvres in one hand and drink in the other, sat on a chair askew and fell to the floor. Everyone turned and laughed. With arms and legs splayed in all directions, the awkward posture of the person was very different and funny. But then the person moaned, and everyone's laughter immediately stopped. Everyone recognized that the dissociate evil event was no longer a transient, inconsequential evil but an enduring, consequential evil. The people around the fallen person no longer laughed but rushed to be of assistance.

Kidding comes naturally to kids—whether they are humans, apes, wolves, or tigers. They spar with each other, feign attacks, and fake each other out in all kinds of ways. A good feint first appears to be an attack, and it gets one's adrenalin going, tightening up the muscles in the body, giving the individual the emotion of fear. But once the mind recognizes that the attack is feigned, the individual relaxes, the muscles "let go," and the individual feels pleasure in the peaceful, associative state. Especially for the young, this flood of adrenalin, followed by a muscular release, is highly emotional and enjoyable. Kids love to be scared on Halloween, on roller-coasters, and in horror films. The crucial thing is that the scary, tense state of fear is short. Then, the individual recognizes and enjoys the threat as inconsequential. Fear and enjoyment are closely related. In fear, the individual mentally holds on to the possible negative effects of a dissociation, and laughter comes from the release of that tension. Older people have a muted sense of fear and humor because, from many remembered past experiences, they already have a good idea of what is coming. The magnitude of the dissociation is less, the surprise is less, the tension is less, and the enjoyment is less. Instead, older people find deep inner joy in recognizing the order and rightness of a situation.

In contrast, the longer a tension is without relief, the more likely that a would-be joke turns into a dissociative aggravation. By continuing too long, it loses its surprise, spontaneity, and unpredictability, and it is no longer enjoyable. A puppy or a kitten will bite peers or human playmates, and its feigned seriousness is enjoyable for a time. But there is a limit to how hard or how long they may bite, for biting disturbs the normal status of an individual, and it is an evil. As long as the dissociative evil is temporary and inconsequential, it can be greatly enjoyed. However, when the bite or the tease is too intense or too long, the play is no longer fun but is resented and knocked aside.

Serious people with a perfectionistic, essentialistic philosophic outlook have difficulty with humor. People who push themselves to live and think according to some ideal norm, will projectively judge all minor dissociations from the norm to be wrong, rather than humorous. These ideologues consider all variations from their idealistic norm or intellectual program to be consequential and morally wrong. By claiming that their moral standard and program is right, and others are wrong, they box themselves into a mental logistic that has a strong, separating boundary. For them, all dissociations from their idealized position are judged significantly evil and worthy of strong, negative, public condemnations. To them, frivolity is a distraction from reaching what is virtuous and righteousness. They focus on the current moment, and they insist that every part of every action should be performed in the ideal, perfect way. If an action or part of an action is not according to their moral expectations, they consider it against God's law, for they presumptively consider their standards and expectations to be identical with God's. Rational stoicism and medieval religious rules condemned paying any attention to emotions and humor. Those strictly rational values are deeply imbedded in the lives of many serious, religious people even today. Focused on 3rd level rationality, essentialists regularly condemn the role of emotions in human actions. They fail to appreciate how in a progressive logistical way, 2nd level emotions, when properly moderated, can support and invigorate 3rd level rationally right actions unto 4th level advancements.

People with a "good sense of humor" recognize that not everything needs be taken so seriously and or be so perfect. They recognize that it is not the current intrarelational moment that is most important, but the extended interrelational consequences of an action. Mature interrelationalists do not hone into every aspect of a rational argument or law but look at the overall effect of the matter interrelationally. They see a bigger picture, and they realize that there are many ways to go from here to there. People can reach the same goal via paths that may be somewhat dissociated from current social norms. They are not negatively judgmental or fearful of change but relish those changes that are interrelationally somewhat askew, different, and new without going too far dissociatively and becoming threatening to the integrity of an existing individual or system. Nonetheless, some dissociative remarks "are not funny" because they "go too far." When the dissociative pain is not quickly let go, it causes enduring, personal pain by being emotionally consequential rather than inconsequential.

IPhil is a rational philosophy that has room for making mistakes, as long as they are logistically inconsequential. In fact, IPhil understands how we can learn from making mistakes that are temporarily, dissociatively evil and painful. Whether a child is a human, or a giraffe, or a fawn, etc., we enjoy watching a youngster taking a few steps and falling on its butt, then getting up again and after a few steps falling again. In all areas of life, we learn to succeed by making mistakes and standing up and trying to take a few more steps—as long as our failures are small and inconsequential.

For an idealistic, "know-it-all" person on the 2nd level, there is only one right way to bake an apple pie—the traditional, orthodox way. A 3rd level associative person will try different recipes for an apple pie, experiencing different tastes and pleasures. That is why single-minded idealists are regularly told to "lighten up," or "loosen up." Idealistic, unchanging norms and religious laws can be heavy burdens for people to bear.

Perfectionism constricts people's creative freedom and imposes the heavy burden of constant scrutiny and condemnation over the smallest things. 2nd level perfectionists and idealists find it very difficult to change, for change would require a fundamental shift from being idealistically self-centered to becoming open to variations that are logistically inconsequential and more enjoyable. Tolerance and enjoyment of variations is a 3rd level treasured virtue, and through logically inverse variations, significant, evolutionary changes are possible.

IPhil has a beloved niche for humor as well as the notion of inconsequential evil. IPhil shows how there is room for and interrelational value in correctable mistakes and inconsequential evil in our world. IPhil is opposed to logistically consequential dissociative evils, but tolerant of inconsequential dissociative evil, allowing far greater freedom and creativity in our world. Thus, IPhil recognizes that the Almighty has a "good sense of humor" in allowing many of our minor inconsequential imperfections to become inverse opportunities for greater improvement.

Despite and through periods of dissociative, painful, intermittent evil, our potential energy is always pushing us to advance through dissociations to new, better associative opportunities. Sometimes, a good quick joke can teach a lesson faster than a half-hour lecture. We know that humor is good medicine, especially for those who are struggling with severe problems. Through music and humor, an injured individual can switch his mental focus from the current painful evils of his injuries to the possibility of future, enjoyable, advanced harmony. In some people, this view may require a metanoia or paradigmatic change of mind and heart from what is inconsequentially evil now to what can be associatively better in the future.

`There are two forms of bad jokes. First, when the teller considers something as dissociative and therefore funny, but the hearer does not find it sufficiently dissociative and thus not funny, responding, "I just don't get it." Teenage girls are constantly giggling over the slightest variation from the social norm, while older people do not find it sufficiently variant to find it funny. Second, when the teller of the joke considers something as inconsequentially dissociative and therefore funny, but the hearer finds it too dissociative and therefore painfully consequential and not funny. For example, one person may laugh at how another's date is fat, but the hearer is severely insulted. Then the speaker runs after the offended hearer, saying, "I really didn't mean it," or defensively responding, "That person can't take a joke." That is why laughing and crying are so close to each other, and why it takes real wisdom for a comical person to walk between those two extremes.

Thank God

When a good event happens, many people say, "Thank God." When something bad happens, most people look for someone else to blame, usually a sinful human or the devil. What does IPhil say about this? IPhil recognizes that the foundation of the

Almighty is Primal Absolute Potential Energy, which is directed toward the greater good and unto associative interrelational fulfillment. The beauty we see in created things originated from the well-ordered logistic in everything. IPhil recognizes that in establishing a greater good there necessarily is some intermediary state of dissociation, which is painful and considered evil. However, IPhil does not demonize this evil but mentally prays and physically works for advancing that dissociative evil unto a more interrelational good. In this way, IPhil finds philosophic foundation for the common practice of thanking God for the enduring foundational potential from which a new associative good comes. IPhil does not blame but recognizes free individuals—from the atomic to the human level—as the efficient cause of intermediate physical dissociations, which IPhil recognizes as *opportunities* for advancing to a physical state of greater interrelational association or good. In this way, IPhil finds philosophic foundation for always thanking God for the enduring potential foundation found in *all* interrelational things in our world. This includes striving to locate the source of dissociative, kinetic evils in our world, so that corrective measures can be taken unto greater good in the advancing physical order of our world. This practice arises naturally from the potential and kinetic aspects of everything energetic in our world. In this way, our received energetic nature pushes us to carry our dissociative crosses daily in hope unto progressive, evolutionary, gloriously associative states in the future. IPhil progressively views many dissociative material evils in our world are opportunities for mentally and physically advancing associatively in the formation of a better interrelational world. Because of this, interrelationalists thank God for the good things that have already happened, *and* in confident hope thank God for the material evils in life, since they are opportunities for good already in potential energy form unto greater interrelationships and love in our lives.

Trial and Error

The explorative process of "trial and error" involves much dissociative evil, for this mode of learning has many dissociative failures before one or a few successes are found. In this process, each failed trial is viewed experimentally as processively inconsequential. In this experimental process, an experimenter intuits metaphysically that there are possibilities of success in the experiment. This is what is medieval Jewish grammarians called the "prophetic present," where a satisfactory future physical event solution is already potentially present and known to exist metaphysically in a system. Each failure brings the experimenter one step closer to finding the associative solution. Infants are driven by the instinctive intuition to stand and walk, even as they inconsequently fall on their butts again and again. IPhil locates the impetus behind the process of trial-and-error in the forward leaning of the potential energy of the experiment, as the experimenter works through the set of metaphysical possibilities in a situation.

In the trial-and-error process, there are many more inconsequential, dissociative failures than there are possible, consequential, associative successes. Striving to go beyond the known logistic in a logical evolutionary way, experimenters are quite blind in their pursuit of finding a progenitor that can lead to an evolution in the intellectual and/or physical order being studied or experimented with. When an innovative advancement is found, the release of the deep inner tension of metaphysical possibility and physical actuality is experienced as happiness and joy. Eureka! I've found it.

Why are there so few successful attempts at evolutionary advancement in a system, while there are so many failed interrelational possibilities within the investigated system? Recall that constructive operators can produce many terms within a logically consistent system. However, in the evolutionary formation of any new logically consistent system, there is need for just one logically inverse progenitor term to generate an entire, more advanced, evolved system. Gödel's incompleteness theory establishes that for every logically consistent system there *can* be one statement that is beyond that antecedent system. The trial-and-error method seeks to find that singular, inverse, catalytic progenitor, which can operate both within the old system and beyond it to produce a new, interrelationally advanced, logically consistent, enduring, evolved system. A single, physical, kinetic, 2nd level event is evolutionarily more advanced logistically than the entire, antecedent system. The singular discovery of just one evolved progenitor or "germ" can generate a whole new interrelational system in the physical order. Once the 2nd level intrarelational progenitor is physically actualized, no other new progenitors are evolutionarily necessary, although logistically possible. Once discovered, the initial progenitor often can be the guide for analogously evolving other progenitors.

Death

One of the greatest evils is physical death, and religions strive to provide a smooth bridge from this life to a better afterlife. Physical death occurs in the 4th stage of one's individual logistic. Here there is total "letting go" of the physical integration of one's body, resulting in physical death. While this happens in the physical domain, IPhil recognizes that the metaphysical logistic of the individual endures, and the individual's logistic is now open. Consequently, there can be a possible 5th stage in which the enduring metaphysical logistic *can* receive new energy from another source. That potential energy can come from the Almighty as holder of the Primal Absolute Potential Energy (PAPE) that serves as the existential foundation of the situation. In this scenario, physical death can become a transforming "inconsequential evil." Here, the re-energizing of an individual's logistic by the free act of the all-potential Almighty can push each individual's logistic in a logical evolutionary way to a higher interrelational state.

Using an evolutionary logical inverse in his own way, Paul wrote, "When the corruptible form takes on incorruptibility and the mortal immortality, then will the sayings of Scripture be fulfilled: 'Death is swallowed up in victory. O death, where is your victory? O death, where is your sting?'"[20] In most religious traditions, religious persons believe in the possibility of one's logistic-soul being energized in an advanced, evolved afterlife. From that afterlife, the spiritual individual is able "to laugh at death," seeing physical death as a temporary, "inconsequential evil," leading to an enduring, consequential, glorified, eternal life. I will write more about the logistic-soul and the possibility of an afterlife in a later chapter.

---

20. 1 Cor 15:54–55.

Part II
Beginnings

Chapter 7

The Almighty as Primal Absolute Potential Energy

> Models of God through History. The Almighty Not Infinite but Absolute. I. Primal Absolute Potential Energy (PAPE). Knowledge in PAPE. The Origin and Nature of Time. Quantized Time and Energy. Time Always Goes Forward. Origin of True Freedom. Who Made God? II. Primal Absolute Energetic Reality (PAER). III. Primal Absolute Interrelational System (PAIS). Trinitarian Concerns. Arius's Objection Resolved. Freedom Advanced. Logistical Entanglement. IV. The Absolute as Panentheistic. Nothing before Something. World Religions' Models of God & IPhil.

The most common essentialistic description of God is tri-omni (omnipotent, omniscient, and omnibenevolent). Neoplatonic in origin, this description usually imagines God as a perfect and unchangeable Being, the foundation of all that has existential being in our world. In contrast, IPhil describes the Almighty as the singular potentially energetic reality antecedent to and cause of the singular kinetic big bang from which everything in our universe came. In contrast to the perfect, static, essentialistic view God as absolute being, IPhil's potential energy description of God as the Almighty is developmental and progressive, involving both being and becoming.

Since energy in our world has an enduring fourfold interrelational logistic, and since the logical evolution process carries over logistical characteristics of the antecedent system into the subsequent system, the potential energy antecedent to the big bang must also have the fourfold logistic of energy. It will be shown that in its antecedent state, the interrelational form of the Almighty logistically advances through four energetic stages as {None, then One, then Three, and then a Panentheistic Many}. Through these progressive, interrelational stages, nevertheless, the Almighty remains logistically the same absolute, being-becoming, energetic reality. Absolute potential energy's developmental energetic process is analogous to the four major stages of human development: {child, adolescent, adult, and elderly}, through which each human retains the same logistic. Consequently, this important and long chapter is divided into four major parts.

The idealistic, perfect, Neoplatonic model of God as the Good or as Being does not change, so, it lacks the dimension of time. Consequently, the idealistic Neoplatonic tradition cannot rationally explain how a perfect, timeless God could create and sustain an imperfect world having the dimension of time. In the first major section of this chapter, IPhil establishes the first stages of a progressive energetic model of the Almighty as Primal Absolute Potential Energy (PAPE), whose forward leaning has the parameter of time. Subsequently, by the logical evolutionary process, all evolved finite realities will have the dimension of time after the big bang. Furthermore, I will show that the parameter of time in PAPE always goes forward, not in a continuous fashion but in a quantized, step-like manner.

The second section of this chapter focuses on the intrarelational character of energy, establishing a Primal Absolute Energetic Reality (PAER). This is the integral, monotheistic reality to which various religious traditions give different names: God, Brahma, Dao, Almighty, etc. The third section focuses on the interrelational character of energy, establishing the Primal Absolute Interrelational System (PAIS). One of the "spooky" discoveries in atomic physics is entanglement, in which some changes can occur instantaneously at a distance. IPhil shows that this entanglement phenomenon originated in the primal logistic of the Almighty before the logical evolution of the parameter of distance in our extended spatial system. This explains why entanglement is independent of the parameter of distance. Also, IPhil explains how the wave-particle duality originated in the interrelational logistic in the primal metaphysical order of the Almighty. The fourth section of this chapter explains the panentheistic evolution of our finite world from the logically consistent Primal Absolute Interrelational System (PAIS) via a logically evolved absolute-finite progenitor. This chapter concludes with a comparison of IPhil's description of the above stages of the Almighty with various models of God found in different world religions.

Interestingly, IPhil recognizes that the first sin of disobedience against our God-given logistic is procrastination—not advancing according to our received, energizing potential.

Models of God through History

In ancient times, folk cultures were limited to their immediate experiences. They metaphysically intuited the existence of unseen spirits behind the beneficial and malicious physical phenomena they experienced. We should not be too critical of them in this conceptualizing process, for modern scientists similarly project the existence of "dark energy" and "dark matter" to explain effects whose causes they cannot see. Behind surprising, awesome phenomena, the ancients projected imaginary, metaphysical causes, which came and went invisibly like the wind, like breath, like spirits. That is why in the 1st stage of sociological development, primitive religions focused upon local animal spirits and other significant natural spirits, whose actions were quite random and unpredictable. Some of these spirits seemed to be more active near certain locations than others, and these locations became centers of worship, or places avoided by the primitive community. To foster regular, positive, beneficial blessings from these spirits, people gave gifts to them, and fostered respectful taboos in imitation of the gifts and signs of respect they gave important leaders in their tribes. These practices were performed in the hope of turning the actions of different natural spirits to their favor. I will discuss the foundation and rise of spirit-religions more in a later chapter.

As cultures matured and expanded in their 2nd sociological phase of development, community leaders began to organize and become stronger in the face of neighboring opponents. Now unexplained invisible things were seen as major spirits, demigods, and gods, ordered and respected in an imperial manner. Classical mythologies described the ancient gods acting in self-centered ways that benefited themselves—much as local tyrants were often self-protective, self-aggrandizing, benevolent, and revengeful. As kings and emperors lived in palaces on higher ground for the sake of extended observations and for military advantage, the greater gods were projected as living, observing, and acting from mountaintops, in towering storm clouds, or as lights in the heavens. Inversely, commoners lived in the valleys where they worked. They viewed themselves as destined to spend their afterlife in a subterranean, darkened underworld, like the Hellenistic Hades and the Hebrews' Sheol.

The 3rd stage of development of people's understanding of and interrelationship with a singular Almighty began to take place in different parts of the world in the Axial Age around the sixth century BCE, as described by the German philosopher Karl Jasper. In various, unconnected parts of the world, thinkers opposed the fickleness of the many materialistic, self-centered gods of the previous era.[1] In their logically evolved thinking, sages inversely insisted that there were not many but only one spiritual and well-ordered Primal Principle or God. This Almighty operated according to consistent philosophic and moral principles, always rewarding the good and punishing the bad. This spiritual God transcended our material world. Going beyond the Babylonian cosmology as expressed in Genesis 1, the Hebrews located the throne of their God *transcendentally above* the dome of the firmament of the sky. Similarly, Plato located the Good and the world of Ideal Forms transcendentally beyond our imperfect world, and Aristotle located unchanging Being in the farthest unmoving crystalline sphere in the heavens. Inversely going against the polytheism of Hinduism, the psychological tenets of Buddha led to the doctrine of individual no-soul but recognized a single World Soul or transcendental Buddha. The founder of Daoism recognized that the universal Dao (Way) transcended all human utterances. In these ways, in the Axial Age human understanding of the One Almighty advanced in inverse, logical evolutionary ways.

Because the Greeks philosophically believed that everything in the heavens had to be perfect, they incorrectly assumed that all motion of celestial bodies had to be in perfect circles. Using this assumption, Claudius Ptolemy in 140 CE published his star catalog, which described the paths of heavenly bodies as being in epicycles or

---

1. Jaspers, *Origin and Goal of History*.

circles within circles around the earth as the center.[2] This geocentric, earth-centered, Ptolemaic model of the universe became the standard model for 1,500 years. In this model, the Almighty, perfect, unchanging God reigned transcendentally above everything in the universe, with the earth at its center.

Not everyone accepted this geocentric view of the motion of stars in the heavens. Aristarchus of Samos in the third century BCE put forward a heliocentric model in which the earth revolves around the sun.[3] That idea was strongly opposed by Cleanthes, head of the Stoics, who thought it was the duty of the Greeks to indict Aristarchus on the charge of impiety for advocating that model of the universe. This politically backed opposition and religious condemnation of the heliocentric model of Cleanthes was similar to the condemnation of Galileo by the Catholic Church for advocating a similar idea 1,800 years later.

In an effort to find a simpler way to predict the vernal equinox for the establishment of the liturgical date of Easter, Nicolaus Copernicus broke from the Ptolemaic geocentric tradition by writing *De revolutionibus orbium coelestium*, which was published posthumously in 1543 CE. It described the revolutionary theory that earth traveled around the sun in a heliocentric way.[4] He supported this position not because of physical evidence but because this model was mathematically simpler. His theory unfortunately retained Plato's assumption of perfect, circular orbits of the planets. This made Copernicus's heliocentric model as inaccurate as Ptolemy's geocentric model.

Using his limited observations from a newly discovered telescope, Galileo Galilei in the early seventeenth century, strongly advocated the Copernican heliocentric model of the universe in *Dialogue Concerning the Two Chief World Systems*, even though he had promised church officials to present both theories equitably and as theories rather than fact. At the time of the publication of his *Dialogue*, there were as yet insufficient physical observations to back his side of the argument. Elsewhere, however, Tycho Brahe obtained highly accurate astronomical data, and his student Johannes Kepler presented his three famous laws about planetary motion, which stated that the paths of the planets around the sun were not perfect circles but ellipses.[5] The heliocentric model was affirmed after Sir Isaac Newton's discovery of calculus and the universal law of gravitation, for they described accurately the observed motion of the planets in the solar system. Because of the limitations of the telescopes of his day, no parallax of the stars in the heavens was observed as the earth traveled in its annual orbit around the sun. Thus, the stars continued to be viewed as being at a fixed, indeterminate distance. This pushed the heavenly abode of God beyond those stars.

When Hubble demonstrated that galaxies were billions of light years away, many people gave up on the idea that God existed in a specific place beyond the stars of

2. Bagrow, "Origin of Ptolemy's Geographia."
3. Stahl, "Aristarchus of Samos."
4. Rabin, "Nicolaus Copernicus."
5. Baker and Goldstein, "Theological Foundations of Kepler's Astronomy."

heaven. So where is the Almighty? IPhil recognizes that all things in our world are finite and that the Almighty's Primal Absolute Potential Energy in our world is the imminent, potential undergirding of all that is kinetic and physical in our world. While the ancient 2nd level view of God is a heavenly King over all, the 3rd level logical inversion in IPhil locates the Almighty *interiorly*, as the ever-present potential energetic foundation of all. In this way, the Almighty is my and our supportive, potential, metaphysical founder and interrelational partner of what is physical. The Almighty finds its transcendence not in being materially beyond all finite reality, but in the Almighty's total potentiality being beyond the kinetic, material character of all evolved finite reality.

The Almighty Not Infinite but Absolute

In the past, writers have applied the term "infinite" to God. However, relative infinities, ∞, are not precise quantities. Consider the following four infinite numbers:

$A = 1 + 2 + 3 + \ldots$ $B = 2 + 4 + 6 + \ldots$ $C = 1 + 3 + 5 + \ldots$ and $D = A + B + C$.

Each of these different sums is infinite. However, the subtraction of two of these infinite numbers can be either infinite, as in $A - B = C = \infty$, or it can be zero in $A - B - C = 0$. Also, $D = A + B + C$ is infinitely bigger than the other three infinite numbers. Note that every combination of A, B, C, and D is less than "all." That is, the absolute amount of "all" is equal to or greater than all infinite amounts put together. Furthermore, when an absolute amount is reduced by an absolute amount, there is only one numeric answer: 0. This recognition will be important later. Consequently, I have avoided using the term *infinite* because it is an ambiguous term and not suitable for precise argument.

Georg Cantor proposed the concept of "absolute infinity," symbolized by: Ω. He maintained that there can be multiple sets of absolute infinity. IPhil maintains that the absolute of the Almighty contains *all* and is greater than all these absolute infinities put together.

I. Primal Absolute Potential Energy (PAPE)

What I call the Almighty refers to the absolute, singular potential energy antecedent to the singular kinetic energy event commonly called the big bang. Since all potential and kinetic energy of our finite world came from the big bang, IPhil recognizes this Primal Absolute Potential Energy (PAPE) as the ur-stuff of our universe. Also, being energetic, this potential energy has the fourfold logistic of energy, of which PAPE is the 1st phase. Theoretically, there may have been more potential energy than what was express in the kinetic big bang, but that inactivated potential energy could not be physically observed in our evolving universe, and being inactive, it properly should not be called energy. Later, I will present an argument that all antecedent potential energy in the Almighty was expressed in the big bang. That is, in the big bang the Almighty gave us his absolute all.

Knowledge in PAPE

Some people consider knowledge to be the exclusive ability of conscious humans. However, our hearts change their rate of beating from information gleaned unconsciously from other physical parts of our bodies. The gravitational attraction between the earth and the sun comes from knowledge gleaned unconsciously within that physical system. If there is no knowledge of another, there cannot be any interrelationships with another. Consequently, there must necessarily be knowledge in some form in all energetic systems for them to be interrelational.

The interrelationship of potential energy and kinetic energy causes things to be and to happen. Since kinetic energy arises from potential energy, the antecedent potential energy must have some type of metaphysical knowledge of its current state of energized being. In addition, potential energy must have some awareness of its potentialities of *becoming* that which is different from its current state of energetic *being*. Furthermore, because the logistic of potential energy is fourfold, its progressive development demands four formally different forms of inner, metaphysical knowledge: (1) of potential energy as existent; (2) of potential energy as progressively dissociative with respect to energy's existential base; (3) of possible logistical interrelationships of the actual existent with itself or others as potential, virtual receivers; and (4) of an awareness of its capacity to freely self-initiate its potential from its current potential state into an advanced kinetic state.

In addition, in the possibility of succeeding states of potential energy, there must be a parameter which distinguishes its current forward-leaning existential potential state from any subsequent realized potential states. That parameter is the dimension of time.

The Origin and Nature of Time

Contemporary physicists and cosmologists say that there was no time before the big bang. For them, time is a parameter that is physically measured by the kinetic interactions in material clocks and physical interactions. Interrelational philosophy recognizes energy to be both kinetic and potential. IPhil recognizes that 3rd order, *physical time* began with the kinetic interactions after the big bang, and 2nd order. *Metaphysical time* existed in logistically specified, potential energetic interrelationships before the big bang. Since the parameter of time is founded on difference in change of states of being, there is 1st order *existential* time in the foundational potentiality of potential energy. Even within metaphysical, possible interactions, energetic events are not simultaneous but progressive because of the interrelational logistic of potential energy. Thus, there was *metaphysical time* in the potential energy before the big bang, followed by *physical time* which began with the evolutionary transformation of progressive potential energy into kinetic energy in the big bang.

Energy follows the interrelational logistic. Being interrelational, Primal Absolute Potential Energy (PAPE) has four metaphysical aspects, which are the foundations of all change:

1. Potential energy inherently is ordered both to exist "now" and to be forward leaning toward a subsequent "not-now," future state. This indicates that potential energy is capable of progressive development. That is, potential energy has an inherent *push* or gradient toward a subsequent state of energy, and it has the progressive characteristic of carrying over the antecedent energetic characteristic into a subsequent potential state. In human experience, even when a person is sitting still, without any physical motion, the internal *push* of one's potential energy can be intuitively felt and personally known to be real, carrying forward past human characteristics into a subsequent state. Adolescents, when they are forced to sit still, experientially know their antsy compulsion to move. Here potential energy in an antecedent static state is inherently directed toward some type of self-initiated kinetic change—or not. This demonstrates that freedom on this level can be self-initiated. However, lacking an intrarelational sense of self at this point of development, this progressive urge to be free would not be self-directed but non-self-directed, random, or "by chance."

2. Internally interrelational potential energy is intrarelationally oriented toward being in the same energetic state, as well as oriented toward becoming interrelational with respect to oneself in a different subsequent existential state. Here potential energy is directed toward a possible *virtual change*, that is, toward some movement from the current energetic state of PAPE to a different, dissociative, subsequent state of the same PAPE, which inversely is not the same existential state but a new existential state. This is a progressive change that can produce a new state that is formally distinct from the original state. It draws upon the energy of the original state to go beyond it to produce a new existential state temporally beyond the antecedent. This logically inverse state carries over the potential energy of the antecedent state. Therefore, this progressively new state can be said to be logically evolved from the antecedent existential state. Progress is a form of logical evolution, Thus, it can be argued that the logistical, metaphysical origin of the fourfold evolutionary process is found in the forward leaning of primal potential energy, progressively advancing in a logical evolutionary way from one state of potential energized existence to the next.

3. Potential energy is also directed toward impacting another state in the proximate future in a way that can be repeated subsequently. Analogous to the

systemic evolution of a logically consistent system, progressive development of a logically consistent, integrated individual is an advancement from the current energetic state of the individual in a repeatable, logically consistent sequence of existential states. That is, in moving from a current existential state to a series of subsequent existential states, the energy of PAPE must have its own, forward, repetitional logistic.

4. As the potential energy of PAPE pushes toward fully self-initiating an actual subsequent state, the former state of PAPE must "let go" of the energetic ownership of the previous state, while retaining a carryover remembrance of the antecedent state in its gaining of ownership of the subsequent state of PAPE. In this way, each new, subsequent state receives in a fourfold way the existential potential energy from the "letting go" of the antecedent state of PAPE. By the carryover characteristic found in the progressive, evolved change, the forward-leaning logistic retains both a memory of the previous state, and knowledge of the new state.

In this way, potential energy in its four-stage logistic has *existential knowledge* of its current state, *remembered knowledge* of its past state, and *virtual knowledge* of its immediately possible future state. Also, because of the conservative, unending forward-leaning essence of potential energy, there is intuited also a progressive, projection that is long-term transcendental. That is, potential energy has: (1) knowledge of its current, existential state; (2) memory of the past state from which the energized reality came; (3) projective knowledge of a near future state; and (4) extended knowledge of more possible states in the distant future. In this way, IPhil recognizes that in potential energy, there is fourfold, distinctive knowledge of the present, past, near future, and distant future. Because potential energy exists before its intrarelational formation of an individual, knowledge in this phase is not concisely formed. So, it is not conceptual, rational, or conscious. Unlike formed intrarelational 2nd level knowledge, in its 1st phase potential energy would have been inversely known as unformed, diffuse, and intuited. Being diffused, unformed, and intuited, PAPE's knowledge initially is not well-formed as is conscious knowledge. How knowledge advances from 1st stage unconscious, intuited, thetic knowledge to a 2nd stage into conscious, intrarelational, antithetical knowledge will be discussed later.

It is common in the physical sciences to describe time as the duration between successive, physically observed events. However, IPhil recognizes that not all changes of state are external and externally observable. Some changes of state can be strictly internal and only subjectively knowable. A common experience is the internal decision to not act. Here, silence means "no." Still, in the "no," we nonetheless intuitively sense a duration of self-existence while we internally move toward self-initiating an observable kinetic change—or not.

Quantized Time and Energy

The Greek philosophers Democritus and Leucippus first proposed the atomic hypothesis in the fifth century BCE.[6] They held that when a person divides something like a board in half, then in half again, and then in half again, etc., there comes a point at which the material object can no longer be divided and still maintain its original physical characteristics. That is, physical reality cannot be infinitely divided into indeterminate infinitesimals but only into very small, finite atoms. The Greek word "*a-tom*" literally means "not-divided" or indivisible.

Once we study their argument, it becomes clear that they reached this conclusion by applying an early, undeveloped form of logical evolutionary thinking. (1) To a finite physical reality, (2) they applied the inverse operator of division, which produced a distinct, "smaller," individual result. (3) Then by repeated division, they set up a repeatable logistic, which produced increasingly smaller results. As the number of divisions increases and advances toward an infinite number of divisions, the results become increasingly smaller, tending toward becoming infinitesimally small. Within that repeated division logistic ordered division, it seems that there is no end to this division process, and no-thing can be said to be infinitesimally small. (4) Finally, they broke the symmetry of that logistical operation by inversely affirming that there is a point at which this division process can no longer continue. This breaking of the logistic of the repeated division process resulted in the recognition of a "smallest" result. The existence of a "smallest" size *cannot be logically deduced* from the symmetric logistic of repeated division. However, it can be known through the breaking of that repeatable logistic. This results in the establishment of a smallest atomic amount, which became the progenitor of further conclusions in other physical systems.

Let's look at this argument more closely and applying the logical evolutionary theory to the parameter of time duration or time between successive existential states of PAPE.

1. All potential energy, including the Primal Absolute Potential Energy of PAPE, exists in an existential state of absolute energy.

2. That potential energy is also forward leaning and pushes for a change of state that is different from and subsequent to the current state. This allows for repeated changes of state.

3. An extended duration can be divided, and the new divisions of duration will be smaller. If the operation of division can be applied repeatedly to an extended duration, subsequent divisions will become smaller and smaller. Logically, it would seem there would be no end to this division process. This repeated feedback process indicates that there is no smallest time duration.

---

6. Laertius and Yonge, *Lives and Opinions of Eminent Philosophers*, 332.

4. Being a logically consistent system, however, according to the logical evolutionary theory, there logically *can* be a duration that is inversely "not-smaller" than the antecedent. This would be the smallest possible duration. Thus, by logical evolution it can be said that there *can* be a smallest quantum-duration.

Beyond the logical evolutionary theory, how can it be argued mathematically that there must be a shortest quantum duration in the numerable physical order? Let's go back to the basics. Multiplication is defined as multiple addition. In the definition of multiplication there is a multiplicand and a multiplier that produces a product. Regardless of the magnitude of the multiplicand, in the natural number system the multiplier is a natural number—which from experience we know consists not of fractions but only positive integers. Regardless, the definition of division holds that $C \div B = A$ if and only if $A \times B = C$. The product could be finite only if the multiplier B of the infinitesimal multiplicand A is also finite. *Infinity is not a number in the foundational natural number system*. So, if the multiplicand A is a finite number, and the multiplier B must be of the finite natural number system, the product will always be a finite number. This argument assumes that in numerically making any progressive system, there was a finite—not an infinitesimal—first duration to start with.

Infinite division may be fine when that division is in the theoretical, metaphysical order. However, in the physical order, as Johnny said, "That doesn't go." Since we know as physically real only what can be immediately and physically observed, and the number of divisions that we physically observe is finite—not infinite—then it is impossible for us to physically know and observe an infinite number of divisions, and infinitesimals cannot be a part of the physical world as we know it. What is metaphysically possible is possible in the metaphysical domain of theoretic possibilities. However, what is physically possible is limited by the boundary conditions of a given physical system. Since smaller and smaller infinitesimals of duration cannot be produced or observed in our knowable interrelational physical world, there must be a "smallest physical duration." That is, in our physical world where we can make only a finite number of operations and observations, there must be an indivisible time-quantum in our energetic universe, which I symbolize by Δt.

A similar argument can be applied to the amount of energy transferred from an antecedent to a subsequent state of potential energy. The amount of transferrable energy can metaphysically be divided into smaller and smaller amounts for a time. The number of divisions of that energy, however, cannot be infinite in our observable physical world. Consequently, the size of the amount of energy transferred from one physical state to the next cannot be reduced to an infinitesimal but must be finite. Thus, in our physical world, there must be a "not-smaller" or "smallest" amount of energy transferrable from one energetic state to the next. This "energy-quantum" I symbolize by ΔE. Physical experiments on the atomic level show that energy is quantized.

Because the dimensions of time and energy are distinctly quantized, the subsequently logically evolved dimension of space will be distinctively granular, and logically evolved material things in time and space will be foundationally indivisible or atomic.

Time Always Goes Forward

The potential of Primal Absolute Potential Energy (PAPE) to advance into subsequent states provides the energetic origin of ordinal numbers {1st, 2nd, 3rd, 4th . . .}. Like ordinal numbers, time always goes forward.

However, some scientists argue that time can goes backwards. This possibility is suggested by some equations in physics. For example, according to the Pythagorean theorem, the distance traveled by light at the speed c for the time t in an orthogonal space system from point (0, 0, 0) to point (x, y, z) is: $x^2 + y^2 + z^2 = (ct)^2$. In this equation and similar equations in physics, the time term is t^2. Because of the symmetry of the square-operator with respect to positive and negative numbers, the time variable t in these equations can be replaced with the inverse time variable, -t, without any change in the truth of such equations.

Yet looking at this claim from IPhil's perspective of potential energy, time must advance positively because of the nature of potential energy to move to subsequently possible states. Energetically, can time go backwards from a subsequent state to its original state? Potential energy has both an existential aspect as well as a quantitative aspect. In the fourfold interrelational process of energy exchange, an energetic, existential reality in its 4th phase must "let go" of ownership of its current existential possibility to advance to a subsequent existential actuality. When an energetic reality advances to a subsequent state, its original set of possibilities is reduced by the antecedent possibility. As a result, the set of existential possibilities of potential energy in a subsequent state will be less than the existential possibilities in the antecedent state. It may be possible to go back to the subsequent state with respect to the quantitative potential, but it cannot return to the existential state of its total antecedent potentials. That is because in the advancement from a former state to a subsequent state in PAPE, the former existential potential of the previous state is "let go" in advancing to a subsequent state. The former state can be remembered as past, but the past's potentiality as a potentiality cannot be retained or reclaimed. Once an existential potentiality is actualized, that existential potentiality is no longer a potential part of that potential energy. So, it is impossible to go backwards in time because the existential potential of the future state of reversed time cannot be the full potential of one's actualized past state. If that were the case, the existential potential of the subsequent state would be the same as the existential state of the original state. While there may be memory of an intermittent state, there would be no evidence that its potentials and its time had been reversed. From that reversed situation there would be energetic knowledge that nothing has changed *and* interrelational knowledge that something had changed.

These two bits of knowledges are contradictory. Therefore, this scenario is logically impossible. In time reversion, there would no knowledge that time had reversed or that the individual reality had returned to its former state.

Approaching the question of reversibility of time from another direction, it can be argued that because of the formal reciprocity of interrelational energy, potential energy not only has a virtual knowledge of its future possibilities, but once actualized, the advanced state energy has a memory of its antecedent state. In the advancement of potential energy, memories are known successively and cumulatively. Consequently, to reverse time would mean that the subsequent returned state would simultaneously know its original state and know its return state as both potential and actual. If that were the case, potential energy would simultaneously know its possible future state both as not yet actual *and* as now actual. This is self-contradictory and logical impossible.

IPhil recognizes that the ur-stuff of reality is potential energy with the potential to move forward. A reversal of time would render potential energy both forward leaning and inversely equally backward leaning—making potential energy non-progressive, and that does match our internal experiences.

Origin of True Freedom

Absolute potential energy not only orients that energy toward some type of subsequent state, but it must have the capacity to effect that change. Where does that initiative come from?

Since evidence indicates that the kinetic big bang of our world was singular and absolute, the antecedent pre-bang potential energy must also be singular and absolute. Consequently, to be able to initiate that change, the pre-big-bang, Primal Absolute Potential Energy must be inherently capable of self-initiating a subsequent dynamic state. There is no external "other," other than the Almighty existing in the singular state of absolute potential energy before the big bang. Consequently, its self-initiative to transform primal absolute potential into the kinetic energy of the big bang cannot be under the influence of another, for there is no other. Therefore, this initial self-initiative must be *truly free*. Because of this, the origin of true freedom is found within the potential energy of PAPE, the ur-stuff of all kinetic, physical, and material reality.

Subsequently, the theory of logical evolution indicates that by the carryover process found in logical evolution, every subsequently evolved reality will have within its received potential energy the capacity of free inner self-initiation. However, since the Almighty's potential energy and logistic are at the foundation of all finite individual's logistics, it is possible that the free self-initiatives of the Almighty could *internally* curtail or even totally shut down all evolved individuals' expression of their potential freedom. Some people argue that since God continually empowers our ability to perform free acts, all individual self-initiatives must be determined by God. This is a classic conundrum, and it will be resolved later in this chapter.

Notice that I deliberately avoid using the expression "free will." The notion of "will" is associated with exertion of a human's potential energy toward a consciously known and desired goal. In early physical systems, however, as in atoms, trees, and human infants, scientists observe many self-initiated, free actions that by their statistical distributions are clearly not deliberate but random. A simple example of unconscious self-initiated acts is found within one's physical body while asleep. One's sleeping body is unconsciously aware of the things happening in one's environment, and yet one's body does not consciously respond to them. The heart will change its rate as the temperature in the room changes. The lungs will change their breathing pattern if one's supply of oxygen changes. In these ways, IPhil recognizes that self-initiatives and freedom can originate spontaneously and randomly from the potential energy present in early material individuals. Nonetheless, with the progressive emergence of self-consciousness, actions *can* become deliberately self-initiated. Only then can one's self-initiatives be called acts of "free will." In a later chapter I will say more about the advancement from unconscious randomness to conscious deliberateness in matured species and humans.

Who Made God?

A student asked me: "If God made everything, who made God?" My answer surprised him: "If God made everything, then *God* made God." His head jerked back, and he looked at me askance. I continued: "Let's think about this a little bit. You have been standing in front of me for a few minutes. Who immediately made you as you are?" He answered with doubt in his voice, "God?" I shook my head and asked, "Is there any real evidence that God, your parents, or your teachers gave you the same body *now* that you had just a few minutes ago?" He shook his head, but it was clear he was still confused. Different people and things contributed originally to making your body as a baby, but once your body was formed, it was really *you* who have kept it going and growing. It was the energy/matter you took in as food that pushed your consuming body to grow and be as it is now. However, as you get older, the operations of your body become increasingly difficult and someday both you and I will die. Why is that? Well, all humans are finite; we all have a beginning, and we will all have an end one day. Then everything in our bodies will dissipate into the universe for others to use in their growth process.

In contrast, God has absolute potential energy now. Because that potential energy is forward leaning, and God's interrelational logistic is absolute, God gives absolutely all that energy to himself to establish the next moment of God's energetic existence—absolutely. As a result, the Almighty in the next moment will be energetically the same as in the previous moment. Consequently, God will remain absolute with the same interrelational potential energy as before. This is the principle of conservation of energy as it applies to the Almighty. Total energy is conservative, and likewise, the Primal Absolute Potential Energy of the Almighty is conservative and therefore eternal. In this way, the Almighty as PAPE is constantly being re-actualized in an absolute self-loving way. In this way, it can be said that God is energetically constantly making himself anew. Yet it is the same God because personal identity comes from one's enduring logistic through time, just as you retain the same logistical personal identity while you change materially through time. Thus, the Almighty energetically remains logistically ever the same Almighty.

II. Primal Absolute Energetic Reality (PAER)

In the above discussion about PAPE, energy is in a potential state, *and* it is leaning forward and moving into a subsequent state of potential absolute energy. Still, the logistic of energy is oriented toward becoming more interrelational. The 2nd phase of energetic development is toward the intrarelational formation of one, formed reality, PAER.

Because of the formally distinct stages of the logistic of energy, the logistic of the diffused Primal Absolute Potential Energy of PAPE is oriented toward becoming more interrelational into the intrarelational phase of the integrated, individualized Almighty PAER. Logistically advancing intrarelationally, there must be some intercommunicating knowledge between all diffuse energetic parts of PAPE. The intrarelational character of knowledge in the integrated One, forms in PAER an integrating knowledge center. Only with PAPE's advancement into an intrarelated, energetic One can this integral absolute energetic reality be called the Almighty.

The establishment of PAER is subsequent, and its absolute interrelational characteristics are carried over from the primal energetic characteristics and logistics of PAPE. Unlike the disorderly tyrannical gods of ancient times modeled after human tyrants, who often acted with impunity in inordinate, self-centered self-initiatives, the well-ordered, interrelational logistic of PAPE is carried over and advances PAER in a well-ordered intrarelational way, expressive of the interrelationally well-ordered logistic of energy. In this way, the well-ordered primal logistic of potential energy guides PAER always to self-initiate in an orderly, reciprocal, and just way. The carried-over logistic of energy establishes the eternal law that Greek philosophers intuitively knew when they condemned the disordered actions by the poet's powerful, mythical gods. In a similar way, Jewish prophets intuitively knew that the Lord's physical punishments for human disordered behavior was metaphysically, spiritually right. Here, IPhil provides an explanation why the Almighty, as PAER, was guided and pushed toward right, reciprocal, progressive behavior, through the carryover, reciprocal logistic of interrelational energy received in a developmental, evolutionary way from PAPE. It is the absolute interrelational push of the received

Primal Absolute Potential Energy that keeps PAER from self-initiating anything less than what is absolutely interrelational. How then is PAER free? In PAPE there is the parameter of time, and absolute freedom received from PAPE in the absolute order can be self-initiated as to *when* absolute self-initiatives are taken. How finite self-initiatives can be freely made will be explained later.

Different religions give different names to this singular, primal, intrarelated, absolute, and powerfully energetic One. The most common term for PAER is "God," but other religious and philosophic traditions call it: the Good, the Lord, the Way, Brahman, the Absolute, the Almighty, the Greater Spirit, etc. Each of these titles is associated with a different religious or philosophic worldview. Traditional Deists conceive of God as an all-powerful reality, an Almighty, which started creation moving. Modern scientists maintain there is a primal, integrated algorithm. Its aspects are expressed as equations, which are logically consistent in their integration. However, mathematical equations do not produce physical realities. It is the interrelational *energy* of the variables of those equations which give them fire, spirit, and life.

The logical evolutionary process carries logistical characteristics from the antecedent to the advanced subsequent system in a subordinated way. How can the Almighty be intrarelationally one as PAER and simultaneously energetically diffused like PAPE? Well, look at your own body. All the parts are intrarelationally organized to establish you as an integral, holistic individual. Yet, the atoms that form the molecules of your body have parts that are going in different directions within their individual logistics in ways that are indeterminate and energetically diffused. IPhil claims that by the carryover character of logical evolution, the 2nd level, formed Almighty must retain the diffusion of 1st level potentialities if the Almighty as an integral reality is to be potentially free.

Consequently, intrarelational PAER is not logically deduced from but logically evolved from existential PAPE, where the logistic of PAER is formally distinct from and intrarelationally more advanced then PAPE, *and* the logistic of PAPE is carried over and endures as the existential logistical foundation of PAER.

III. Primal Absolute Interrelational System (PAIS)

The study of energy in thermal devises in our physical world shows that energy logistically can form and support enduring interrelational systems. In the interrelational logistic of potential energy, symmetry in the exchange of energy is possible. That symmetry makes the system reciprocal, logically consistent, conservative, and enduring. Optimally, the amount of energy received *can* be equal to the amount of energy given. However, in our partial and finite world, the amount returned can and usually is less than optimal. Partial exchanges display the need for continued input of energy to keep such systems running, lest they run out of interrelational energy, stop, and die. Such inequities of exchanges in finite systems display the 2nd law of thermodynamics.

However, what happened in the singular PAPE and PAER before the big bang? If the interrelational logistic of the Almighty is absolute, it can only operate in an all-or-nothing, absolute way. Although unlikely, the 2nd law of thermodynamics admits that such an absolutely conservative energetic system is logically *possible*. IPhil considers this possibility seriously. This is our logical evolutionary starting point.

In PAPE, absolute energy is communicated from one existential state of absolute energy to a subsequent existential state of absolute energy. This energetic possibility gives PAPE not only a progressive character through time, it also offers the possibility of an equal, total transfer of absolute energy from one temporal moment to the next. This symmetry establishes conservation of energy in transfers between subsequent, different existential states of intrarelational PAER through time.

If absolute energy is involved in the *intrarelational* formation of PAER as an absolute One, how can there be any energy available to form an *interrelational* absolute system? The answer to this question is found in the logically evolved 3rd level notion of *ownership*, and in the "breaking" of the closed logistical boundary of individual ownership.

IPhil recognizes that in the interrelational logistic, it is possible for the holder of absolute energy to communicate ownership of available interrelational energy to a virtual receiver logically distinct from itself. It is then possible for that energized receiver to reciprocate and return ownership of that absolute energy to its original holder of that absolute energy. This exchange of ownership of absolute potential energy between an all-giver and all-receiver establishes what IPhil calls the Primal Absolute Interrelational System (PAIS). This reciprocal metaphysical change of ownership takes place in time but without any lateral or physical movement. When that reciprocity occurs on an absolute level, there is established an absolute feedback system. That absolute reciprocal system could be ongoing and perpetual in an absolute way. However, because of the freedom in each energized giver, that exchange of ownership does not *have* to happen this way. Either the giver or the receiver could continue to hold on to that absolute energy. However, the absolute potential energy involved in that exchange pushes the energized receivers to be more interrelational. The progressive Spirit of that absolute potential energy pushes each energized individual toward acting act righteously according to the interrelational logistic of absolute energy in as shortest time interval as possible. Optimally, the quantized character of time indicates that the potentiality of interrelational exchanges of absolute energy would push for the absolute exchanges of potential energy to take just one quantum of time in each phase of the interrelational logistic of PAIS. If the exchange were less than absolute, PAIS would not be eternally enduring absolutely. Rather, the internal push in that absolute energy is toward being more relational, pushing interrelational exchanges in four time-quanta.

Consequently, interrelational PAIS is not logically deduced from but logically evolved from integrated PAER, where the logistic of PAIS is formally distinct from and interrelationally more advanced than PAER, *and* the nested logistics of PAPE and PAER are carried over and endure as the logistical foundation of PAIS.

In PAIS, the communication of the absolute potential energy from the all-giver to the all-receiver and from the all-receiver-giver back to the all-giver-receiver is logistically symmetrical in operation. Since the all-communicator from the all-giver to the all-receiver and back is not a static term but a transitional agent, the energizing all-communicator is like the wind (*ruah in Hebrew*) and can be called the all-communicating Spirit. Because the transfers of all-energy from the all-receiver to the all-giver and back are both absolute, the all-communication Spirit going from the all-giver to the all-receiver will be interrelationally identical to the all-communicating Spirit going from the all-receiver-giver to the all-giver-receiver. Thus, in PAIS's four-sided exchange of absolute energy between the two absolute dialogical terms via the all-communicator, there really are only *three* formally distinct absolute dynamic participants, namely two absolute interrelational dialogical terms and one two-directional absolute communicator. Christian theologians call this four-phase, three-participant Primal Absolute Interrelational System (PAIS) the Trinity.

Trinitarian Concerns

This leads to a major conundrum: How can it be said that there are three distinct absolutes in PAIS? To say that three distinct things are simultaneously absolute is logically impossible, for how can each of the three individuals possess all and still be energetically distinct? This logical contradiction is the main reason why many rational people, like Newton, have not believed in the Trinity. Because they are thinking in a materialistic, individualistic, independent, static, physical, 2nd level ways about the Trinity, many thinkers and scientists reject the notion of an absolute Trinity. They recognize that three distinct individuals cannot simultaneously possess absolute or infinite energy and remain formally distinct.

IPhil's model of the Trinity, however, is interactive, requiring absolute, 3rd level, mutually cooperative, and interdependent progressive exchanges. In IPhil's model of the Trinity, these three dynamic parties successively have and then give ownership of absolute energy to the other interactive agents of PAIS in succession. All three participants are always logistically and potentially Almighty, *and* only one interactive agent actually possesses absolute potential energy at a time. That is, the interrelational logistic of IPhil gets around the apparent contradictory impossibility of multiple absolutes by recognizing that the logistic of PAIS is not static but dynamic. Each participant is always potentially capable of acting in an absolute interrelational way, while *only one* interrelational participant term has actual ownership of that absolute energy at a time. This is much as the participants in the game of catch, where only one player has full possession of the ball at any given time. The Eastern Orthodox theologians hold a similarly dynamic model of the Trinity, calling this mode of operation *spiration*. That which is exchanged is the system's 3rd level interrelational energy, while each participant retains its 1st level energized existence and 2nd level intrarelationally logistical identity.

PAPE's model of God is not static but has the parameter of time. Sequentially in each phase of interrelating, only one interrelational party *actually* "owns" or possesses absolute interrelational energy at a time—while all three interactive parties logistically possess the *potential* to possess the absolute energy during *all* stages of the absolute exchange. Throughout the four phases of the exchange, each of the three interrelational dynamic participants in the absolute exchange receives, has, orients, and gives absolute energy at different times, and they do so in a progressive manner, as in the game of catch. This is like a musical round, in which each individual begins his or her round at a different time. It is through the sequence of surrendering ownership that only one participant actually possesses absolute energy at a time, while all retain their absolute logistic through all phases. It's not either/or but both/and.

This requires that in each phase the interactive parties must "let go" of their actual possession of that energy for the sake of empowering the other participants absolutely. This is similar to ownership of the car described above. The parents must actually "let go" of their ownership of that car—even as they are remembered as the previous owner and continue to have the metaphysical potential of receiving ownership of that car back from the youth in the future. The distinction of actual ownership from potential ownership is very important in understanding how there can be three potentially absolute participants in PAIS, while only one participant is the current, actual holder of that absolute energy at any given time. Unlike a static, essentialistic 2nd level idealization in which only one party continues to hold on to that absolute energy, the progressive, 3rd level, systemic exchange continues because each dynamic participant receives and then gives Absolute energy in sequence for the sake of the absolute system/game. While all participants *can possess* absolute energy throughout the exchange, only one dynamic participant *actually possesses* absolute energy at a time.

In ancient times, like possessive emperors, God was thought always to possess everything, even when those possession were "given" to his subjects. The understanding was that even after the 2nd level king "gave" something to a subject to use, it was assumed that those things really remained under the dominion of the king. The subject was only a "steward" of those gifts, which could be reclaimed by the king at any time. Consequently, everything and everyone was always considered to be an intrarelational possession of the king, who in his graciousness had allowed his subjects to tend and use them in the name of their lord. In contrast, 3rd level IPhil recognizes that a progressive exchange of ownership can take place in PAIS when the participants successively "let go" of actual possession of the Primal Absolute Potential Energy (PAPE) so that each recipient really has full ownership of the gift—for a time.

Augustine devoted an entire book, *De Trinitate*, to this topic. He showed how the Trinity was present and active in many things.[7] The first illustration he gave of the Trinity came from the interrelational character of love. He described how in the absolute love process, there is an absolute Lover, an absolute Beloved, and the absolute Love exchanged between them. The Lover gives the Beloved all Love. The Beloved receives that total Love and returns that total Love back to the Lover—who is now the Beloved of the Beloved-now-Lover. His illustration is very similar to PAIS. However, Augustine did not take this illustration as the prototype of the Trinity, probably because the Spirit of Love in the illustration appears to be an unsubstantial, transient, communicating reality, in contrast to the Lover and Beloved terms in that Love exchange who are well-formed and stable. Since the three participants of this illustration were not of the same form, for Augustine this inequality of forms made this model of the Trinity deficient. In modern times, however, we know that matter and energy are but different forms of the same thing. By Einstein's famous equation, $E = mc^2$, we know that static matter and moving waves may not be of the same form, but they arise from and can possess in different ways the same potential energy.

This brings up an important distinction. Unlike most finite systems in which loving individuals are independent and can walk away from such an interrelationship, in PAIS, according to the interrelational logistic of absolute energy, the participants are necessarily interdependent. Each must always be functioning in a phase-dependent way in coordinated cooperation with the other participants. They are inherently interdependent because logically there cannot be an all-communicator or an all-receiver unless there is an all-giver, and vice versa.

In PAIS, each participant must play its part and must "let go" of its absolute energy to another in its proper phase. Metaphorically, it can be said that in "letting go" of all interrelational energy, each individual participant in the Trinitarian (PAIS) needs to "die" to its *actual possession* of transferrable energy in order that the other participants can "live" in an absolute way. The word "sacrifice" comes from *sacra-facere* = holy-making or whole-making. A gift of anything to another for the sake of the improvement of the life of another may be considered a beneficial sacrifice. In PAIS, each participant's "letting go" of absolute energy in its transfer to another participant is the first example of "sacrifice," where the "letting go" of absolute energy is a "dying to self-possession," which here produces something transcendentally whole and holy. Similarly, human lovers say to their beloveds: "I give you my all." To give a part of oneself is to give from one's surplus. IPhil recognizes that committing "one's all" to one's beloved is Godlike and analogously divine.

IPhil recognizes from PAIS that "giving one's all" refers to what is *interrelationally* possible on the 3rd level. IPhil recognizes that in "total" sacrifice, it is all one's *interrelational energy* that is transferred to another party. Still, the giver progressively needs to retain her 1st level existential energy and her 2nd level individual intrarelational

---

7. Augustine, *On the Trinity*, bk. I, ch. 10.

energy so that she can become a potential giver-receiver and a loving exchange is enduring. For this reason, the participant's *active*, energetic "letting go" of one's 3rd level interrelational energy cannot be suicidal on the 1st and 2nd levels of energy. In war memorials, it is realized that the spirit or logistical soul of the soldier is still existing (1st level) and is individually (2nd level) related to the (3rd level) community for (4th level) giving the ultimate gift for a worthy cause. (Suicide is wrong when a person wishes to eliminate one's individual 1st level existence and 2nd level individuality, which is contrary to the logistic of one's received potential energy.)

IPhil finds in PAIS the origin of the greatest commandment: "You shall love the LORD your God with all your heart, with all your soul, with all your mind, and with all your strength." Before the LORD directs us to love God totally in the finite order, the Almighty fulfills that command in the absolute order. Because of its lofty origin, it is very difficult for us to fulfill the greatest commandment, but our potential energy pushes us to do so. The second greatest commandment is the Golden Rule: "You shall love your neighbor as yourself." This directive also was first displayed by the equitable exchanges found in the reciprocal behavior of the participants in the interrelational logistic of PAIS. That is, the second greatest commandment is an advanced, interrelational, covenantal, 3rd level expression of the 2nd level greatest commandment. In this way, as the Almighty has already done to an absolute degree in PAIS, we are asked to act reciprocally toward the Almighty and toward others, analogously according to our received energy, logistic, and abilities.

Arius's Objection Resolved

Arius (250–336 CE) was a Christian priest in Alexandria, Egypt, who emphasized the Father's divinity over the Son and thereby opposed Trinitarian Christology.[8] Christians applied to Jesus this line: "The LORD said to me, 'You are my son; this day I have begotten you.'"[9] Arius reasoned that if Jesus was begotten by God and is God's son, then he came into existence *after* God. That is, "there was a time when he was not." From this reasoning, Arius claimed that Jesus could not be God but only a superior human. This position became very popular. At one time there were more Arian Christians than there were Trinitarian Christians. Mohammed and his Islamic followers object to the doctrine of the Trinity for similar, physically based reasons.

Seeking to establish peace and order within the Roman Empire, Emperor Constantine in 325 CE called the Christian bishops together to settle all disputes among Christians. Arius's theological position was a prime topic. The First Council of Nicaea produced the first draft of the Nicene Creed, which affirmed that Jesus was indeed divine. However, the bishops did not establish their position rationally. Biblical texts could be interpreted either way. Statements in the Nicene Creed were imposed

8. Williams, *Arius*, 98.
9. Ps 2:7.

executively with the emperor's blessings.[10] An amended form of the Nicene Creed is professed in all Orthodox, Roman, and Christian churches today.

The reason I bring up this topic is because IPhil's model of Trinitarian PAIS has the same problem. However, IPhil can provide a solid philosophic argument to resolve this dispute. The key to resolving both Arius's problem and IPhil's conundrum is found in the dual aspects of the absolute energy. IPhil recognizes that absolute energy can exist in two states—as an essentialistic metaphysical possibility and as a progressive physical actuality. Since the metaphysical order deals with possibilities, and the physical order deals with actualities and changes, the metaphysical order takes precedence, just as possibilities and potential energy always precede expressions of kinetic energy and subsequent actualities. In the interrelational logistic of PAIS, when the initial "all-giver" is called the "Father," because of the logistic of absolute energy exchanges, there *simultaneously* exists an "all-receiver" or "Son." On the other hand, IPhil recognizes that *time* exists in the movement of potential energy from a state of metaphysical potency to a physical state of act. Consequently, as Arias said, time must pass for the subsequent transformation of the "virtual receiver" into an energetically actuated all-receiver or Son. Note, however, that all three absolute agents in Trinitarian PAIS existed metaphysically and logistically before the actual exchange of absolute energy in Trinitarian PAIS. Furthermore, because the metaphysical order takes priority over the physical order in the domain of possible existence and knowledge, priority in the creed rightly is given to the metaphysical, logistical order to the subsequent physical order. It is not a matter of either/or but both/and—with simultaneous potentiality preceding subsequent activation.

Freedom Advanced

Earlier, I established that in the 2nd stage of logistical development of absolute potential energy, PAER is absolute, and there were no other energized individuals. Consequently, the self-initiative of PAER's potential energy would not be influenced by another and therefore would be truly free of all outside influences. However, by the logical evolutionary process, in PAIS, it is possible that there *can* be another individual that is a receiver of absolute interrelational energy that is inversely not the giver of that absolute potential energy. Here, if the all-giver retained any ownership or control of the energy interrelationally given to the all-receiver in PAIS, then the all-receiver's ability to self-initiate could not be truly reciprocal toward the all-giver-potential-receiver. Consequently, for logical consistency and enduring reciprocity, in the absolute interrelational logistic, the all-giver totally "lets go" of all interrelational energy to the all-receiver. In this way, the freedom received by the all-receiver must symmetrically also be totally free. The all-giver's act of total sacrifice of interrelational

---

10. Schaff and Wace, *Select Library of the Nicene and Post-Nicene Fathers*.

energy, including its capacity to act in total freedom, enables the all-receiver to become the totally free all-receiver-giver.

In "letting go" of the ownership of that absolute energy, the all-giver "lets go" of its *control* of the self-initiative inherent in the all-receiver-giver's energy. In this way, the all-receiver is empowered to self-initiate with the same total freedom that the all-receiver initially had. Furthermore, by logical evolution into the finite order, that surrender of individual control by the all-giver to the all-receiver in PAIS would be carried over into the evolutionarily establishment of all subsequent finite receivers. That is, by logical evolution, the total relinquishing of freedom to the all-receiver would be carried over to all finite receiver-givers, whether they are minerals, animals, or humans.

Logistical Entanglement

In the field of physics, quantum entanglement is a phenomenon that does not make sense physically.[11] Quantum entanglement was first identified in a 1935 paper by Einstein, Podolsky, and Rosen. Einstein described this as "spooky action at a distance." It is commonly known as the EPR paradox. Here is an example of quantum entanglement.

In the reciprocity of energetic exchanges, when a physical measurement recognizes that one member of a mutually generated pair has a definite value (e.g., clockwise spin), by conservation of energy, the other generated member of the entangled pair is always found to have the inverse value (counterclockwise spin). Equations describing systems indicate that when the spin of one member is reversed (that is, made to go counterclockwise from going clockwise), the spin of the other entangled member of the pair will *instantaneously* likewise reverse (changing from going counter clockwise to going clockwise)—*even if these particles are a large distance apart.*

The physical problem here is this: How can this type of change happen at a distance instantaneously? Many experiments have verified the existence of this phenomenon. All other physical changes occur through some force particle traveling at the speed of light or less. How can the changes of the parameters of an entangled, separated pair of particles occur *instantaneously*? Let me describe a physical experiment that shows quantum entanglement

When an electron in an atom is stimulated by some external energy, by the conservation laws of physics, that electron will send off not one but two photons in opposite directions at the speed of light, c. Photons have an internal spin which can be influenced by a magnetic field. The spin of photon A going in one direction will have the opposite spin of photon B going in the opposite direction. That these two photons have opposite spins can be experimentally verified. Now, let's change the experiment slightly. As photon A travels away from the original electron, it can be made to pass through a magnetic field which flips the direction of the spin of photon A. At this

11. Francis, "Quantum Entanglement."

point, a person would expect both photons to have the same internal spin. However, experimental measurement of the spin of photon B going in the opposite direction shows that the spin of photon B has *also* flipped—thereby maintaining the conservation of spin in that experiment. Here is the issue: How did photon B know that it should flip its spin orientation?

Look at the mechanics of the experiment. Since the flipping of photon A took place at a particular distance from the original electron, and if the information of the flip of photon A travels at the maximum speed of c, while photon B is also traveling at the speed of light in the opposite direction, information about the flipping of the spin of photon A could never catch up to photon B, signaling it to flip. There is no physical way for photon B to know that the spin of photon A has flipped. Yet experimental evidence shows that it does. The entanglement of linked pairs of particles has been observed over large distances. How can this be?

IPhil explains this phenomenon by recognizing that the phenomenon of entanglement first arises in the Primal Absolute Interrelational System (PAIS) *before* the evolution of space and *before* the establishment of the parameter of distance. Consequently, a change due to pair-entanglement is not communicated through space by means of a communication-particle traveling through space, because the phenomenon of entanglement first existed in the absolute logistic before the logical evolution of space and the parameter of distance. In PAIS, the moment when PAER self-initiates an interrelational exchange with a virtual receiver, the dialogical operations of the all-giver and all-receiver are established metaphysically, and they are reciprocally locked together—logistically. That is, reciprocal entanglement is a *metaphysical* aspect of the interactive logistic of energy before any energy is *physically* transferred via an energy-communicator in space. There cannot be a giver without a receiver—virtually or real—just as there cannot be a receiver without a giver. Change through entanglement is a metaphysical, directional, *qualitative* relational change; it is not a physical, *quantitative*, energetic change.

Similarly, when the all-receiver receives that potential energy and is transformed from an all-receiver into an all-receiver-potential-giver, the original all-giver *immediately, logistically* is transformed metaphysically into an all-giver-potential-receiver via the interrelational *logistic* of potential energy and not via an *energy communicator*. In entanglement the change is not caused by the transfer of *energy via a communicator* but by *information via their shared metaphysical logistic*. In atomic entanglement changes, the *energy* of those entangled particles does not change. Rather, what changes is the dialogical *sign* of those entangled particles. Thus, the phenomenon of entanglement is not *efficiently caused* by the transfer of energy but *formally caused* by the transfer of information within their shared, *pre-spatial* interrelational logistic.

Entanglement is also a part of the Dao, which expresses itself as the dialogical yin/yang, which are simultaneously driven forward cyclically by the inner dynamism of the *qi* of the Dao. Note that the yin begins to appear at the height of the yang phase of process, and immediately there is established the beginning yang within the main yin phase of that process. This occurs without any physical communication, only logistical connection between them. Daoism recognizes that this asymmetrical, cyclic process existed before the material, spatial creation of our finite world. In this way, IPhil is able to explain *why* that is the Way of the universe.

Unfortunately, many vocal scientists are materialists and believe that all changes occur through efficient causality and *only* through efficient causality. Entanglement is a mystery to them because it breaks the closed system of their materialistic thought. Philosophic analysis of entanglement shows that over and above efficient causality, the observed changes that occur in entanglement are caused by formal causality. Entanglement, thus, demonstrates the formal reality of the logistical, metaphysical order—in addition to the materialistic laws of the physical order.

Because experimental verification of knowledge of entanglement over a physical distance cannot be explained by a kinetic, physical expression of energy, entanglement provides evidence of the real existence of a logistical causality and the metaphysical order. IPhil recognizes that the non-spatial phenomenon of entanglement also provides evidence of the existence of a pre-spatial interrelational system, existing metaphysically before the physical big bang. Then, after the big bang, this logistical phenomenon by logical evolution is found not only in electromagnetic systems but in all logically evolved systems, even in the human order.

When I was a teenager, one night the mother in the neighboring house suddenly sat up from a sound sleep and screamed, "My son has been shot." He was a soldier in Vietnam on the other side of the world. Later, it was verified that her son was indeed shot in Vietnam at exactly the same time his mother in Marshfield, Wisconsin, awoke with the knowledge of her son being shot. Identical twins regularly report knowing if something significant happens in the life of their identical twin. It would seem that logistical entanglement can be experienced by people who have an intense interrelationship with others who logistically are "on the same wavelength."

IV. The Absolute as Panentheistic

Because of the perfect reciprocity of the absolute terms of the Primal Absolute Interrelational System (PAIS), this is the first *logically consistent* system. Furthermore, the push of this system's potential energy to be *more interrelational* is logistically possible such that logical and energetic evolution of this logistic absolute system can advance into the logistic of the first non-absolute system.

Yet, the expanding characteristic of evolving finite mathematical and physical systems presents a problem when we apply the logical evolution theory to the Primal Absolute Interrelational System (PAIS). Being absolute, it seems that the Primal Absolute Interrelational System (PAIS) cannot become any larger. So how can an absolute interrelational system evolve? Not outwardly but inwardly, as in the intrarelational

2nd phase of the interrelational logistic of energy. Furthermore, the logical inverse of absolute sharing is not-absolute, that is, finite. However, how can a new finite term be generated in a system in which all logistical operations are absolute? The answer is found in making the new non-absolute, receiving progenitor energetically "in" and partially "of" the all-receiver of PAIS. When that is the case, all evolved finite terms and systems in IPhil can properly be described as "pan*en*theistic" in the Almighty. The "*en*" in the word *panentheistic* is very important because evolved finite receiving elements and systems are partially "in" the Almighty energetically, especially "in" the all-receiver. In addition, by the carryover character of logical evolution, the interrelational logistic of the Almighty is foundationally "of" the more interrelational, evolved logistic of each finite receiver.

Panentheism is different from pantheism insofar as pantheism recognizes that all things are absolute or divine. Since God is absolute, rationalists commonly assume that all things are or can be directly, deductively generated from God. This means they would have the same interrelational nature as God, and they are thereby in some way be divine. In contrast, because logically evolved systems are generated via a logically inverse progenitor, elements within a logically evolved system are not logically deduced from the antecedent system, so they are not pantheistic. Rather, since the logical inverse of the progenitor is not-absolute but finite, a finite system that is logically evolved from the absolute system would be "in" PAIS and therefore panentheistic.

The first finite receiver in the all-receiver in PAIS is not logically deduced from but logically evolved from the all-receiver in PAIS, where the logistic of the first finite receiver is formally distinct from and interrelationally more advanced from PAIS, *and* the nested logistics of PAPE, PAER, and the all-receiver in PAIS are carried over as the logistical foundation of all finite receivers in our evolving finite world.

Nothing before Something

Gottfried Leibnitz was one of the first individuals to ask: Why is there something rather than nothing?[12] In cosmology, theoretical scientists now recognize that the initial state of reality after the big bang was raw, unformed energy. In time that energy formed into nuclear particles, like quarks, electrons, protons, neutrons, neutrinos, etc. It is from these particles that all macroscopic things developed. Looking back at the potential energy before the big bang, the metaphysics of IPhil recognizes that the 1st existential stage of reality is one of diffused Primal Absolute Potential Energy (PAPE). It is in potential energy's intrarelational 2nd stage that energy became distinct, integrated, and individualized. Energy is initially not a thing. By logical evolution, all material things emerge from potential energy and from the absolute energy of the Almighty, as in $M = E / c^2$.

---

12. Copleston, *History of Philosophy*.

Before energy's 2nd stage intrarelational formation of substantial things, the 1st stage of the Primal Absolute Potential Energy of PAPE did not yet have an integrated form, and it was "no-thing." Only in the 2nd stage of the logistical development of interrelational energy does the intrarelational phase of energy produce individual "somethings," which are capable of being externally observed and interactive in their 3rd stage of development in reciprocal energetic systems. This explains how and why integral 2nd level "somethings" came from 1st level potentially energetic "no-things." Individuals who are materialistic are prone to call potential energy a "thing." However, the fourfold logistic of energy explains why that is wrong.

Why is there energy? The logistic of energy is cyclic, making it self-sustaining, conservative, without beginning or end—while space and matter both have a temporal beginning and ending. Youngsters can sometimes rationally drive parents crazy by keep asking to every explanation given by a parent, "Why? Why? Why?" In the mental, metaphysical order there is no final answer to that repeated question. However, by logical evolution, the correct, parent answer is, "That is the way it is." Primal energy and knowledge is simply 1st level existential; it is antecedent to all 2nd level conceptual ideas and 3rd level rational explanations.

IPhil basically agrees with the intuited ancient myth that the 1st state of reality was that of chaos. This chaos consisted of a myriad of energetic elements interacting in random ways. However, we now know that being energetic, chaos is oriented toward becoming interrelationally ordered—freely capable of energetically establishing intrarelational individuals and then interrelational systems, which are capable of evolving other things on logically distinct levels of existence.

Related to this is another question: Why did God create the world? Medieval scholastics quoted the scriptural statement "God is Love," noting that "Love tends to go outside of itself." But why is God love? IPhil recognizes that the ur-stuff of reality is Primal Absolute Potential Energy (PAPE), which pushes potential energy to become more interrelational. Being free, PAER, or the Lord, pushes to be more interrelational. Potential energy inherently pushes it to be progressively more interrelational and more lovingly. This is how IPhil explains why the Lord as PAIS can be systemically called Love.

The potential character of energy, the dissociative aspect of interrelationships, the evolution of logically consistent systems through progenitors, and self-initiative freedom are all means by which the energetic Almighty is able to increase the interrelational joy and glory of all in All. This is IPhil's explanation of why and how our finite, physical world energetically came from the Almighty, why logical evolution is panentheistically in God, and why the logistic of the Almighty is the foundation of all the logistics of what energetically has logically evolved in our world.

Albert Einstein wrote: "The most incomprehensible thing about the universe is that it is comprehensible." Contrariwise, IPhil observes that our world is energetic. Energy is interrelational and systemic. Thus, interrelational terms must have at least unconscious knowledge of their systems. Evolution naturally arises from enduring and logically consistent systems. Therefore, the evolved systems in our world increasingly know their world. That is why humans are able to comprehend increasingly our energized universe. Thus, the most *awesome* thing about the universe of which we are a part is that it is comprehensible by humans.

A classical paradox says the only thing that is constant in our world is change. How, then, without logical contradiction, can that there be *constancy* in the midst of *universal change* in our world? This paradox comes from 2nd level, antithetical thinking, which pits the concept of absolute change against absolute constancy. IPhil provides an easy 3rd level both/and explanation of this paradox. Everything that exists in our world is energetic, and energy exists in two states: potential and kinetic. On every level of existence, the potential side of energy has an *unchanging* 3rd level interrelational logistic, *and* the 4th aspect of that *unchanging* logistic of potential energy is its capacity to freely self-initiate kinetic *changes*. The answer is not 2nd level dualistic in an either/or way but 3rd level dialogical in a both/and way.

World Religions' Models of God & IPhil

A survey of the religions of our world shows that they have different religious descriptions of God. From IPhil's perspective, these easily fit into the four formally different categories found in the fourfold logistic of absolute potential energy.

1. *God is Spirit*. Primitive conceptions of the spirit world arose from observations of surprising, significant, and mysteriously influential things in local nature, as found in folk religions, like the Lakota religion, Wicca, and New Age. Spiritualities concerning Gaia express non-institutionalized awareness of the sacredness of all things. There is an awareness not only of the spiritual powers of the finite things in our world but also of an intuitive awareness of a single spiritual power that is overall. However, this overarching spiritual absolute is not intellectually well grasped. Experience shows that this overarching Spirit *could* influence one's daily life—but not frequently. Recognition of this universal spiritual reality is intuited, vague, traditional, and only rarely experienced. But when it is, it is surprising and awesome. This vague, potential, spiritually great reality is analogous to nebulous, energetic, almighty PAPE.

2. *God is One*. Judaism and Islam view God in an imperial way. They identify God analogously with a humanlike lords, judges, leaders, and kings. Their monotheistic, essentialist view of the world staunchly maintains that God is One, that God has a special relationship with a chosen people, and God is known through the experiences of local seers and prophets. This autocratic conception of the Almighty is like the intrarelational individuality of the Almighty found in PAER.

3. *God is Triune.* How can there be three interdependent absolute agents in one God? It can happen in the same way that there are many interdependent parts in the distinct organs in one human body. The interdependent human parts in our human bodies are human. There are three interdependent Absolute interactive systemic elements internal to the corporate Almighty. The interdependent divine/absolute parts have the same interrelational logistic as the intrarelated, absolute One. As the finite interdependent parts in the human body can interact independently in their own way within the logistic of the whole, so too there are interdependent yet free absolute dialogical participants in the Primal Absolute Interrelational System (PAIS), cooperatively establishing the interrelated Triune God. Hinduism and Christianity recognize God as individually triune and still communally One.

While many may consider these religions as polytheistic, they advance a logical, evolutionary, integrated view of God. Beyond the static, 2nd level, unchanging view of God as totally intrarelational and self-centered, the 3rd level interactive view of the Almighty is dynamic and interrelational. Because of the interrelational character of the logistic of energy, the individually independent and systemically interdependent participants in the Almighty can act interrelationally at different times in different ways for the spiritual advancement of our world. Their Trinitarian models of God are analogous to IPhil's description of the triune Primal Absolute Interrelational System (PAIS). Within this absolute logically consistent, interdependent logistic, our finite world logically evolved in their image.

4. *God is ineffable.* Buddhism, Daoism, and apophatic ascetics in the West focus on the prime absolute reality as transcendently unknowable. The writings of Pseudo-Dionysus and the *Cloud of Unknowing* first introduced this view of God to Western Christianity.[13] Their *via negativa* focuses inversely upon the ways God is not, and they shy away from making any rational descriptions or judgments about God. They "let go" of every imaginary or rational conception of the absolute, and only hold in faith a mystical recognition of the foundation of everything as existing beyond the limitations of the human mind. This contemplative tradition focuses on the transcendental, ineffable character of the Almighty as found in the final stage of total self-surrendering devotion to the absolute Almighty.

IPhil recognizes that getting to know anything is a fourfold process. All processes are logically evolutionary insofar as each stage retains the logistic of the antecedent step as its foundation, and in freed way each succeeding phase has an integrity of its own, while potentially oriented toward a more interrelational step unto the completion and closure of the process. In getting to know a *horse*, a person first senses an existing horse and unifies all these sensations into a sensual image. Then seeing many

---

13. LeClercq and Luibheid, *Pseudo-Dionysius*.

similar horses, the knower abstracts what is common in all images and comes up with a static metaphysical concept and essentialistic definition of a horse that endures through time and space. Yet, because that abstraction process deletes much information about that horse and all horses, the individual is not able to have a holistically true knowledge of both this horse and of all other horses. That is, an exterior individual's knowledge of this and all horses is incomplete because there are evolutionary free self-initiatives by horses that are not logistically predictable. In a similar way, knowledge of the Almighty analogously will be fourfold. It is philosophically inadequate and erroneous to claim full knowledge of God or anything by referring to a 2nd level essentialistic definition of the Almighty.

Consider this: When I mention the name "Abraham Lincoln," what image comes to mind? Do you picture Abraham Lincoln as a baby, or as a scrawny pioneer, or as a struggling lawyer, or as the sixteenth president of the United States of America, or when he was shot in the Ford Theatre, or as he is sculpted in the Lincoln Memorial? Each of these different images of the same person is true—partially.

All humans grow through diverse stages and images, all fauna pass through stages, all flora pass through stages, all stars have their life cycles, and all molecules go through stages of formation, development, interaction, and decay. Through the analogous carryover character of logical evolution, IPhil recognizes that before all other finite realities, the primal analogate, the Almighty, energetically also went through the four stages of the interrelationship of energy.

We need to "let go" of the assumption that there is only one *right* model of God. God is each of them and none of them. We need to break away from the assumption there is one, true, 2nd level, essentialistic, ideal description of God or anything in our world. A fully mature, 3rd level knower, wanting comprehensive knowledge of Abraham or of the Almighty, will embrace all four developmental stages of the Almighty's energetic existence, intrarelationship, inner absolute interrelationships, and panentheistic imminence in and transcendence to all finite reality. We need to embrace lovingly the diverse, partial truthfulness of all the above models God.

Chapter 8

Logical Evolution of Space and Mathematics

> First Finite System. Quantum Distance. Maximal Speed of Energy Transfer, c. 3-Dimensional Space. Lines in Space. Number Systems. Fourfold Expansion of Space: Expansion of Space by Receiving; Inflationary Expansion; Reciprocal Systemic Expansion; Interior Expansion. Dark Energy. Why Was the Big Bang so Big?

The previous chapter established that the logical evolution of the logically consistent Primal Absolute Interrelation System (PAIS) could occur through the inverse operator "not-absolute," such that there could be a progenitor receiver that is not an absolute receiver but rather a not-absolute or finite receiver logistically *in* the all-receiver. The absolute logistic of PAIS requires that the all-receiver receive all-energy. This condition does not ban the logical possibility that there *can* be finite, not-all receivers in a new logically system, as long as those receivers are energetically *in* the all-receiver. By the carryover process of the logical evolutionary theory, this makes not the absolute energy but the interrelational logistic of the all-receiver-giver *imminent* within all evolved and generated finite receivers. Being interrelational in finite ways, their finite logistic (or nature) is logically distinct from and interrelationally beyond the *transcendent* absolute logistic (or nature) of PAIS. Also, energetically the all-receiver-giver transcends all its evolved and generated finite receiver-givers. When this happens, it can be said that the all-receiver and PAIS are *panentheistic*—where all subsequently evolved and generated finite receiver-givers *participate* in the logistics of the intrarelated Almighty.

First Finite System

In the antecedent absolute system, absolute potential energy pushes the interrelated participants of PAIS to exchange their energy totally. This makes the reciprocal exchange of the terms of PAIS symmetrical, conservative, and unending. Furthermore, being logically consistent, PAIS *can* logically evolve a finite receiver, which can be the progenitor of the first finite system.

This progenitor through its received interrelational energy can freely project and establish the second finite receiver-giver. However, communication of empowering energy from the absolute-finite, partial progenitor receiver to this second, finite, partial receiver-giver *can* also be not-total but partial. That is, the first finite receiver-giver logistically *can* retain some of its energy and freely give only part of its energy to its receiver. Likewise, the second finite receiver-giver logistically *can* freely give only part of its received energy back to the first receiver-giver, retaining part of its received energy. Consequently, the potential energy exchange in this first finite system *can* be, but need not be, total. That is, it *can* be partial. Here, the possibility of partial exchanges of energy between the first two interrelational terms of the first finite system provides the origin of the second law of thermodynamics. The more total the energy exchange is, the longer the subsequent, reciprocal system will endure. The less total the exchange is, the faster the energy of the reciprocal system will wane—unless some additional energy is inserted in that system from without.

Recall that in PAIS, because of the absolute character of its interrelational logistic, it was logistically demanded that the dialogical primal interrelational terms exchange totally the absolute amount of energy received. Nonetheless, each of the phases of that interrelational logistic had to take at least a quantum of time. The freedom of the participants of PAIS is found in the possibility that a participant can be slow in expediting the exchange of that absolute energy. Nonetheless, each participant in that Absolute logistic is pushed by its received absolute energy and interrelational logistic toward *total absolute sacrifice* of its received energy as fast as possible. In the first finite system also, it is possible for the giver and receiver-giver to transfer its received energy in a fast or slow way. In other words, in the first finite system, in the first time period, it is possible that some energy can be retained by both participants in an exchange cycle. This energy retention makes the first finite system significantly different from the antecedent absolute system. Since both dialogical terms in the first finite system *can* retain or continue to own some received energy in the first finite system, there is need for a new parameter to keep these partially energized parties apart during their interrelational process.

What is this new parameter that is necessary to keep energy-retaining finite dialogical terms distinct in the first finite interrelational system? Looking at our physical world, experience shows that the parameter that keeps physically interactive things distinct is the parameter of *distance*. Thus, by the carryover character of evolution, it can be said analogously that in the logical evolution of the first finite system, a new interrelational parameter of distance is logically necessary. Furthermore, because of the symmetrical character of the reciprocal exchange of partial energy between givers and receivers in first finite interrelational system, the new parameter of distance between two exchanging terms must be *bidirectional*. Consequently, the subsequently evolved finite system will be interrelationally more complex than the antecedent PAPE with its parameter of *unidirectional* time.

Because only some of the energy of the second receiver is received from the first receiver-giver and some of that received energy is returned back to its giver, it is possible that the second receiver-giver give use some its unused potential to a potential third receiver. That third receiver must logistically be quantumly spatially separated from the first and second receiver, producing an expanding "spatial field." I call this expanding field, started from the first finite progenitor the "proto-spatial system," and I call each receiving finite term in this energetically expanding system a "spacepoint." The energizing communicators that travels through the space between spacepoints I call "spaceons."

The progenitor finite receiver existed antecedent to the generation of a second point and the proto-spatial system. Thus, the absolute-finite progenitor of the proto-spatial system initially had no dissociative spatial dimension. Likewise, all generated new spacepoints in this spatial field of themselves do not have the parameter of distance. Only their shared interrelational proto-spatial logistic does. Knowledge of the parameter of distance comes only from the energy-communicator's going between spacepoints, much as knowledge of the parameter of time comes from an energy-communicator's going between successive potential energy states of a progressive receiver-giver. Thus, the parameters of space and time are not substantial interrelational terms but logistically defined, interrelational mediators between terms.

Quantum Distance

Recall that in PAPE, in the empowering transfer of potential energy from one state of PAPE to the next state of PAPE, two subsequent existential absolute states cannot exist at the same temporal instant. If the transfer took place in no-time, then the initial potential state and the subsequent existential state of PAPE would be the same, there would be no change, and the potential of the antecedent state would not be real. I also showed that two distinct, successive absolute states cannot be separated by time durations which are infinitesimally small because time divisions can only be divided a natural number of times in our physical world, and infinitesimals cannot be multiplied an infinite number of times in our finite physical world to produce an observable finite time duration. Rather, there must be a smallest, quantized time duration, which I indicate with the symbol Δt.

Democritus 2,500 years ago argued that an actual, physical object cannot be unendingly divided laterally. In our physical world, there must be an ultimate quantum size that is indivisible. Similarly, IPhil argues that the parameter of distance can be divided only a natural number of times in our physical world down to a minimum, quantum size. That minimal quantum-distance I identify with the symbol Δx. This minimal quantum-distance (Δx) will likewise be the minimal distance between all spacepoints subsequently generated in the entire proto-spatial system.

Maximal Speed of Energy Transfer, c

Energy-communicators traveling between neighboring spacepoints a quantum-distance (Δx) apart could take the minimal quantum-time, Δt. This means that, in the first logically evolved finite spatial system, the maximal speed of the energy communication between neighboring spacepoints a minimal quantum distance apart is, $\Delta x / \Delta t = c$. This c is a specific finite number. With the interrelational logistic of the proto-spatial system as its foundation, the maximum speed of energy-communication throughout the generated proto-spatial system will be the same number, c. In addition, by logical evolution, c will be the maximum speed of communicating energy in every subsequently evolved system which has the proto-spatial system as a carryover foundational part of its evolved logistic.

One of these future logically evolved systems is the electromagnetic system, and the energy-communicator in that system is electromagnetic radiation or light. The speed of light in empty space has been very accurately determined to be: c = 299,792,458 meters/second. By the principles of logical evolution, the maximum speed of energy communication in all evolved systems, like the energy-communicating gravitons in the gravitational system, will also be c. In this way, IPhil establishes the logical evolutionary explanation of why the speed of light through quantized space is the constant, c, which was one of the foundation stones of Einstein's theories of special and general relativity.

3-Dimensional Space

Newton proposed that there was nothing between things in space. Einstein, however, proposed a different view of space. He said we should imagine space as a fabric which is not visible to us but is nonetheless there. Space can be approximated in the classroom with a flexible rubber sheet, pulled taut over an area. When a lead ball is placed in the middle of this sheet, it makes a depression. This illustrates how the fabric of space is warped around objects which have a large mass. This model worked well for his general theory of relativity.

IPhil's model of the fabric of space is similar to a rubber sheet. However, IPhil maintains that space is connected not through anything physical but through the metaphysical interrelational logistic of space. It is not continuous but quantized. On the macroscopic level a rubber sheet appears continuous, but on a molecular level, a rubber sheet consists of molecules of rubber kept apart yet held together at a definite distance by electromagnetic forces. IPhil similarly pictures space to be like a malleable crystal consisting of spacepoints kept apart and together by the actions of spaceons moving between spacepoints according to the quantized interrelational logistic of the proto-spatial system. That is, in the proto-spatial system, its spacepoints are held in place by energized spaceons going back and forth between spacepoints. They are kept roughly a quantum-distance apart by the phases of the wavelike logistic of the spaceons. In this way, IPhil's model of quantized proto-space is midway between Newton's empty-space model and Einstein's continuous membrane model.

The spacepoints in the proto-spatial system cannot be directly observed by electromagnetic light because spacepoints and spaceons are of a logistical system that is formally distinct from the subsequently evolved electromagnetic systems. Because of the primitive character of the proto-spatial system, spacepoints are logistically not able to receive or reflect electromagnetic waves. Electrically charged particles can only observe electromagnetically things of the logical evolved electromagnetic system—while in the gravitational system they cannot observe mass that has no electrical charge.

Because spaceons going between spacepoints are energized, they have a substantial, 2nd stage, particle phase, and the logistical requirement that spaceons be a quantum distance apart applies to them in this particle phase. Thus, spaceons in their 2nd intrarelated, individualized phase will likewise be pushed and pulled to be a quantum distance apart. Consequently, in the energetic generation of neighboring spacepoints in the most primitive interrelational system, there are four energized participants that will be logistically pushed apart by the logistical requirement that all interactive parties be at least a quantum distance apart. Putting all this together, the quantum distances between two energized spacepoint particles and the two particle-phased spaceons can be pictured as the four corners of a tetrahedron. A tetrahedron is a 3-dimensional figure. Here is the origin of the 3-dimensional space we observe in our physical world.

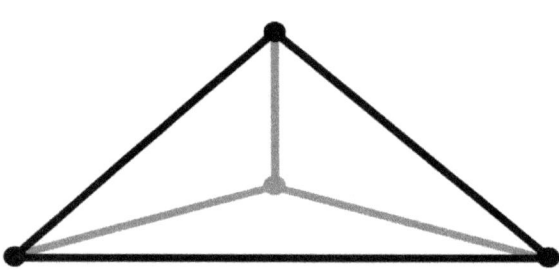

4-point, 3-dimensional Quantum Tetrahedron

Made up of two spacepoints and the particle phase of the spaceons traveling back and forth between them, the quantized interrelational proto-spatial logistic produces a tetrahedron, which is a three-dimensional figure! That is, in a stable, ongoing interaction between two spacepoints via spaceons, there must be three formally distinct distance-parameters keeping the four interactive, energized participants at least a quantum distance apart. So, by the interrelational logistic of energy, the logically evolved quantized proto-spatial system between every pair of neighboring spacepoints shows itself to be 3-dimensional. Here is the quantized, energetic interrelational origin of the formally distinct, three dimensions of the space of our physical world.

Because of the reciprocal character of the interrelational logistic, those space communicators can go back and forth between neighboring spacepoints, making the parameter of distance in all three dimensions of our 3-dimensional spatial world *bidirectional*. This contrasts with the parameter time, which originated in the antecedent absolute system and which is *unidirectional*. It is common to speak of living in a 4-dimensional world. However, IPhil shows why and how in proto-space and in all evolved systems the time parameter is unidirectional, and the three distance parameters of space are bidirectional. That time is not bidirectional will have an impact in the evolution of the fourth fundamental force, the weak force. But more about that later.

Lines in Space

At this stage of logical development, there is no reason why the extension of these three distance dimensions should be *linear*, much less *orthogonal* or at right angles to each other. Linearity and orthogonality arise from various symmetries found in the interrelational logistic of the proto-spatial system.

Thus far in the development of the geometry of this energized spatial system, I have only talked about the first two neighboring points. As this spatial system develops, spaceons traveling to new spacepoints will generate communication lines. However, because the quantum distance requirement demands only that a new spacepoint be at least a quantum distance from the generating spacepoint, it is possible for the expansion to go in random directions in this proto-spatial system. The path of energy advancement then would not be in straight but broken lines. Is there any reason why the expansion of the proto-spatial system should occur in straight lines?

Euclidean geometry assumes that between two distinct points there will always be another point on the straight line between the points. In the following illustration, point F is recognized to be *off* the straight line between points E and G, while point H is *on* that straight line, if EF + FG > EH + HG, which is the shortest broken line between E and G.

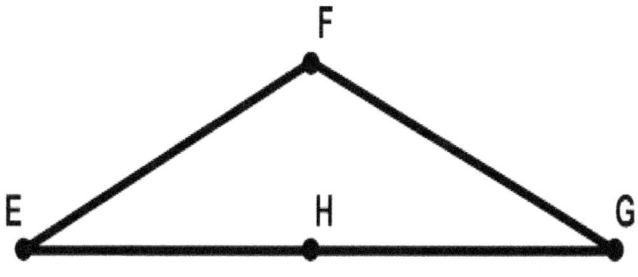

Illustrations of a longer broken line EFG and a shortest straight line EHG

PART II: BEGINNINGS

However, that is not true in an energetically quantized universe. In contrast to Euclidean geometry, which defines a straight line as the shortest distance *between* two points, IPhil defines a straight line as the set of points that energetically expresses maximal quantum extension *from* two points. In IPhil's proto-spatial system, lines of spacepoints are generated by the push of received potential energy to be more interrelational, or as interrelationally expansive as logistically possible.

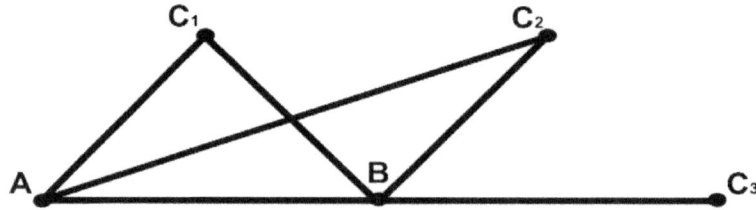

Illustrations of different angles, showing that a straight angle produces a most extended straight line

The above illustration indicates how a straight line is understood in the quantized spatial system of IPhil. Establish two distinct points A and B a quantum distance apart. In the quantized proto-spatial system, there can be no spacepoints between A and B because of energy's logistical requirement that spacepoints be least a quantum distance apart. Now consider several other points, C_N, each a quantum distance from B. As the angle ABC_N increases, the distance AC_N gets longer and longer, reaching its maximum length when $AB + BC_3 = AC_3$. Here, the longest distance, AC_3, occurs when angle ABC_3 is a straight angle or 180°. This is true of *any* two reference points at least a quantum distance apart.

It is because of the fourfold progressive character of energy that the generation of a straight line occurs in four interrelational stages.

1. There exists a point.

2. There exists another point at least a quantum distance from the first point.

3. There can be a finite extension of a linear ray from the first point, A, through the second point, B, such that the distance of another point, G, is: $AB + BG = AG$, thereby establishing AG as a straight line through A and B because potential energy pushes the establishment subsequent possibilities to be more interrelational, which occurs in straight lines.

4. That extension can be repeated to establish an indefinitely extended straight line—if there is enough quantized energy to generate further spacepoints.

Applying the above description of how to produce a straight line in all three distinct dimensions of interrelational quantized proto-spatial system, this system can be extended in three formally different directions. In addition, since energy's interrelational logistic is symmetric, applying symmetry to the logistic of the 3-dimensional, quantized proto-spatial system, these three formally distinct dimensions then can become orthogonal, or at right angles to each other. Because the potential energy of the proto-spatial system pushes toward increasing interrelationships, the logistic of absolute energy will push to generate a proto-spatial system that is quantized, reciprocal, linear, and orthogonally 3-dimensional: a "quantized Euclidean space."

While 2nd level, intrarelational, metaphysical concepts are individualistically ideal and perfect, 3rd level interrelational physical, free, quantized events occur in a quantized space, in which interactions are not perfectly precise but statistically normative. Thus, the potential and kinetic sides of energy, ground and form deterministic, singular, ideal, potential interrelationships in an idealized 3rd level perfect space, with free, normative, quantized, kinetic interactions in a 4th level quantitatively imperfect space. This results in two formally different forms of knowledge—one of generalized, metaphysical, potential descriptions of intrarelational concepts that are deterministically ideal, and the other of physical descriptions of logistically defined, interrelational events that are free and statistically normative.

Number Systems

Observations show that our world in many ways is mathematical in its structures and interactions. If our world came from the absolute Almighty, how did the mathematics of our physical, material world originate in a finite manner? Some pundits may say that axioms of mathematical systems are just assumed by mathematicians, and mathematics has nothing to do with energetic reality. However, if there is no energy, there would be no physical world, no metaphysical world, no mathematics, and no mathematicians.

IPhil locates the metaphysical origin of mathematics in the interrelational logistic of the potential energy of PAPE, then in PAER, then in PAIS, and then in the logical evolution in the proto-spatial system.

The logistic of energy indicates that Primal Absolute Potential Energy (PAPE) existed initially in an unformed, diffused state. Yet, energy is forward leaning, and although diffused, PAPE as a whole can exist in successive, absolute, existential states that can be numbered ordinally: 1st, 2nd, 3rd, 4th . . .

1. Regardless of its diffused state, each subsequent Absolute state is energetically equal to its prior absolute state. In this way, the equality operator, =, is identified. From this, comes the law of identity: $A_1 = A_2 = A_3 = \ldots$

2. Once intrarelated as individual PAER, the intrarelated energetic state has an integrated, nameable, identifiable quantity of energy, which can be the first cardinal number: 1. The receiver's remembered antecedent, virtual state, and there

its potential state was energetically "0" energy owned. From this dissociative comparison, there arises the inverse existential operator, "not," because the existential absolute energy of 1 is "not equal," ≠, to the absence of energy, 0. 1 ≠ 0. Also, "greater than" and "lesser than" operators can be defined, such that 1 > 0, and 0 < 1. Here is the origin of the natural binary numerical system. The binary system has well defined terms which have no fuzziness to them because they are of the absolute system. In current electronic communication systems, the binary system shows the clearest information transfer.

3. When interrelational transfer of that absolute energy to another receiver takes place, the energy transfer to another term establishes the 1st phase of the Primal Absolute Interrelational System (PAIS). Here is established the constructive operator of addition, +. With respect to the receiver's energy, the successive reciprocal stages of the receiver (2) from a giver (1) is: $1_1 + 0_1 = 0_2 + 1_2$. Energy is interrelational in a reciprocal way, leading to the formation of a system that corporately contain both giver-receiver and receiver-giver. In this way, the absolute reciprocity of PAIS establishes the commutative law: A + B = B + A, where one and only one state can be absolute, 1, making the other state 0. In addition, each movement of energy in PAIS has *three* energetic parties. These can be identified as the two dialogical terms (A & C), and the energy communicator (B) between them. Within a single cycle of the Primal Absolute Interrelational System (PAIS), there arises the distributive law is A + (B + C) = C + (B + A). Here also can be established the inverse fourth-stage operation of subtraction, -, with respect to itself in giving to another: $1_1 - 1_1 = 0_2$. In this way, IPhil finds foundational principles of arithmetic in the systemic phases of the energetic PAIS antecedent to the big bang.

4. The reciprocity in the Primal Absolute Interrelational System (PAIS) makes it logically consistent and allows for the logical evolution and advancement of the mathematical relationships found in the absolute order into the non-absolute, finite order.

The complete logical evolution of the numeric system of the absolute order into the finite order is not given here. Nonetheless, I have given enough information and directives such that the axioms of the quantized Natural Number Systems and all subsequently more complex number systems can be logically evolved from the above binominal Primal Absolute Interrelational Number System. In addition, because of the carryover character of the premises of antecedent, logically consistent systems into the subsequently evolved systems, it is expected that there will be analogous similarities between the axioms of the Primal Absolute Number System as well as all subsequently evolved and more interrelational, finite mathematical systems. In this way, IPhil predicts that all logically consistent mathematical systems can be advanced not by deduction but by logical evolution into an evolutionary Tree of Mathematics.

But let's move to something that is less theoretical and more practical. Consider the following examples:

2 apples + 3 apples	= 5 apples	(1)
2 oranges + 3 oranges	= 5 oranges	(2)
2 apples + 3 oranges	≠ 5 apples	(3)
	≠ 5 oranges	(4)
	= 5 fruits	(5)
And abstractly, 2 + 3	= 5	(6)

Why don't 2 apples + 3 oranges equal 5 apples or 5 oranges, while they do equal 5 fruits? Experientially, it is obvious that they don't. However, in theoretical, metaphysical philosophy, why don't they? The progenitor or first term in an evolved system establishes the nature or logistic of all interactive members of the system generated by and from that progenitor. Consequently, all generated members in a mathematical system will have the same logistic or nature. In the cases of equations (1) and (2), the + or "and" operator in the mathematical logistics of these system requires that all its participative elements have the same logistic or nature. Since apples and oranges are terms with different biological natures or logistics—arising from different progenitors—the rules of arithmetic do not apply to their intermingling, as seen in equations (3) and (4). Nonetheless, because they have a common evolutionary ancestry in the branching Tree of Life, apples and oranges have the same fruit evolutionary heritage and the same "fruit" logistic, and it explains why statement (5) is true.

In the transformation of potential energy into kinetic energy, the kinetic energy retains the logistic of the particular form of potential energy from which it came. It is only in this way that the amount of potential energy and the kinetic energy in a particular system can be legitimately added.

IPhil recognizes that statement (6) is an abstraction in which the labels have been removed. How is that justified? IPhil recognizes that mathematical logistical interrelationships initiated in the spatial system are carried over into every system subsequently evolved from the spatial system where the distinct members are "energized spacepoints." In subsequently evolved systems, those energized spacepoints take on the nature or logistic of the newly evolved system. Regardless of how the energy of a distinct spacepoint is evolved—into an apple, an orange, proton, or whatever—the same logically evolved arithmetic relationship will be true. Consequently, by logical analogy, the names of the different forms of energy/matter can be dropped if all quantized terms have the same logistic and of the same logically evolved order

Idealistic, essentialistic thinkers assume that there is one and only one right answer to a question. IPhil recognizes that everything that exists is energetic, and energy has four metaphysical aspects and four physical phases in its logistic. Consider the question: How much are 2 + 2? There is more than one answer to that question.

1. 2 + 2 = 2, when the first 2 is 2 liters of nitrogen and the second 2 is 2 liters of oxygen, which when added together produce 2 liters of mixed gas.

2. 2 + 2 = 4, when all the numbered items are totally, enduringly dissociative from one another, as are the spacepoints in the proto-spatial system. In a binary numbering system there is no 2, and this equation becomes 10 + 10 = 100.

3. 2 + 2 = many, when the first 2 are 2 female rabbits and the second 2 are male rabbits, producing many little rabbits.

4. 2 + 2 = 1, when the first 2 is the first 2 stages of an exchange, and the second 2 is the second 2 stages of an exchange, producing 1 logically consistent system, from which advanced interrelational systems can evolve.

Which answer is right depends upon the interrelational context currently considered. As a result, 2 + 2 are usually 4 because that arithmetic expression is usually applied to sets/systems in which the terms remain linearly distinct and have the same logistic.

The ancient mathematician philosopher Pythagoras, after recognizing musical harmony from strings of definite ratios, said that number governs all forms and ideas. With the invention of mega-computers, some people today say that all reality is numbers. In contrast, IPhil recognizes that number arises from the 2nd level dissociative aspect of quantized physical reality. Physical forms are quantized and numeric, while metaphysical ideas are sequential and numeral. From potential energy comes interrelational dissociations, from which come the logistics of the distinct numbered terms of mathematics and separations in geometric space. In IPhil, numbers are but a few of the many aspects of the energy of things.

Fourfold Expansion of Space

Cosmic microwave background (CMB) experiments have provided physical evidence for the theoretically predicted big bang. Unfortunately, visual observations of the expansion of the universe did not precisely agree with theoretical projections for the big bang. To resolve this discrepancy, Alan Guth in 1980 proposed the inflation hypothesis, which maintains that there was a very rapid inflationary epoch lasting about 10^{-36} seconds between 10^{-33} and 10^{-32} seconds after the big bang.[1] Following this inflationary period, the universe continued to expand but at a much slower rate. This computer model matches the data. However, no one has been able explain satisfactorily why that brief inflationary period happened when it did, or why it came to an end after a very short time.

---

1. Guth, *Inflationary Universe*.

Also, recent observations show that the rate of expansion of the universe is not constant but is slowly increasing. The current hypothetical explanation for this phenomenon is that there is some type of "dark energy" causing that acceleration. Its nature, however, is not understood, so it is rightly called *dark*. According to data from the Planck Mission and based on the standard model of cosmology, the total mass-energy content of the universe is currently recognized to be: 4.9 percent ordinary matter, 26.8 percent dark matter, and 68.3 percent dark energy.[2] This leaves the question: Why is so much matter and energy in our universe dark?[3]

IPhil recognizes that space is not nothing. Space is more than a mathematic invention. Space also was not created by material realities moving relative to each other in space. Space needed to have existed potentially *before* material realities could move relative to one another in space. Above, IPhil described how the logically evolved quantized proto-spatial system began with the Almighty's energizing of a single, finite receiver in the all-receiver. This energized finite receiver became the pre-spatial progenitor from which was interrelationally generated the quantized, 3-dimensional proto-spatial field which became our universe. The logistic of this quantized spatial field is maintained by spaceons going between energetically generated neighboring spacepoints, keeping them at least a quantum distance apart. Like the fourfold potential energy that made this spatial system, the logistic of the proto-spatial system is fourfold. Metaphysically this suggests the possibility of four different types of expansions of the original, quantum-separated, proto-spatial system.

1. Receiving Expansion

IPhil recognizes that time is quantized, and a change from the initial state of potential energy to a state of kinetic energy takes time—at least a quantum of time. Before becoming metaphysically capable of producing an expanding spatial system, there needs to be a quantum of time for the initial, virtual, possible, progenitor spacepoint to receive the potential energy needed to generate another spacepoint. Here the energetic expansion is internally non-laterally *within* the progenitor spacepoint. In its first, receiving, childlike phase of spatial generation, the progenitor of the proto-spatial system must receive, integrate, and orient the received energy before "letting go" of some of its energy to generate another spacepoint externally. In this 1st phase of spatial generation, the progenitor spacepoint as yet occupies no space. Its energetic generation in the all-receiver as the progenitor receiving spacepoint would take time—approximately four quanta of time. This 1st internal "expansion" of the progenitor of the proto-spatial system is like a receiving child, which is generally quiet as it gathers its internal strength to act externally in the world.

---

2. European Space Agency, "Planck."
3. Peebles and Ratra, "Cosmological Constant."

2. Inflationary Expansion

When the progenitor spacepoint's receiving phase is completed, the intrarelated energy in the integrated progenitor receiver spacepoint can exteriorly use its received potential energy to generate another spacepoint. That energetic transmission results in energetically establishing a second spacepoint at least a quantum distance apart, then a third, then a fourth, etc., where each is at least a quantum distance apart in only a few quanta of time. Since a sender does not yet reciprocally receive from a newly established point, the reciprocal spatial logistic is not yet physically established, and the speed of communication in the outward formation of space is subsequently near the speed of c, the maximum speed of light in our currently fully established spatial system. Newly generated spacepoints here receive a tremendous amount of energy, and their giving of energy freely, randomly causes spatial extension in all directions. Furthermore, in this phase, promulgations of energy will be interrelationally maximized by going in straight lines. This is how the principles of IPhil proposes that the inflationary expansion of our universe took place at speeds approaching the speed of light.

By analogy, this 2nd inflationary phase may be described as an adolescent phase, during which the expansion of energized space individualistically occurs as fast as possible in all possible directions from each generated spacepoint. This is how IPhil provides a philosophic explanation for the rapid inflationary period hypothesized by Alan Guth about 10^{-33} or 10^{-32} seconds after the very short 1st short, quiet, receiving phase of the singular progenitor spacepoint, the "seed" of the big bang. But why did this inflationary period soon come to an end?

3. Reciprocal Systemic Expansion

In reciprocal interrelationships of logically consistent systems, energy first moves from a spacepoint giver to a spacepoint receiver and then returns from the receiver-giver to the giver-receiver. Forming a reciprocal, logically consistent system, the logistic of the spatial system now forces the speed of energetic promulgation to be much less than in the simply outward inflationary period. The reciprocal exchange between neighboring spacepoints (or evolved elementary particles) would take eight time-quanta or more. Consequently, in its 3rd phase, the generating spatial system would expand more slowly. Here, in the expanding, reciprocal spatial system, the kinetic energy density and the temperature will progressively drop. These slower, reciprocal activities are much like the slower, developmental concerns of an adult.

4. Interior Expansion

Expansion of the proto-spatial system occurs through the movement of energy-communicating spaceons between spacepoints. These spaceons go through their own four-phase interrelational process. This fourfold process gives each spaceon a wave-like characteristic. Also, in the 2nd phase of that process, the intrarelational character of its energy briefly gives each spaceon a particle-like characteristic. It is this 2nd intrarelational particle-like stage that justifies giving the energy-communicator an integrated, individual identity, which I call "spaceon."

Normally, all the energy that goes into forming the spaceon particle phase is "let go" in the spaceon's transfer of all its energy to a receiving spacepoint. However, because it can be partial in its operations, it is inversely possible for a self-initiating spaceon to "not let go" of all its received energy to its receiver. When a spaceon does "not let go" of at least a quantum of energy, that energy-holding spaceon can be transformed into a spacepoint.

When the energized spaceon traveling between two spacepoints a quantum distance apart is transformed into a new spacepoint, that presents a logistical problem. The interrelational logistic of the quantum spatial system demands that all energized spacepoints be at least a quantum distance apart. When a new, spaceon-made-spacepoint elbows its way in-between its original neighbors, the quantum-distanced *logistic* of proto-space will push those former neighbors apart from each other to make room for the newly formed spacepoint. This results in a stretching of the fabric of space at that point. It also causes the quantumly separated neighboring spacepoints to accelerate away from each other. This breaks the symmetry of the well-ordered logistic of the quantized proto-spatial system at this point. Then, in a logical evolutionary way, this symmetry-breaking can transform the antecedent symmetrical Euclidean proto-space into an asymmetrical, non-Euclidean warped-space at that point.

The transformation of a transient spaceon into an enduring new spacepoint between neighboring spacepoints is a logistical inverse. Logistically ordered to remain a quantum-distance apart, neighboring spacepoints are normally static with respect to each other. Now, however, the introduction of a new spacepoint between them forces neighboring points apart with respect to the antecedent proto-spatial system. This introduces something totally new—acceleration. Through the principle of equivalence, Einstein showed that inertial acceleration and gravitational acceleration are equivalent. In the next chapter, I will show how this acceleration within the proto-spatial system logically evolves into the gravitation system. This logical evolution through acceleration also introduces the new parameter of mass and explains the development of gravitational force in space. But more about this in the next chapter. Finally, this may be viewed as the elderly phase because it leads to the death of the spatial system. This will be explained in the closing chapter on the "big collapse," which is logistically opposite the conception phase of the "big bang."

In summary, IPhil finds that there are four types of spatial expansion arising from the four phases of the interrelational logistic of potential energy. By the principles of logical evolution, these stages are analogous to the four phases of human life: (1) Initially, like an infant, the energy-receiving and interiorly integrating progenitor interiorly becomes the "cosmic seed" of the big bang. (2) Like a rapid adolescent growth spurt, the energized progenitor spacepoint sends out spaceons and establishes distant spacepoints as fast as possible, explaining the very fast inflationary period of spatial expansion. (3) While the edge of the expanding proto-spatial system continues to expand rapidly, the systemic orientation of the logistic of space, in an adult-like way, reciprocally slows the expansions of the proto-spatial system in order to make the proto-spatial system logically consistent and enduring, capable of generating subatomic and atomic particles. (4) Finally, the breaking of the symmetry of proto-space by the transformation of some transient spaceons into spacepoints introduces accelerative expansion to the interior of the spatial system. By that logical inversion from static interrelationship of neighboring spacepoint, logical evolution *can* occur within the spatial system, resulting in systemic acceleration, gravitation, and the mass of our material world. Exactly how all four forces of physics logically evolve by the breaking of the symmetries of the four dimensions of the proto-spatial logistic will be covered in the next four chapters.

Dark Energy

Astronomers have observed that galaxies are not speeding away from each other at the same rate. They are speeding away from each other at an increasing rate. Cosmologists have named the cause of this acceleration "dark energy," because the cause of this acceleration is still unknown to them. What kind of energy is the cause of the increasing internal expansion of space?

IPhil proposes that the acceleration of "dark energy" can be explained by the random transformation of the energizing spaceons into enduring spacepoints, forcing the gradual expansion of the logistic of the quantized proto-spatial system. However, the pushing energy that causes this type of expansion of space does not come from the new spacepoints themselves but from the *logistical requirement* that all spacepoints be a quantum distance apart. From this point of view, the "dark energy" for spatial expansion comes not from any physical body but from the quantum character of the interrelational logistic of the proto-spatial system.

The energy originating from the interrelational logistic of the spatial system is a new form of energy—quantized logistic energy. The 4th level "letting go" of quantized spatial spacing results in the logistic force of "dark energy" as a reaction to breaking of the symmetry of the proto-spatial system. This logistic force is akin to the 3rd level interrelational logistical transformations of dialogical terms in entanglement. It is also akin to the 2nd level intrarelational logistical transformation of inward potential energy into outward kinetic energy. This is akin to the logistical transformation of existential energy into forward-leaning potential energy. In these ways, IPhil recognizes that within the fourfold logistic of energy, there are four levels of metaphysical logistical forces antecedent to the four physical forces of our world.

Energy is interrelational. Since interrelationships have four, formally distinct aspects, there are four formally distinct types of potential energy. (1) The existential aspect of energy gives all receivers enduring progressive existence. (2) The intrarelational aspect of energy makes intrarelational receivers into individual particles with their own interior individual logistical identity. (3) The interrelational aspect of energy produces systems with an enduring, outgoing communal logistics, as in the generation of the proto-spatial system, in which potential energy pushes elements toward becoming an enduring, logically consistent system. (4) The outgoing aspect of energy gives members of a system the ability to break the logistical boundaries of a system and the capacity logically to evolve a more advanced interrelational system.

IPhil finds that "dark energy" is not a form of 2nd level individualized energy. Rather, it is a form of 3rd level "field energy," where the space's communal logistic pushes the members of that system toward remaining at a quantized distance within the logistic of that system. The communal logistic of the spatial system expresses the "natural law" of the proto-spatial system, striving to keep that system well-ordered. Thus, it is not the energy, mass, or any other characteristic of the substantial spacepoints that causes the accelerative expansion of space. Rather, it is the spatial system's interrelational *logistic*, maintained by inter-term energy communicators, that causes an accelerative expansion when transient spaceons are transformed into stable spacepoints within the quantized logistic of the proto-spatial system. In a fourfold way, the new spacepoint is the *material cause* grounding that change, the logistic of quantized proto-space is the *formal cause*, the transformation of the spaceon into a spacepoint is the *final cause*, and the energy-communicators of the communal logistic of the proto-spatial system are the *efficient causes* of that change.

In an enclosed gas, when one part is heated, that heat quickly spreads through the gas, until the entire gas is at equilibrium. Similarly, "dark energy" may arise initially in one place in space, but through the reciprocity of proto-space expansion, it will quickly spread through the entirety of space. While originating locally on the quantum level, the effects of "dark energy" will quickly spread, because of the tendency of proto-space toward reciprocal equilibrium throughout the spatial system—as IPhil predicts.

A crucial step in the scientific process is experimental verification. IPhil claims that the source of dark energy is the quantum distance demand of the spatial logistic of the parameter of distance between spacepoints in the evolved proto-spatial system. IPhil predicts that energy-communicating spaceons can become spacepoints between neighboring spacepoints, forcing the quantized spatial system to expand. Then, as space expands, there will be more spacepoint possibilities that spaceons can randomly become new spacepoints between them, producing more dark energy, increasing the rate of expansion of the universe. Telescopes today can observe stellar bodies at different times since the big bang. These observations can tell whether the rate of the universe is increasing as the spatial size of our physical universe increases. Verification by observation that the rates of expansion of our universe correlates with the expansion of space through time would add credence—or discredit—to the quantized model of spatial expansion logically evolved in IPhil.

Why Was the Big Bang so Big?

To understand why the big bang was so big, it is important to go back to the antecedent absolute order. In establishing the all-giver (PAER), the Primal Absolute Potential Energy (PAPE) pushed the all-giving (PAER) toward interrelationally giving of absolutely all potential energy to an intrarelating all-receiver. This action empowers the virtual all-receiver to become a formally distinct all-receiver-giver capable of reciprocating the actions of the all-giver, thereby establishing (PAIS) as an absolute reciprocal, logically consistent system. In all these successive stages, absolute energy is moved forward in time and in the sequential absolute empowerment of all participants in PAIS.

Then, by breaking the absolute logistic of PAIS, an absolute-finite progenitor can be evolved, capable of generating the finite proto-spatial system. How much energy would be given to the first finite receiver? Although the amount freely given by PAER to the finite progenitor receiver *can* be only a single quantum, the absolute interrelational energy of PAPE will push PAER toward establishing maximal interrelationships in PAIS, as well as maximal interrelationships between all finite receivers panentheistically in the all-receiver. That is, because absolute potential energy pushes toward maximal interrelationships, it is possible and probable that *maximal* potential energy was given to the progenitor of the proto-spatial systems within the all-receiver-giver of PAIS. In this scenario, PAER, as the all-receiver-giver gives absolutely all-energy not only to the all-receiver in PAIS, but also, to achieve maximal interrelationships within the finite order, to the absolute-finite progenitor panentheistically in the all-receiver-giver. It is in this way that it can be said that the absolute-finite progenitor was initially absolute. This is IPhil's explanation of why energetically the big bang was so big.

In the 1st phase of the evolutionary absolute-finite progenitor's logistic, IPhil recognizes the likelihood that the absolute-finite progenitor receives the same absolute amount of potential energy as given to the all-receiver. God has given humans the greatest commandment: You shall love the Lord your God with all your heart, with all your soul, with all your mind, and with all your strength. Where did this come from? IPhil puts its logistical, evolutionary origin in the absolute interrelational logistic of PAIS. As PAER has given absolutely All potential energy to the all-receiver in PAIS, there is a carryover of that same total gift and reciprocal command into the finite order via the absolute finite progenitor. In this way, the Almighty has first given His all to us finite creatures, before asking us to reciprocate totally. By the carryover characteristic of logical evolution, every subsequently evolved system would likewise be urged by the received potential energy toward giving to others of its own kind a maximal amount of its received potential energy for the sake of maximizing interrelationships within our expanding, evolving universe.

Chapter 9

Logical Evolution of Gravitation

Four Fundamental Forces of Physics. Limits Have Limits. Logical Evolution of Universal Gravitational Formula. Dark Matter.

Four Fundamental Forces of Physics

Experiments in physics indicate that there are four fundamental forces in nature. These forces cause interactions between distinct substantial things via force-carrying particles traveling through their intermediate space or "field." In the following four chapters, I will show how the four forces of physics arise via logical evolution in their own distinctive ways by various symmetry-breakings of the four dimensions of the proto-spatial system. These collectively make up what I call the "evolutionary field theory."

The four fundamental forces of physics have been observed to have the following characteristics:

- Gravitational interactions are long-range and the weakest of the four. Still, gravitation is the dynamic that pulls and holds together the largest, galactic bodies and systems in the universe. The gravitational force operates between objects of mass. Gravitational force and mass are never negative but only positive. The graviton is the massless force particle that conveys the gravitational force at the maximum speed c.

- Electromagnetic interactions also are long-range, but they are very strong. This force operates between electrically charged particles at the maximum speed, c, by a massless particle called the photon. Electromagnetic particles can carry either positive or negative electrical charges, although there are some particles with mass that carry no electrical charge. Antithetically, there is a negative, repulsive force between like-charged particles, but a positively attractive force between unlike-charged particles.

- Strong interactions are caused by a very strong, attracting force between elemental particles called "quarks," which combine to form neutrons, protons, and other nuclear particles.[1] Their strong force-carrying particle is called a gluon; it is of

---

1. Nave, *Hyperphysics*.

a class of particles called bosons. This synthetic force is short ranged because gluons have mass and travel at a speed less than c.

- Weak interactions are caused by a repulsive weak force, which manifests itself primarily in the expulsion of particles from atomic nuclei in radioactivity. This force is conveyed by bosons that are variously electrically charged {W^-, Z^0, and W^+}. Since these bosons have mass, they travel at a speed less than c, and this force is short ranged.

Various attempts have been made to connect these four fundamental forces. Einstein coined the term "unified field theory" (UFT), but he was unsuccessful in his attempt to unify the general theory of relativity with electromagnetism. The "theory of everything"[2] and the "grand unified theory" were other proposed unification schemes closely related to the unified field theory, but they were unsuccessful also.

The dominant approach for establishing a unified field style theory has been to create inductively a model of a spatial energy field that has a specific mathematical algorithm, from which theoreticians strive mathematically to logically *deduce* the four forces of physics by considering their emergence from that general field formulation in different physical situations, especially as the temperature of the universe dropped as it expanded after the big bang. This approach assumes that *all* the laws of physics, describing everything in our universe, can be derived by logical deduction in a deterministic way from a single, unified, logically consistent algorithm or set of axioms. However, Gödel's incompleteness theory points out that all logically consistent axiomatic systems are incomplete, and there *can* be more true statements, which cannot be logically derived from a given set of logically consistent axioms. In other word, theoretically, a strictly deductive unified field theory will be incomplete, and it cannot completely express all mathematical and physical relationships in our evolving, logically consistent universe.

However, there was one successful interconnecting of two fundamental forces. In 1963 Sheldon Glashow proposed that the weak nuclear force and electromagnetic force could be logically connected within a unified electroweak theory.[3] In 1967 Pakistani Abdus Salam and American Steven Weinberg independently revised Glashow's theory by having the W-particle and Z-particle arise through spontaneous breaking of one of the symmetries of the electromagnetic field.[4] From IPhil's perspective, it is noteworthy that Salam and Weinberg's approach was successful in connecting electromagnetism and the weak force not through logical deduction but through "symmetry breaking," where particles broke through the symmetrical boundaries of the logistically consistent electromagnetic system to manifest the logistic of the weak force.

---

2. Dongen, *Einstein's Unification*.
3. Weinberg, "High-Energy Behavior."
4. Glashow, in the section on the proposed unified electroweak force in *From Alchemy to Quarks*.

From IPhil's perspective, the symmetry-breaking of the electroweak theory producing the weak force indicates logical evolution.

The progressive character of the parameters of these formally distinct forces—mass, electrical charge, color charge, and weak-force bosons—cannot be logically deduced from one another. However, the logical evolution of mathematical systems formally does produce new and different parameters in its symmetry-breakings. Thus, the progressive advancement of parameters is another indication that these forces emerged not deductively but in a logical evolutionary way from a primal spatial field or system. IPhil proposes that all four physical forces have emerged through logical evolution from the logistic of the proto-spatial system through various logical inverses of the four dimensions of the proto-spatial system. That is why I call this approach the "evolutionary field theory," which I will develop over the course of four chapters. I will explain how there are only four fundamental forces in our universe, arising from the symmetry-breaking of the four dimensions of proto-spatial system.

Unfortunately, the mathematics associated with this process quickly becomes complicated. Consequently, what I have presented in these four chapters are but preliminary stepping-stones toward a full mathematical development of the evolutionary field theory. If the reader wishes to skip over my discussions of some of the technical, mathematical details of the logical evolution of the four forces of physics, I suggest turning to the summary of the evolutionary field theory at the end of chapter 12.

Isaac Newton had to invent calculus to derive the elliptical orbits of planets from the universal gravitation formula he hypothesized. Similarly, I need to use first-year calculus to evolve that formula. I know of no other way to explain how the classical universal law of gravitation can be logically evolved except by mathematically applying calculus to the proto-spatial system. The current chapter will show how Newton's universal law of gravitation can be logically evolved from—not deduced from—the breaking of the symmetry of the linear, 1st dimension of the quantized proto-spatial system. But there is a problem in using calculus to do that within IPhil.

Limits Have Limits

Newton needed calculus to apply the gravitational formula to concrete situations. The name "calculus" came from Gottfried Wilhelm Leibnitz, who, it is said, invented this branch of mathematics slightly before Isaac Newton did.[5] Infinitesimal calculus is a necessary mathematical tool in the field of mechanics. But there is a problem here: IPhil recognizes that space is quantized, and that means there are lower quantum limits with respect to time, space, and energy in the development of equations within the foundational proto-spatial system. In contrast, infinitesimal calculus assumes the existence of what are called "mathematical limits," in which variables can theoretically be made infinitesimally small. It is thus a serious question to ask whether the mathematical concept

5. Child and Gerhardt, *Early Mathematical Manuscripts of Leibniz*.

of infinitesimal calculus can be applied to the symmetry-breaking of the quantized proto-spatial system, in which there are no infinitesimals. To solve this problem, it is necessary to delve into the theoretical foundations of mathematical limits.

The classical Greek paradox of Achilles and the Tortoise shows that many Greeks were philosophically perplexed by the notion of limits. In this paradox, Achilles races against a tortoise. Because the tortoise is slow, it is given a head start in which it travels a certain distance. Then Achilles starts to run, traveling the distance the tortoise just traveled. Meanwhile, the tortoise moves ahead a distance. So, Achilles runs to where the tortoise just was. However, the tortoise by then would have moved ahead a little more, and by the time Achilles moves to that spot, the tortoise would have moved ahead further, and so on. Following this line of argument, it appears that Achilles could not catch up to, much less pass, the tortoise. However, by observations we know that Achilles can and does pass the tortoise, but why? Today we know that the solution to this paradox is found in the realization that the time taken by Achilles to move to the next spot gets shorter and shorter in each movement. Adding together the decreasing time intervals in each movement, even if there are an *infinite* number of moves, Achilles would catch up to the tortoise in a *finite* length of time. The reason for this is that the sum of an infinite series of decreasing numbers need not be infinite but *can* be finite.

$$\tfrac{1}{2} + \tfrac{1}{4} + 1/8 + 1/16 + 1/32 + \ldots = 1$$

From this, it becomes clear that a "limit" is a number which is logistically more advanced than all rational numbers.

Although the Greeks were not able to establish formally that the sum of an infinite number of decreasing terms can be finite, they did have an intuitive understanding of the notion of limits, especially regarding their calculations about circles. They recognized that the perimeters of *inscribed* regular polygons in a circle constantly increased as the number of sides in the polygons increased. Also, the perimeters of *inscribing* regular polygons around a circle constantly decreased as the number of the sides of those external polygons increased. From these experiences, they intuited that the circumference of a circle had to be a fixed number between those two perimeters. That is, as the number of sides of the inner and outer polygons increased, the length of the perimeters around the circle would converge to the length of the circumference of a circle, which is not a rational number but is called a "transcendental" number. If the diameter of a circle is 1, the circumference would be a number, named π. The Greeks showed that this ratio, π, is the same number for every circle, regardless of its diameter.

PART II: BEGINNINGS

The realization that an infinite series of finite terms could produce a finite number played an important role in the development of mathematics. However, a formal definition of limits didn't come until the seventeenth century.[6]

The traditional expression of a limit is:

$$\lim_{x \to c} f(x) = L$$

In 1821 Augustin-Louis Cauchy formalized the (ε, δ)-definition of limit. This definition uses ε to represent a small positive number such that f(x) becomes arbitrarily closer to the limit L when |f(x) - L| < ε for all (decreasing) values of ε. The phrase "as x approaches c" refers to values of x whose distance from c is less than some positive number δ where $0 < |x - c| < \delta$. The (ε, δ)-definition of limit of a function $f(x)$ can be expressed in four steps.

1. If for every number in an open interval containing c, except possibly at the number c itself,
2. as *x approaches c*, that is, when $0 < |x - a| < \delta$, and
3. for any ε, regardless how small, there is a $\delta > 0$ such that $|f(x) - L| < \varepsilon$,
4. then the limit of the function $f(x)$ is L.

Experience has shown the value and the accuracy of calculus. However, there is a major philosophical problem with the application of calculus to a quantized space. The above definition assumes that the values of ε and δ can get smaller and smaller or infinitesimally small. However, in a quantized space there is a finite, minimal quantum-distance between all terms. Thus, in quantized proto-space, mathematical limits have a quantum limit. So how is it possible to use limits in the proto-spatial system?

Because proto-space is quantized, there is a definite, minimal, quantized distance between spacepoints, corresponding to their numerical values. In other words, in our finite, quantized, energetic world, there is a real, existential lower limit to the values of δ in line (2) of the above definition. That means there is a finite (not infinitesimal) quantum limit to the difference between the numeric values of the variables of the energized proto-spatial system. Smaller divisions than a quantum distance simply *do not physically exist*. So, the determination of a mathematical limit with a dimension less than one quantum would not exist, and an infinitesimal value of the mathematical limit of a function less than one-quantum in size is also metaphysically not possible in our physical, quantized world. As a result, below the quantum level, the value of the function would be physically *indeterminate*, but still it would be *approximate*. Thus, in the proto-spatial system, calculus is true down to the level of a quantum size, and it is approximate below that. Similarly, physical experience shows that all physical

---

6. Bradley and Sandifer, *Cauchy's Cours d'Analyse*.

observations are limited by the imprecision of making an instrument. The Heisenberg uncertainty principle establishes that there is always theoretical uncertainty in dimensions of less than one quantum wavelength. In physical experiments, such minute variations are interrelationally uncertain and functionally insignificant. The disassociate difference between what is quantumly actual and what is infinitesimally mathematical is systemically *inconsequential*.

Because of the *actual* quantum limitations of proto-space, the use of calculus in any evolutionary derivation of the universal gravitational formula can be considered physically true down to the quantum level, and indeterminate, uncertain and inconsequential below that.

Emmy Noether proved a theorem showing that any differentiable symmetry of the action of a physical system has a corresponding conservation law. Because the earliest logistical symmetry of a parameter is that of the shortest quantum-time interval, Δt, in the exchange of quantized energy, ΔE, between dialogical terms in the proto-spatial system, the first conservation law is that of conservation of energy through time. IPhil recognizes that in the domain of our universe's spatial system—down to the level of quantum distance—there is interrelational consistency. But in distances less than one quantum length, there need not be systemic closure or logical consistency. This is significant. For it means that in distances between spacepoints less than a linear quantum and in a period less than a time quantum, the conservation law of energy and *all* subsequently logically evolved conservation laws can be broken.[7]

It is easy to recognize this fact in the Primal Absolute Interrelational System (PAIS) in time intervals shorter than that of one complete interrelational cycle. During the 1st, 2nd, and 3rd phases of the exchange of absolute energy, the reciprocity of the energy is not yet complete, so it cannot be said that during those three phases the system is logically consistent or energetically conservative. In the interrelational cycle of the primal exchange of absolute energy in PAIS does not display reciprocal conservation of energy until it is received back by the all giver-receiver. Thus, conservation of energy and the solid foundation of calculus are valid down to the level of a single quantum of time, space, and energy. Below that, findings are not only uncertain but physical variations are interrelationally inconsequential.

Logical Evolution of Universal Gravitational Formula

Before proceeding to evolve the gravitational formula logistically, there is an evolutionary corollary that needs to be highlighted. In each successive evolved system, the progenitor that makes the initial actions of establishing an evolved system will have a particular inverse value in the antecedent system. That value of the progenitor, however, is not logically determined by the logistic of the antecedent system. Its breaking of the symmetry of that system can freely be from anywhere within the domain of that

---

7. Kosmann-Schwarzbach, *Noether Theorems*.

logistic. The initial value of the progenitor can be of any energetic value for that parameter from the antecedent system. Consequently, in the new, non-derivable, logically evolved system which the progenitor generates, the bridging value of the progenitor at the symmetry-breaking establishes the initial standard measure in the new, evolved system. Because of this, all generated measurements in the new, evolved system will be multiples of the value carried over into the new, evolved system by the progenitor. For that reason, in the logical evolutionary process there is need for a constant of proportionality in the new system, which translates the standard values of the logistical variables in the antecedent system carried over into the standard values of the variables in the logistic of the evolved system. This constant of proportionality can rationally be made 1 in humanly evolved metaphysical systems for the sake of simplicity. However, physical evolution happens by chance, and the constant of proportionality in all evolved physical system usually will not be 1 but another number established by the progenitor of that physical system. Furthermore, because the physical constants of the various, physically evolved systems in our world result from the free, 4th level self-initiatives of progenitors, the physical constants of the successively evolved systems in our world will be energetically random. For this reason, there is no logically determined mathematical connection between the constants of proportionality of evolved physical systems. I will apply this corollary later.

In the previous chapter, I established that interrelationally the potential energy of a spaceon would normally cause a lateral extension in the proto-spatial system of the big bang. I showed that this lateral extension energetically took place in straight lines to maximize the dimension of that proto-spatial system. In that straight-line extension of the proto-spatial system over an extended duration, t, the spaceons travel at maximum speed c. They would normally traverse a multiple quantized distance, d, such that

$$d = c\,t$$

In its intrarelational, 2nd particle phase, however, if the spaceon energy communicator does not freely "let go" of a quantum of its received energy, the energy retained by the spaceon in its 2nd intrarelational particle phase will turn it into a new, enduring spacepoint between the original neighboring spacepoints. Because the logistic of the spatial system demands a quantum distance between neighboring spacepoints, the newly generated spacepoint needs to elbow its way between neighboring spacepoints. This elbowing-in by the new spacepoint causes the antecedent neighboring spacepoints to move *physically* with respect to their original *metaphysically static* places in the proto-spatial system. This results in the emergence of a new phenomenon—*acceleration*—with respect to the antecedent, metaphysical, ideal proto-spatial system. This action breaks the quantized symmetry of the perfectly symmetrical logistic of the antecedent proto-spatial system. Because acceleration is a *not-constant* change in speed, acceleration is an inverse operator, capable of producing the acceleration of an energized spacepoint and of logically evolving a new system in which acceleration is the new generative operator. In this chapter, only 1-dimentional, linear acceleration in 3-dimensional proto-space will be discussed.

One of Albert Einstein's famous thought experiments established the principle of equivalence, linking gravitation and linear acceleration.[8] He proposed two closed elevators with an observer in each. One elevator is left static on the surface of the earth; the other is accelerated through space at a rate equal to g. Here, all material things in both elevators would drop at an accelerated rate of 1 g. A bullet shot through the elevator would travel on a parabolic path, regardless of whether the elevator is in place on earth or accelerating through space at the accelerative rate of g. In this way, it is seen that the equations of the motion of the projectiles will be the same for an elevator in the gravitational field of 1 g, as it is in the elevator accelerating at the rate of 1g. In the following illustration of two elevators, the first is in the gravitational field of earth, and the second elevator is accelerating in space at the rate of 1 g.

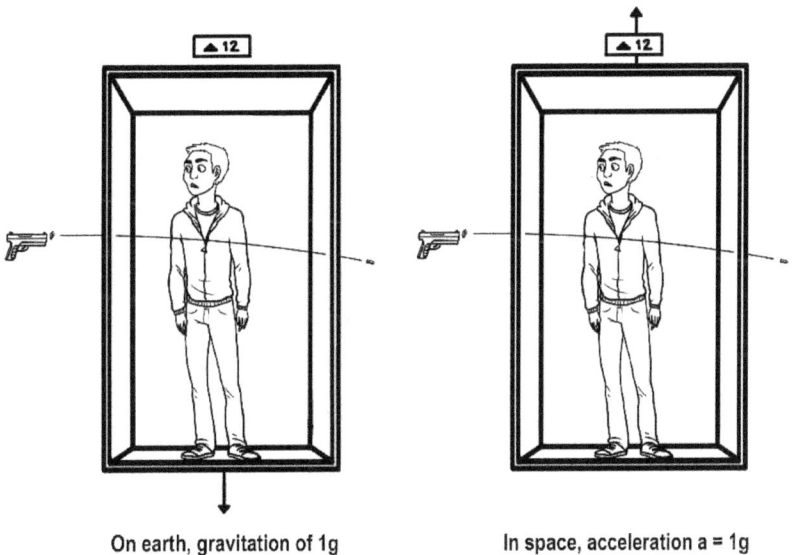

On earth, gravitation of 1g   In space, acceleration a = 1g

Illustration of relative equivalence of acceleration and gravity

In the quantized proto-spatial system, with respect to each other, spacepoints have no velocity, v = 0. The speed of any motion in that situation would be defined by the formula v = d / t. Vectors arise because energized communicators in a 1-dimensional spatial system push terms to maximize interrelationships and therefore produce maximized extension in straight lines. In addition, by conditions established in the antecedent system and communicated to this subsequent system, the velocity v of terms cannot

---

8. Einstein, *How I Constructed the Theory of Relativity*, 17–19.

be greater than the maximum speed of energy-communication, c, or v < c. For small intervals, Δ t, where d and v are vectored quantities in the same direction is:

$$\Delta d = v \Delta t$$

This equation is true down to the level of the quantum distance which separates spacepoints in this spatial system.

Initially, the quantum distance between neighboring spacepoints is fixed, and their velocity with respect to each other is zero, v = 0. However, after the generation of the new spacepoint, their velocity with respect to each other will increase a finite amount. Here, a new quantity, *acceleration*, can be defined in the following way:

$$\Delta v \equiv a \Delta t \qquad (1)$$

where a is held at a *constant* acceleration as the velocity v increases from 0 to Δ v in time interval Δ t.

By integral calculus, it can be established that the distance traveled during this constant acceleration in the antecedent, regular spatial system of equation (1) is:

$$d = \tfrac{1}{2} a t^2, \text{ where a is } constant \qquad (2)$$

However, in the short time duration, Δt, when the neighboring points are pushed away from each other, the acceleration *can* be variable, that is, "not-constant." As insignificant as it may seem, *here* is where the inverse acceleration operator breaks the logical boundary of the antecedent system of equations (1) and (2), when the acceleration of the spacepoint separating from the introduced new spacepoint *can* be variable rather than constant. That new acceleration breaks the symmetry of (1) and (2), and that variable acceleration through space and/or time would be written "a (d, t)."

In this interrelational system's energized space, logistically every interaction between spacepoints is reciprocal. The original spacepoint cannot physically respond reciprocally unless it has knowledge of the accelerating action of the second spacepoint. The original reference point cannot know the action of an accelerating first point unless the time (t') that the second point is accelerating away from the original point in the above equation is less than or equal to the time (t) of the spaceon's communication of the information at the speed c about that change.

That is, in equation (2)

$$t \geq t'$$

When the initial distance between the separating spacepoint is d_0,

$$t \geq d_0 / c \qquad (3)$$

With this, I have laid enough stepping-stones to define a new variable that can be logically evolved, but not derived. With this new accelerative parameter, I can inversely break the boundary condition of (1) and (2) such that the acceleration is not constant but *variable*.

Analogously, I can define a new variable, M, using the terms of the antecedent spatial system, modeled in a logically evolutionary way after equation (2), namely,

$$M \equiv \tfrac{1}{2}\, a\,(d, t)\, t^2, \text{ where a is not constant but } variable \quad (4)$$

Also, this new variable must conform to condition (3) throughout the new logically evolved system, in which the logistic is reciprocal and symmetrical. That is, this same quality must be recognized logistically and simultaneously in the original spacepoint because of logistical entanglement.

As indicated above, the value of t in equation (4) is the minimum value of t possible in (3). That is,

$$t = d_0 / c \quad (5)$$

In other words, communication of energy in this new, evolved system would occur by means of an interrelational energizer, which is characteristic of this system and is traveling at the evolutionarily inherited rate of speed c. Combining equations (4) & (5), we have

$$M = \tfrac{1}{2}\, a\,(d, t)\, (d_0 / c)^2$$

Notice that in the expansion of space the acceleration, a (d, t) of spacepoints from each other is positive, and $(d_0 / c)^2$ is always positive. So, M must always be positive.

Transposing this equation, we have:

$$a\,(d, t) = 2 c^2 M / d_0^{\,2} \quad (6)$$

Here we find that the maximal speed of communication by the energy-communicator of this evolved system produces a reciprocal acceleration in the receiver. Furthermore, notice that in the right side of this equation this acceleration is not a function of time but only of space. That is, at any given moment, this acceleration is a function only of the distance between accelerating, neighboring space points.

In addition, since the logistic of the interrelational spatial system is reciprocal, and since the accelerated receiver has the new property of M, by logistic symmetry the energy-giver must have a similar quality, which I will call "m." With the new variable m, which is affected by the acceleration, a new function can be defined called "force" or "F."

$$F \equiv m\, a \quad (7)$$

Combining equations (6) and (7) we find

$$F = 2\,c^2\,M\,m\,/\,d_0^{\,2} \qquad (8)$$

Previously, I indicated that in the logical evolution of one system into the next, the standard measurements in the antecedent system would not normally be equal to the standard measurements in the non-derived new system. Rather, in the subsequent, logically evolved system, the standard of measurement will be proportionate to the standard established by the progenitor of that evolved system. When the constant of proportionality is G:

$$F_g = G\,\frac{M\,m}{d^2}$$

By the principle of equivalence, the above equation can be identified with the universal gravitational formula. Now, we also can identify that the function M defined above is the "mass" of one accelerating energized spacepoint, and m is the "mass" of the accelerating energized receiving spacepoint, which is separated by the distance, d.

However, that leaves one significant concern. In the above derivation, the foundational vectored acceleration of the spacepoints caused by the intruding new spacepoint is *not toward* each other but expansively *away* from each other. However, experience shows that the direction of the gravitational force of one mass is always *toward* the mass of the other. How can this discrepancy be resolved? Newton's third law of motion is: Every action has an equal and opposite reaction—which is an expression of conservation of energy. In the above scenario, the action is the push of an intruding spaceon-spacepoint causing the neighboring spacepoints to accelerate away from each other. Inversely, the equal and opposite reaction will be toward each other. The equal and opposite reaction is the gravitation force of the logistic of the proto-spatial system counteracting against the intrusive force of the new spacepoint against the quantized logistic of the proto-space. That is, the *cause* of the acceleration is the intruding new spacepoint, while the reactive *effect* of the gravitational force is from the proto-spatial system's *logistic*, which internally seeks to maintain its antecedent, ideal, equally spaced logistic. That is why gravitation is not attributed to the separating cause of the accelerating spacepoint pushing outward, but to an inverse, inward reaction of the logistic of the spatial field seeking to balance that distortion. Thus, gravitation is a characteristic not of the accelerating spacepoints but of the resisting logistic of the *spatial field between* the accelerating spacepoints in the proto-spatial system.

The introduction of spacepoints within the equally spaced proto-spatial system will cause the original symmetrical proto-spatial system to expand and warp. In this way, IPhil links gravitational force, mass, and the warping of the quantized

Euclidean geometry of proto-space, transforming it into an evolved, quantized, non-Euclidean spatial system.

Dark Matter

The term "dark matter" has been applied to some form of matter that is not observed directly by electromagnetic radiation or light. Rather, it is known secondarily through increased gravitational effects in rotating galaxies. Observations can determine the rates of rotation of stellar systems, as well as the masses of the stars going around these galaxies. The rate of rotations of these galaxies indicates that there is more mass in these galaxies than what is observed. Where is that dark matter? Also, it has recently been found that in dispersed galaxies with only few, smaller celestial bodies, there is no observable influence of dark matter.

Even if the temperature of a body with mass is very low, it can be seen optically by observing infra-red radiation. However, no such evidence has been found. If 26.8 percent of all the mass-energy of the universe is dark matter, why haven't we had so much as a glimpse of it?

Before answering this question, another important question needs to be asked: Is the proto-spatial system described in the previous chapter real? If it is real, why haven't we observed the proto-spatial system? In the famous Michelson-Morley experiment, performed in 1887, Albert Michelson and Edward Morley attempted to detect the earth's relative motion through a proposed a luminiferous ether, which was believed to be the medium of the electromagnetic waves of light.[9] This ether would have been either stationary or moving with respect to motion of the earth moving around the sun. The negative result of this highly sensitive experiment was the first strong evidence against the presence of ether in space as the spatial medium for the promulgation of light. Because this experiment demonstrated that there was no physical ether medium in which light traveled, many scientists have assumed that this meant there is no objective space. Yet, Einstein's general theory of relativity considers space to be like a stretchable fabric that can be warped in the presence of large masses. Is there any philosophical explanation that space is like a stretchable fabric that is influenced by the effects of mass and gravitation, but is unobservable by light or electromagnetic observations?

IPhil proposes that dark matter is the result of an energetic carryover from the spacepoints that anchor the proto-spatial field from which the gravitational system evolved. In the formation of the proto-spatial system, all spacepoints receive their energy from neighboring spacepoints via energy-communicating spaceons. That interrelational energy can be called "interspatial energy" because its logistic is ordered toward the generation and maintenance of the proto-spatial system of spacepoints. The receiving spacepoints go through a four-stage process before sending that "interspatial

---

9. Michelson and Morley, "On the Relative Motion of the Earth."

energy" via a spaceon to another spacepoint. In that process, spacepoints receive, hold, logistically aim, and then send the received "interspatial energy" toward another spacepoint via a spaceon. Each phase of this interrelational process takes at least a quantum of time. Consequently, in the process of maintaining quantum space according to the logically evolved proto-spatial logistic, every spacepoint intrarelationally holds its received "interspatial energy" in one place for a short time. From the perspective of the evolved gravitation system, in that interrelational process, the briefly held energy of the spacepoints takes on the character of inertial mass. The mass value of that received interspatial energy is according to Einstein's famous formula: $E = mc^2$. When spacepoints interrelationally possess their received "interspatial energy" in one place for a short time before sending it out again, the spacepoints have an inertial mass, $m = E / c^2$, where E is the existential potential energy held by the spacepoint before passing it on to a neighboring spacepoint.

In this way, IPhil proposes that the mass of "dark matter" is the inertial mass of the interspatial energy held by all spacepoints in the 2nd phase of their individual logistic in the formation of the proto-spatial system. This energetic mass is "dark" because spacepoints in the antecedent proto-spatial system cannot be electromagnetically seen because it does not have the logistic of the electromagnetic system, and it therefore can neither receive nor send back any electromagnetic radiation. Consequently, we are not able to know logistically the retained energy/mass of spacepoints electromagnetically via light. However, we can know of this phenomenon indirectly via observation of associated gravitational phenomena upon those particles which do have electromagnetic characteristics.

Spacepoints in the proto-spatial system are normally evenly dispersed, so no gravitation gradient or effect would be observed. However, when space is warped due to the presence of a large celestial body of mass, like a galaxy, the space around that galaxy will be warped, producing a gradient of spacepoints in that non-Euclidean geometry, producing a gravitational gradient from the spacepoints of that system. IPhil would propose that the amount of "dark matter" within a galaxy from the spacepoints of its space will be proportional to the warping of the space due to the mass of the stars of that celestial system. Similarly, smaller galaxies should manifest less gravitational effects from dark matter. A Yale University-led study in 2018 established that galaxy NGC 1052-DF2 has few celestial components, separated at great distances. Observations of this galaxy show that the orbits of its celestial components show no dark matter gravitational effects.

A crucial step in the scientific process is experimental verification. IPhil claims that the source of dark matter is the inertial energy of spacepoints, which are the dialogical terms of the space as generated according to the quantized spatial logistic, as logically generated in IPhil. The more space is warped around a massive stellar object, the greater is the space density around that object, more spatial terms there are in that warped space, and the greater will be the mass-density of the momentary

LOGICAL EVOLUTION OF GRAVITATION

holding of the energy of the antecedent proto-spatial system. Observational verification should be able to confirm—or not—that there is a correlation between the warping of space around large celestial bodies and the magnitude of dark matter increasing the gravitational pull in a given large stellar system. Verification by observation of this phenomenon would add credence—or discredit—to the quantized model of space formation logically evolved in IPhil.

Chapter 10

Logical Evolution of Electromagnetism

Gyroscope and Vector Cross Product. IPhil's Model of Spatial Spin. Logical Evolution of Electromagnetic Force. More Work Needs to Be Done. Quantum Electrodynamics (QED). Why More Matter than Antimatter.

I first need to give some preliminary descriptions of circular motion from the gravitational system before I can apply them to subsequently evolved electromagnetic systems.

In the macroscopic order of the gravitational system, transverse motion can be imposed upon a rigid rod anchored at one point in space, producing circular motion. Here can be defined angular velocity, angular acceleration, and the rotational force of torque. Simple examples of transverse acceleration are the lever, a teeter-totter, and a torque wrench. Transverse motion is at a right angle to the 1st dimension, traditionally called the x-axis. Circular motion expresses a 2nd dimension, which traditionally is called the y-axis. In this way, transverse motion occurs on an xy-plane defined by those two axes. When an object is pushed in the xy-plane, at a fixed distance from the center of the xy-plane, the motion will be perfectly circular. The z-axis is perpendicular to that circling, and it becomes the fixed axis of rotation, while the dynamic rotational motions and accelerations are in the 2-dimensional xy-plane.

Gyroscopes and Vector Cross Products

Consider the following common high school laboratory experiment. Sit upon a stool that can easily spin on its base. Next, while holding the axle of a bicycle wheel with both hands, have someone rotate the wheel rapidly so that the top of the wheel spins away from you. Then, as you would turn a forward-pointing screwdriver, gently twist the axle of the spinning wheel clockwise. Suddenly, you and your stool will spin to the right. This feels very strange because your hands are twisting the axle of the spinning wheel with respect to an axis pointing out in front of you, parallel to the ground, but the consequent spinning of your body on the stool is with respect to the axis of the stool that is perpendicular to the ground. A more common experience is when you ride a bicycle no-handed. If you remain upright, the moving bike goes straight, but if you lean to the right with respect to the ground, the front wheel will automatically turn itself to the right. Your leaning is with respect to the plane of the earth, but the turning of the bike is to the right with respect to an axis vertical to the earth. When you lean to the left with respect to the ground, the front wheel of the bike will turn to the left. It happens so naturally we fail to appreciate that the tilting of the body is with respect to a line parallel to the ground, and the turning of the bike to right or left is really at right angles to the lean.

Because the bicycle wheel is a rigid body, exerting a right-hand force on the axle causes the upper part of the wheel to accelerate a little bit to the right, and the lower part of the wheel accelerates a little bit to the left. As the tire turns, those increments of transverse change are added onto the previous change, thereby adding many small increments to the previous, producing a large transverse change. In this way, the torque to the right on the top of the spinning wheel will progressively turn into an integral acceleration to the right in the front of the wheel, and the torque to the left on the bottom of the spinning wheel will progressively turn into an integral acceleration to the left on the back of the wheel. These combine to put a clockwise torque on the wheel's axle perpendicular to the ground. In the laboratory chair, the integrated torque of the wheel travels from the wheel, through the rigid spokes, and is communicated down your rigid arms, causing the stool on which your body is seated to spin clockwise. The harder the holder twists the horizontal axle of the spinning wheel clockwise, the faster the person will spin clockwise around a vertical axis. If the wheel is spun in the opposite direction, the chair will turn in the other direction.

This feature about torque can be mathematically expressed as a *vector cross product*, A x B = C. If a spin vector A is spun with a spin vector B, the net result will be a spin vector C which is at right angles to the first two. This rotation can be mathematically derived from classical Newtonian mechanics. In the antecedent proto-spatial system, the commutative law, A x B = B x A, is true. However, as illustrated by the chair example, in spinning systems, when the spin vector is reversed, the resultant precession will be in the opposite direction and inversely,

$$A \times B = - B \times A$$

In classical mechanics, spin phenomena have an important boundary condition, namely, a spinning system must be physically rigid, as in the case when the rim of the wheel is rigidly held in place by means of spokes connecting the outer wheel to the axle in a rigid way. Without the rigid structure of a spinning system, the acceleration imposed by the hands on the axle of the wheel could not be transferred to and integrated in the spinning parts of the wheel. The rigidity of physical bodies is necessary for precession to occur within a physical body. Now it can be asked: Is it possible to "break" this rigid body boundary condition on precession and still have some type of precession?

PART II: BEGINNINGS

IPhil's Model of Spatial Spin

IPhil recognizes that in the proto-spatial system, energy-communicators or spaceons travel back and forth in a minimum quantum of time between spacepoints, which are located at a minimum quantum distance apart. In that back and forth movement, the spaceon is traveling at the maximum velocity, c. Because of the symmetries of the proto-spatial system, the parameter of distance is bidirectional.

As a spaceon goes from the first spacepoint to the second spacepoint, another spaceon could at the same time go from the second spacepoint to the first. Consider two spaceons leaving opposite spacepoints at the same time. Initially, they emerge from their space point a quantum distance apart. As they move forward in opposite directions, because of the intrarelational character of their 2nd phase, those spaceons temporarily take on the complementary character of an energetic particle. Because of this, these two spaceons logistically need to keep a quantum distance apart from each other. Throughout their movements to the opposite spacepoint, the center of their quantum-distance interaction will be the midway point between those two spaceons. As a result, the paths of those two spaceons will take the shape of semi-circles, centered midway between the two original spacepoints. Consequently, the paths of these coupled, opposing quantumly separated spaceons will *break from the symmetry* of a bidirectional straight line, moving rather in a perfect circle—with no thing in the center—only a quantized interrelational field.

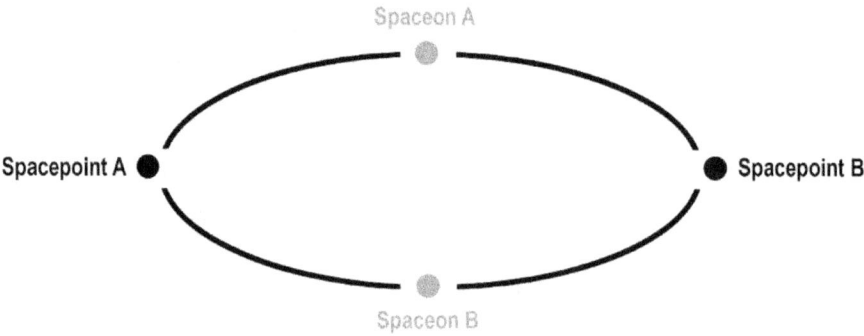

Illustration of spaceon circulating between two quantum-separated spacepoints

This model of spaceon-exchanges between two neighboring spacepoints, takes place on a 2-dimensional plane of rotation, anchored on the neighboring spacepoints in quantized space. As spaceons go back and forth between quantum-spaced neighboring spacepoints in the maintenance of the quantum lattice of the proto-spatial field, the resulting path of these spaceons on a given plane is circular, producing an energized spaceon "spin." Here is the origin of the new parameter of "spin" for the energy-communicators in this new symmetry-breaking system.

Previously, it was shown that all energy in our world cannot be divided indefinitely, but it must be quantized. Consequently, by the carryover characteristic of local evolution, the rotational energy of this "spin" would also be quantized. Also, since this spin can go in opposite directions, this quantum-spin in proto-space could be either positive or negative. As there can be multiples of quantized energy, the *spin* of these exchanging spaceons can have multiple, quantized, energetically positive and negative values.

Furthermore, because quantized, interrelational proto-space is 3-dimensional, the plane of rotation of spaceons between two anchoring spacepoints *can* be randomly oriented in space. So, spin effects from quantum spin from neighboring pairs of spacepoints would generally cancel each other out. Also, pairs of spacepoints in quantized space are not isolated. They are parts of a 3-dimensional quantum matrix of spacepoints aligned symmetrically, as in a pliable crystal. Because the quantum-distance condition applies not only to spacepoints on one layer but also to the quantum distances of neighboring layers of spacepoints, the plane of rotation of spaceons on one layer will tend to push the random planes of circulation of spaceons on neighboring layers toward the same orientation.

Because of the layered quantum condition of the spatial system, if the plane of circulation of the spaceons on one layer is changed for any reason, that twisting of the plane of rotating motion will have a transverse effect, changing the plane of circulation of the spaceons on neighboring layers of space. Because of the circular movement of these spaceons, pressure (1) from one layer of spacepoint will have a precession impact (2) upon the orientation of the spaceon rotations of neighboring layers. That is, twisting on one layer of the proto-spatial system will cause both a linear acceleration and a transverse acceleration perpendicular to the initial twist. This indicates that the forces associated with this evolved system will be of two kinds—one involving linear acceleration, and the other involving transverse acceleration.

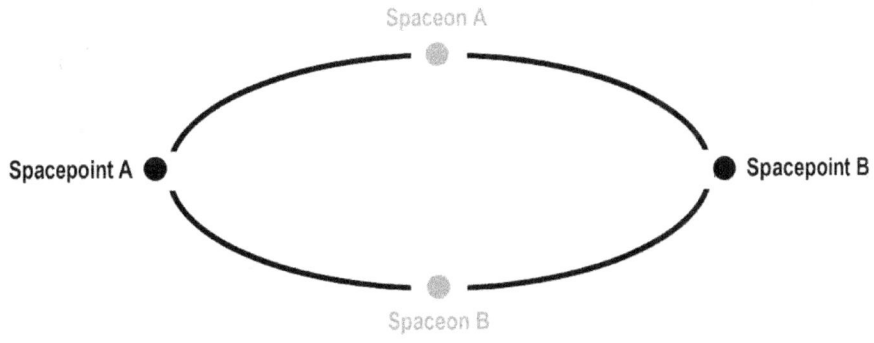

Illustration of spaceon circulating between two quantum-separated spacepoints

In the original equilibrium state of the proto-spatial system, there is no movement or twisting. If there is linear acceleration of the whole matrix, the force that is produced is only gravitation. When a part of the matrix is twisted, more transverse effect will be experienced on one part of the spatial "crystal" and less transverse effect on another part of the spatial "crystal." When that happens, two effects occur together: linear acceleration *and* torque on the spatial system. This causes the orthogonal symmetry of the spatial system to break. This orthogonal symmetry-breaking will cause the logical evolution of a new logistic, a new force, and a new operational parameter not found in the previous spatial system and gravitational system. Experimentation indicates that this new parameter is electrical charge, for macroscopically electrons are known to have both electrical charge and spin. Transverse motion of these spinning systems will have a magnetic style effect. By the logical evolutionary theory, the logistic of the gravitational system is carried over into this newly evolved electromagnetic system. Thus, these new electrically charged particles will have mass. However, there may be some particles that do not logically evolve beyond the logistic of the gravitational system, and they would have mass but zero electric charge and spin.

The above precession effect occurs via the circular motion of spaceons traveling between spacepoints, and this effect will travel at the speed, c, between spacepoints. Since the potential energy of the spaceons is what causes these electrical and magnetic phenomena, their logistic will have a fourfold logistic and be wavelike. Thus, this new circular phenomenon will be communicated through space via a new energy-communicator that has a wavelike logistic. In its 2nd logistical stage, this wavelike phenomenon will have a transient intrarelational particle phase, which is named a photon. Photons are the energy-communicators in this system with magnetic spin and no systemically enduring mass.

In summary, IPhil's model presented here identifies *transverse symmetry breaking* of the antecedent orthogonal proto-spatial field as it causes the logical evolutionary origin the new force of electromagnetism, which travels through space at the speed c via fourfold electromagnetic waves.

This model, however, is only a general picture. Now the classical equations of electromagnetism need to be logically evolved through symmetry-breaking. In the previous chapter, I showed mathematically how the linear, 1-dimensional symmetry-breaking acceleration of a *spacepoint* produced the universal gravitational formula in a logical evolutionary way. Now the question becomes: Can 2-dimensional symmetry-breaking of the exchange of *spaceons* between spacepoints in the proto-spatial field produce the classical equations of electromagnetism in a logical evolutionary way?

Logical Evolution of Electromagnetic Force

Some preliminary thoughts: The above quantum-spatial model indicates the possibility of two different types of interrelated accelerations in two dimensions: linear and transverse. Mass and gravitation arise from 1-dimensional, linear acceleration, producing the universal gravitational formula. By the principles of logical evolution, it can be expected that in the new system, linear acceleration should produce a new parameter (electrical charge) and a new (electrostatic) force analogous to and logically beyond those of the antecedent gravitational system, whose logistic would be carried over into the next evolved system in regard to its linear accelerations.

With respect to transverse acceleration, the formulation of the logical evolutionary magnetic force should show precession effects similar to but evolutionarily more advanced than precession equations found in the gravitational system. Furthermore, since logical evolution occurs through a logical inverse, it is expected that, as the signed number system of positive and negative numbers logically evolved from the strictly positive numbers of the natural number system, the formulation of the electromagnetic force will similarly have a new parameter that has both negative and positive values. Going beyond the symmetrical commutative law of addition in the linear gravitational system, the transverse effect of "spin" should produce an asymmetrical commutative law of the vector cross product in rotating electromagnetic systems.

The universal gravitational formula of classical Newtonian mechanics is:

$$F_g = G \frac{m_1 m_2}{d^2}$$

Classical electromagnetism has two foundational equations. The first equation is that of the electrostatic force and is called Coulomb's law of electrostatic repulsion. It describes the repelling force between two like-charged electrical particles at rest or in parallel motion this way.

$$F_e = \frac{-1}{4\pi\varepsilon_0} \frac{q_1 q_2}{r^2} = -K \frac{q_1 q_2}{r^2}$$

Here F_e is the electrostatic force between the two charges q_1 and q_2. The force and acceleration of the charges will be proportional to their respective electrical charges, where ε_0 is a constant called the permittivity of free space. The constant of proportionality, $1/4\pi\varepsilon_0$, links the standards of the gravitational system with the standard of the electrostatic system. It is immediately apparent that the gravitational equation and the electrostatic equation take on analogous logistical forms, with different interrelational parameters. This is expected in logical evolution. As found in the logical evolution of the gravitational force, this electrostatic force is not caused by the electrically charged particles themselves but by the logistic of the quantized spatial field between two objects with their own electrical changes.

Furthermore, since logical evolution occurs through an inverse operator, it is understandable that while the 1st level gravitational force of the spatial field between objects with mass will reactively always be positive or attracting, inversely, the 2nd level electrostatic force between like-electrical-charges (-/-) will always reactively be negative, while the force between unlike-electrical-charges (-/+) will be reactively positive.

The magnetic force pertains to transverse motion and involves a vector cross product, indicating a precession effect. Magnetic fields can be easily observed when direct electrical current, as from a battery, is forced into circular motion by a coil. The magnetic field produced by the coil of wire will repel one pole of the magnets and attract the other. By reversing the direction of the current, the pole that was repelled is now attracted, and the pole that was attracted is now repelled.

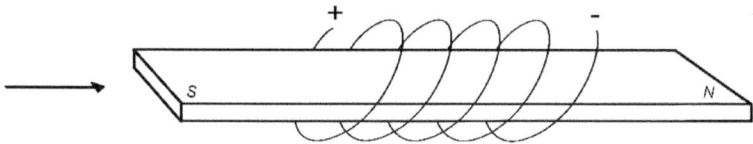

Illustration of a coil and magnet

This electromagnetic phenomenon is very important in the process of generating electricity in turbines spun by the force of dammed water or expanding steam.

The magnetic force $\mathbf{F}_m$ on charge q_1 exerted by q_2 is:

$$Fm = \frac{\mu_o}{4\pi} \frac{q_1 q_2}{r^2} \mathbf{v}_1 \times (\mathbf{v}_2 \times \check{r})$$

where $\mathbf{v}_1$ = velocity of charge q_1; $\mathbf{v}_2$ = velocity of charge q_2; $\check{r}$ = unit vector between q_1 and q_2, and μ_o = permeability constant of space. Mathematically, it can be shown that the quantitative relationship of the constants of proportionality in these two equations are linked to the constant speed c of electromagnetic radiation (light) in a vacuum:

$$C^2 = \frac{1}{\varepsilon_0 \mu_o}$$

The above magnetic force equation shows a vector cross product, which is similar to that found in the operation of a gyroscope, except the parameter is electric charge rather than mass.

Linear, lateral accelerations produce gravitational effects, which will have an effect upon the Euclidian geometry of protospace, evolving it into warped Riemann geometry. In a similar way, IPhil proposes that the transverse accelerations causing electromagnetism will affect that Riemann space, logically evolving it into a torsion Cartan geometry.[1]

---

1. Alpha Institute for Advanced Studies, "Beyond Einstein," 55–59.

More Work Needs to Be Done

Since electromagnetism deals with rotating systems, the logical evolution and mathematical analyses of quantized electrostatic and electromagnetic forces in rotating, quantized systems very quickly become complicated. Unfortunately, I have not yet been able to develop fully the mathematics that establishes the above classical force equations for the electromagnetic system via the logical evolutionary theorem. Nonetheless, because the electromagnetic force equations are highly analogous to the universal gravitation formula, the logical evolutionary process points toward the possibility of generating the electromagnetic equations via logical evolutionary breaking of the symmetries of 2-dimensional protospace. That is, IPhil predicts that through the process of logical evolution, the above classical equations of electromagnetism above *can* be logically evolved—not derived—from two dimensional accelerations in the proto-spatial field. This is an important IPhil prediction. I hope that its establishment will verify the principles of logical evolution as applied to the first two dimensions of the primal, quantized, proto-spatial field.

Fortunately, many theoretical scientists have already done much successful work in this direction. My contributions to this effort are only to provide the logical evolutionary theory, the Interrelational Logistic, the metaphysical character of the absolute potential energy antecedent to the big bang, and the quantum-level model illustrating how 2-dimensional spin occurs within the quantized proto-spatial system. I hope these will guide the mathematical formulation of the logical evolutionary bridge between the logistics of quantum proto-space, the gravitational logistic, the electromagnetic logistic, as well as the two fundamental forces described in the next two chapters.

Quantum Electrodynamics (QED)

This matter does not end with the evolution of the above classical equations of electromagnetism. Because electrons can be accelerated to speeds very close to the speed of light, full formulations must include quantum effects as impacted by the theory of general relativity.

It was the British scientist Paul Dirac who presented the first formulation of a quantum theory regarding the interaction of radiation and matter, by marrying quantum mechanics and relativity to electromagnetic phenomena.[2] In 1928, building on 2x2 spin matrices, he proposed the Dirac equation as the relativistic equation of motion for the wave function of the electron. His equations also contributed to explaining the origin of quantum spin as a relativistic phenomenon. This theory was named *quantum electrodynamics*, or QED. Those equations have been able to describe accurately the quantum behavior of electrons and other charged particles at very high speeds.[3]

---

2. Dirac, "Quantum Theory of the Electron."
3. Feynman, *Strange Theory of Light and Matter*.

The symmetry in Dirac's equations also predicted the existence of a new particle, which had the mass of an electron but a positive charge. This new particle is called a positron. This particle was discovered in photographic emulsions in the laboratory in 1932. Subsequently, a whole matching field of antimatter has been discovered.[4] Being antithetical, the collision of a positron with an electron—or any particle with its antiparticle—results in the disintegration of both particles, with their intrarelated mass transformed into electromagnetic radiation.

Why More Matter Than Antimatter

The symmetries of QED indicate that there should be an equal amount of matter and antimatter in our universe. However, experimental results show that in our physical world, there is much more matter than antimatter. This raises a perplexing question: Why is there more matter than antimatter in our universe? For one thing, the mixing of equal amounts of matter and antimatter in our universe would lead to their mutual destruction—with no particles and only radiation remaining. Fortunately, that has not happened.

Why is there so much more matter than antimatter in our universe, contrary to what these equations seem to predict? IPhil's answer to this question rests on the difference between metaphysical possibility, as expressed in equations, and what is physical actuality, as energetically experienced kinetically in experiments. It is important to stop identifying actuality with what is potentially, metaphysically, possible according to essentialistic equations. Rather, energized actualities in our physical world are kinetically developmental and evolutionary. Here, each successive material realty is built upon what physically happened antecedently, starting with the first energized progenitor of that logistical system. Mathematical algorithms and equations in the mind, on paper, or in computers do not produce reality—*energy* does. In politics and in criminal investigations, an important rule is: "Follow the money." Similarly, in physical reality, a scientist needs to follow the energy. In the metaphysical order, potential energy and its logical evolution contain all theoretical, mathematical *possibilities*. However, in the physical order, *actual* evolution takes place through an energized, self-initiating progenitor, which needs to be expressed physically before any interaction can take place with another receiver with the same communal logistic.

Whether a progenitor is a piece of matter or antimatter is of no real interrelational consequence. In a new, evolved order, whatever the type of intrarelationally formed matter the progenitor may initially receive, on the 1st level, that intrarelational 2nd level progenitor on the interrelational 3rd level will systemically generate on the 4th level its own kind. In this way, the progenitor interrelationally will establish a 3rd level system with new particles of its own kind. So, if the progenitor initially expresses itself as "matter" rather than "antimatter," that progenitor will generate more terms of its own kind,

4. Close, *Antimatter*.

namely, "matter," rather than "antimatter." So, if the progenitor has the theoretical option of being a piece of "matter" rather than "antimatter," the progenitor will logistically generate reciprocal receivers that can establish a system of "matter" rather than "antimatter." Subsequent generations will do the same. As a result, by the rules of logistical development of a logically consistent system, all the descendants of the "matter" progenitor will *physically tend* to generate a universal system of "matter" rather than "antimatter."

Nonetheless, because the energy of the progenitor has "antimatter" as a metaphysical possibility within the quantized electromagnetic logistic, an energized particle in that generated system *can*, and occasionally does, randomly generate antithetical "antimatter," at least for a short time. Being metaphysically possible, "antimatter" *can* and probably will occasionally be generated in an environment where most particles are "matter." However, the generation of an antimatter particle will quickly interact with its diametrically opposite matter, destroying both. As a result, no antimatter system can endure in a universe where the interactive participants are both matter and antimatter. IPhil recognizes that amid dissociative actions in an interrelational system of matter, if its interaction with antithetical antimatter is brief, that interaction is inconsequential to the enduring existence and logical consistency of that system of matter. Even when some possible antimatter is generated, because the progenitor and its progeny had the leading edge, they would quickly physically eliminate antimatter from our universe. In this way, IPhil explains why, even though the symmetrical equations of QED specify equal *metaphysical possibilities* of being "matter" or "antimatter," the *actual physical realities* generated in our physical world by the progenitor of "matter" will tend to produce systems of "matter," rather than "antimatter." This is what IPhil predicts using logical evolutionary principles, and this is what is observed.

In conclusion, the empirical development of quantum electrodynamics (QED) has involved the cooperative work of the world's greatest scientists. It is highly accurate and has served as the template for subsequent quantum field theories. However, *why* is that metaphysical equation formulation true? Phil has presented a spatial model and a logical evolutionary way for evolving these equations through the breaking of the symmetry present in two dimensions of the proto-spatial system on the quantum level. This leads to the proposal that the breaking of all three dimensions of the proto-spatial system leads to the logical evolution of the strong forces. What would be the associated quantized model in proto-space for describing the strong force?

Chapter 11

Logical Evolution of Strong Force

Historical Background. Standard Model of Particle Physics.
IPhil's Model for the Strong Force.

Historical Background

After World War II and the successful production of the atomic bomb, financial resources became available to explore matter and energy on the atomic level for their quantum characteristics. By then it was clear that energy waves have particle characteristics, and matter has wave characteristics. In explaining the photoelectric effect, Albert Einstein recognized that the energy, E, of a light-communicating photon particle in a field between sender and receiver is proportional to the photon's frequency, f, and inversely proportional to the wavelength, λ, of its wave. That is, $E = h f$ or $E = h c / \lambda$. Consequently, material particles used in exploring very small things must have wavelengths that are very small, and this demands that their kinetic energy and therefore their speeds must be very high. This insight gave birth to a series of particle accelerators, which raised the speed of particles closer and closer to the maximum speed of communication, c.

Ernest Rutherford established that atoms had very small nuclei surrounded by sparse fields of electrons.[1] Further experimentation showed that the nuclei of atoms were made up of protons and neutrons tightly bound together. All protons have a positive electric charge, and neutrons have no electric charge, while electrons have a negative electrical charge. The electromagnetic force between like-charged particles is repulsive; yet the nuclei of atoms with multiple positively charged protons were tightly bound together. This led to the hypothesis that in the nuclei of atoms, there had to be a new force stronger than the electromagnetic force. This new force was prosaically called the "strong force." On an atomic scale, this force is 102 times stronger than the electromagnetic force, 106 times greater than the weak force, and 1039 times stronger than the gravitational force.

Multiple experiments at higher energy, and theoretical studies on an atomic scale, soon led to the discovery that protons and neutrons were not fundamental particles but were made up of constituent particles, which are called "quarks." These quarks are

---

1. Rutherford, *Scattering of Alpha and Beta Particles*.

interdependent, constitutive parts. They do not exist independently on their own, so they are analogous to the formally distinct, interdependent, interactive, constitutive parts of PAIS. Quarks are only recognized secondarily in their interactions with other quarks. Strong interactions are observed primarily in two areas: it is the force which externally bonds protons and neutrons together to form the nuclei of atoms, and it is also the force which internally bonds quarks together to produce protons, neutrons, and other hadron particles.

Analogous to the interdependent, constitutive absolute parts which arise in the 3rd stage of metaphysical development of absolute energy in PAIS, quarks arise in the 3rd stage of metaphysical development of the 3-dimensional spatial system. Quarks have electrical charges in multiples of ±⅓ quantum, another indication that quarks cannot exist alone but only in conjunction with other quarks. A force-carrying boson particle called a gluon communicates the energy of the strong force between quarks. On a relatively large scale of 10^{-15} meter, protons and neutrons are bound strongly together in the nuclei of atoms, a bonding referred to as the "nuclear force." On a slightly smaller scale of the radius of a nucleon, gluons convey the strong force that binds together the quarks in protons, neutrons, and other similar hadron particles.

Murray Gell-Mann and George Zweig independently proposed the quark model in 1964.[2] Quarks have six global symmetries called "flavors" in three opposing pairs (3 x 2 = 6). They have been given the whimsical names: up and down, charm and strange, top and bottom. In 1995, Fermilab observed the first quark, a "top" quark. All six flavors have since been observed in accelerator experiments. Quarks have several intrinsic properties, including: mass, electrical charge, spin, and color charge. Oscar Greenberg introduced the notion of "color charge" in 1964 to explain how quarks could coexist in hadrons in otherwise identical quantum states without violating the Pauli exclusion principle.[3] The Pauli exclusion principle affirms that no two atomic and nuclear particles can occupy the same nuclear state. IPhil proposes that the Pauli exclusion principle is analogous to and is logically evolved from the recognition that no two participants in PAIS can have absolute potential energy at the same time. Also, no two spacepoints can occupy the same place at the same time in the logically evolved protospatial system. The term "color charge" was chosen because this property of quarks has three values, which combine in ways analogous to the ways in which the three primary colors of red, yellow, and blue interact on the macroscopic level to produce white light. Quarks carry three types of "color charge," and antiquarks carry three types of "anti-color" charge to identify 3 x 2 = 6 forms of quarks. *Chroma* is Greek for "color," so it is fitting that the study of color-charged quarks and their transient gluon energy-communicators is called quantum chromodynamics (QCD).

Due to a phenomenon known as "color confinement," quarks always appear in groups of three (3). This is similar, though somewhat different, from electric charge,

2. Zweig, "SU(3) Model for Strong Interaction Symmetry."
3. Kaplan, *Pauli Exclusion Principle*.

which in the antecedent electromagnetic system always exists in groups of two (2): positive and negative. However, "electric charges" can be observed separately, while the "color charges" of quarks cannot be observed separately. The reason is that as quarks separate, their energy-communicating particles (gluons) are quickly transformed into new quarks. This is analogous to how energy-communicating spaceons can become spacepoints between other spacepoints. Consequently, quarks may be theoretically established on the quantum level, but they can never be physically separated enough to be externally observed as individual, independent particles.

Standard Model of Particle Physics

In the early years of high-energy particle experimentation, scientists observed a menagerie of new particles, and it took some time to sort them out and organize this nuclear particle "zoo." Gradually scientists developed the standard model in the mid- and late twentieth century. Because ideas came from many scientists by multiple experimental discoveries and theoretical advances, this model has been described as a tapestry woven by many hands. Experimental confirmation of the existence of quarks finalized the current formulation of the standard model in the mid-1970s. It incorporates all the particles associated with the strong force. Then, in 2012, CERN experimentally verified the Higgs boson, which is the foundational particle associated with the mass of nuclear particles. Since then, scientists have experimented with particles on all energy levels, and no new particles have been found. Most scientists now consider the standard model of elementary particles to be complete.

The standard model has been most helpful in putting order into all the discoveries and theoretical insights about the quantized ordering of our physical world. The standard model is believed to be logically self-consistent. However, it has several puzzles. For example: Why are the particles arranged in the 4 x 4 hierarchy in which they are displayed? If this quantized order arose in an evolutionary and subsequently generational way, that hierarchy of particles would metaphysically emerge with the evolution of the equations of the strong force beyond the equations of the antecedent gravitational and electromagnetic logistic. The hierarchy of strong-force particles would normally arise metaphysically in a well-ordered, quantum-energized levels in the logically evolved logistic/equations of the strong force. Nonetheless, in the physical order, because of the free self-initiatives inherent in their potential energy, physical variations in the quantization of their physical appearances would be possible, and empirical conclusions would have to be statistical. A colored graphic display of the standard model of elementary particles can be found in *Wikipedia*.

IPhil's Model for the Strong Force

What does IPhil have to offer to this topic that is new? A 3-dimensional model arising from within the quantized proto-spatial field. Recall that the logical evolution of the gravitational force arose from the breaking of 1-dimensional, quantized symmetry of the proto-spatial system. This logical evolution produced the new operation of linear acceleration in the proto-spatial system, the new parameter of mass, the new force of gravitation, and the new energy-communicating wave-particle: graviton. Then, IPhil proposed that electromagnetism logically evolved from the breaking of 2-dimensional symmetries, by the linear and transverse accelerations of energy-communicating spaceon in the proto-spatial field between spacepoints a quantum distance apart. Because of the quantum distance requirement of all energized bodies in the proto-spatial system, the reciprocal, antithetical motion of spaceon energy-communicators, traversing the quantum distance between spacepoints, produces a circular motion or "quantized spin" by those spaceons. This circular motion is limited to a 2-dimensional plane in 3-dimensional space. This circular motion has no physical center, for its shape is the consequence of the quantum-restraints of the logistic of the quantized proto-spatial system. This 2-dimensional circular motion breaks the 1-dimensional acceleration of spaceons pushed outward by inserting spaceon-spacepoints.

By extrapolation, IPhil proposes that the strong force is the third physical force to evolve logically from the proto-spatial field. This is done by the breaking of planar symmetry of the electromagnetic system. In electromagnetism, that plane of rotation is initially randomly oriented, and the orientation of that plane can be altered by an external electromagnetic field. Breaking that plane-symmetry of electromagnetism internally allows spaceons to circulate naturally between neighboring spacepoints not in a singular plane but in a *spherical* fashion. Because of the logistical requirement that movement of spaceons always be a quantum distance apart, this spherical action has no physical center. This natural spherical motion of spaceons between neighboring spacepoints explains why the strong force does not come from a single spacepoint but from the quantized requirement of its logistic in forming or changing the spherical shell of possibilities between spacepoints.

The reciprocal character of a spaceons formation of a shell of possibilities between two spacepoint allows the logistic of this 3-dimensional interaction and its consequent strong force to endure. However, to produce a complete, energetically quantized shell requires that spaceons go back and forth many times at the speed of c in various random directions between neighboring spacepoints. Consequently, unlike electromagnetism, were the establishment of the circular path would only take one time-quantum, the establishment of a spherical shell would require many time-quanta, and from an external perspective this interaction holistically would travel at less than the speed of light, c. IPhil proposes that the many symmetries in this model give the energized quarks of the strong force the many quantum variations as indicated in the standard model.

Energetically, IPhil recognizes that breaking one degree of freedom within the quantized proto-spatial logistic in the logical evolution of the gravitational forces requires only a small amount of energy, so the gravitational force is the weakest of all forces. Breaking two degrees of freedom would take much more energy, and the reactionary electromagnetic force would be much greater. Further, it would take much more energy to break three degrees of freedom, so the resulting reactive strong force would be significantly stronger than the other two. This matches observations.

Finally, going back to the Primal Absolute Interrelational System (PAIS), its dialogical terms logistically can have two interrelational values—all-giving and all-receiving—but not at the same time! The two interactive terms in PAIS simultaneously cannot both have the interrelational value of "all-giving" or both have the interrelational value of "all-receiving." Logistically, when one interactive term has the interrelational variable value of "all-giving," the other term logistically can only have the interrelational value of "all-receiving." This may be called the "Absolute System's Exclusion Principle." Then, in the logically evolved, proto-spatial system, because of the carryover feature of logical evolution, no two distinct, interrelational, spatial points can have the same four-dimensional spatial coordinates (x, y, z, t). This may be called the "Spatial System's Exclusion Principle." Going further, IPhil proposes that in the logical evolutionary establishment of atomic systems, this same carryover process provides the logistical foundation for "Pauli's Exclusion Principle," where no two interactive terms in an enduring, logically consistent atomic system can have exactly the same matrix of interrelational values.

A full mathematical treatment of this model and the logical evolution of the strong force beyond the gravitational and electromagnetic models would be very complicated. It still needs to be worked out. Unlike the unified field theory, which proposes that the strong force can be logically deduced from a single algorithm, the evolutionary field theory presents a new path by which the strong force and the other fundamental forces of physics arise via logical evolution through the successive breaking of the bidirectional symmetries of the quantized, 3-dimensional proto-spatial system. But what of the 4th parameter of time, which originated before the establishment of the proto-spatial system and before the big bang within the forward leaning of Primal Absolute Potential Energy? While the parameter of time is the first systemic parameter in the *metaphysical* order, it is the last parameter to trigger logical evolution in the *physical* order. Furthermore, how can there be any breaking of the always-forward-moving logistical symmetry of the quantized, dimension of time?

Chapter 12

Logical Evolution of Weak Force and Summary Reflections

Historical Background. Parity Breaking and Weak Force. Weak Force and IPhil. Summary of the Four Forces of Physics. Concluding Remarks.

Historical Background

The discovery of the weak force can be traced back to the French scientist Henri Becquerel, who first discovered radioactivity in 1896 while he was working on phosphorescent materials, which glow in the dark after exposure to light. It seemed to him that the new radiation was like the then recently discovered X-rays. Subsequent research of Marie Curie, Paul Willard, Pierre and Ernest Rutherford, and others indicated that there were several different forms of decay.

Researchers found that elements with the same chemical properties did not have a single atomic weight. Their varied atomic weights were called isotopes. Some isotopes with higher atomic weights, like uranium, were radioactive, expelling alpha particles. Alpha particles were found to consist of two protons and two neutrons, causing a radioactive substance to drop two atomic numbers and four atomic weights on the chart of atomic elements.

In addition, there are two types of beta emissions. When a negatively charged electron is emitted from a nucleus it is called a beta⁻ emission, and when a positron is emitted from a nucleus, it is called a beta⁺ emission. Beta decay allows an atom to achieve the optimal ratio of protons and neutrons within a nucleus. Scientists concluded that these outward emissions were mediated through a different force, which they called the "weak force," because the outward force which expelled particles from the nucleus of atoms is 106 times weaker than the inward "strong force" which holds nuclei and nuclear hadrons together. In addition, when an alpha or a beta particle is emitted from a nucleus, the remaining nucleus is usually left in an electrically excited state. When it moves to a lower energy state, it undergoes what is called gamma decay, emitting high-energy, high frequency, gamma radiation.

The weak force is the fourth of the four fundamental forces of physics. It is responsible for the radioactive decay of subatomic particles. More importantly, it initiates the process known as hydrogen fusion, which provides the radiant energy that is generated when gravity turns large accumulated masses into shining stars. The weak force is also responsible for the aging of stars and their eventual demise.

The distinctive weak force was originally observed in the radioactive decay of different subatomic particles called fermions. These particles, like electrons and positrons, have spin quantum numbers of half integers. This class of particles is called fermions after Enrico Fermi, who presented a theory in 1930 describing how weak interactions occur by means of fermions.[1] This weak force is carried by force-communicating boson particles with different electrical charges, W^+, Z^0 and W^-. The electrical charges on the weak force communicating bosons indicate a close affinity to the electromagnetic force logistic, which describes the behavior of negative, zero, and positive electrically changed atomic particles. Gravitons and photons, the carriers of the gravitational force and electromagnetic force, are without mass, and their force field declines gradually without end. In contrast, weak-force carrying bosons—like strong force carrying gluons—are quite heavy, making their range of interaction short. Thus, the energy-communicator of the weak force is also like the energy-communicator of the strong force, since both have mass and travel at less than the maximum communicator speed of c. From an IPhil perspective, these analogous properties indicate some type of logical evolutionary connection with the three antecedent systems. But how?

Parity Breaking and Weak Force

The laws of nature were long thought to be symmetric in all ways, remaining the same even when they were reflected in a mirror. This mirror-symmetrical rule was called "parity conservation." Parity conservation was assumed to be a universal law because parity was found in the gravitational force, the electromagnetic force, and the strong force. However, in the mid-1950s, Chen Ning Yang and Tsung-Dao Lee suggested that weak interactions might violate this law, and in 1957 Chien-Shiung Wu and collaborators verified experimentally that weak interactions do indeed violate parity.[2]

In 1963, the American physicist Sheldon Glashow proposed that the weak nuclear force and magnetism could arise from a partially unified electroweak theory. In 1967, Abdus Salam and Steven Weinberg independently revised Glashow's theory by suggesting that the masses of the W particles and the Z particle arise through spontaneous symmetry-breaking with the Higgs mechanism. Salam and Weinberg's theory was experimentally confirmed with the observation of Z and W at CERN in 1983.

Because of the many similarities between the electromagnetic force and the weak force, it has been common for reductionist physicists to lump these two forces together into a single "electroweak force." However, from a logical evolutionary point of view, the fact that the gravitational force, the electromagnetic force, and the

1. Cooper, *Enrico Fermi and the Revolutions of Modern Physics*.
2. Hammond, *Chien-Shiung Wu*.

strong force are bounded by the rules of parity while the weak force breaks the rule of parity, means that the weak force must be logistically distinct from those three antecedent fundamental forces.

Weak Force and IPhil

The term "weak force" is confusing because it is not the weakest force—the gravitational force is. Rather, it is called "weak" because it is weaker than the strong force, which hold particles within the nucleus of an atom, while the weak force expels some particles out of the nucleus of some atoms. The strong force is intrarelationally self-oriented toward associatively establishing enduring corporate nuclei, while the weak force inversely is associated with the disintegration of some large corporate atomic nuclei. Logical evolution occurs through symmetry-breaking in the production of a logical inverse. In the first three force, the parameters of distance are bilateral, but in the fourth force the parameter of time is *not bilateral*, but unidirectional. This breaks the symmetry of the first three forces and logically evolved the fourth weak force. While the third, strong force physically builds integrated atoms, the fourth weak physical force is inversely oriented toward disintegration of these atoms, indicating that logical evolution is somehow happening here. Because the process of disintegration is partial, it requires less energy than the integration of whole particles and systems. So, the disintegrating 4th force will be less and weaker than the 3rd order integrating strong force. Also, because the parameter of this 4th weak force is closely associated with the self-initiative which arises from potential energy, the manifestations of the weak force will not be regular in its actions, but it will display free, random self-initiatives of particles emerging from the strongly bound substances—as observed in radioactive decay.

Unlike the back-and-forth, bidirectional symmetries of the three dimensions of the proto-spatial system, the fourth dimension of the proto-spatial system—time—breaks that symmetry by going forward only unidirectionally and not going backwards. Here we have the breaking of the carryover symmetry of the logistic in the fourth dimension of time. The first three fundamental forces display parity, and the breaking of parity indicates logical evolution beyond them. Much of the mathematical work establishing it in this way has already been done, but more needs to be done to frame it in a logical evolutionary manner with the mathematics of the three antecedent fundamental forces.

Beyond establishing a mathematical formulation of the weak force *that* is true to our *physical observations*, IPhil provides consistent philosophical reasons *why* the equations of all four forces of physics are evolutionarily true to the logistical nature of the Primal Absolute Potential Energy of the Almighty before and in the big bang.

PART II: BEGINNINGS

Summary of the Four Forces of Physics

Why are there four, and only four, fundamental forces of physics? In previous chapters it was shown that potential energy is forward leaning and therefore has the parameter of time. It was also shown that time is unidirectional and is quantized. This only-forward characteristic of time in the logical evolutionary process is carried over into all subsequently evolved systems. Also, it was shown that in the first finite interrelational system, the separation of energized, enduring, participant spacepoints requires a new parameter—distance—which is reciprocally bidirectional and quantized. This quantized distance was maintained by the quantized logistic of the spaceon that goes between two neighboring spacepoints. Furthermore, the quantum separation of two spacepoints and the quantum separation of two intermediate spaceons in their 2nd phase as intrarelated particles, produced the 3-dimensional, quantized spatial system of our universe. With the evolutionary carryover of the unidirectional parameter of quantized time, and the bidirectional parameter of our 3-dimensional space, all energetic interactions in our physical world will be expressed in terms of a logically consistent, enduring, 4-dimensional proto-spatial system with the dimensions {x, y, z, t}. Furthermore, the progressive symmetry-breaking of these four dimensions resulted in the logical evolution of four fundamentally different forces. These forces are not *of* the 2nd level distinct, individual terms of the interrelationship but logistically *between* 3rd level systemic terms.

In the following discussion, it will become clear (in accord with the carryover character of logical evolution) how the successive evolution of the four fundamental forces of physics is analogous to the four stages of energy exchanges, as in the game of catch, and the fourfold stages of human life.

1. Gravitational force is analogously like a child, constantly striving to put things in its mouth. Its *thesis* is weak and is always receiving, like the 1st phase of the interrelational logistic. It is an example of the quiet, gentle, receiving, dark *yin* side of *qi*, the foundational potential energy of the primal Dao.

   In the proto-spatial system, generated spacepoints must logistically be at least a quantum distance apart. When a spaceon going between two neighboring spacepoints holds on to at least a quantum of its communicable energy, it becomes a new spacepoint, and the initial symmetry of the proto-space between neighboring spacepoints is broken. Where the spaceon is enduringly transformed into a new spacepoint between neighboring spacepoints, those neighboring spacepoints are pushed apart by the new spacepoint, producing a linear, 1-dimensional acceleration within that proto-spatial field. By the principle of equivalence, this acceleration produces the reactive logistical force of gravitation. In the gravitational system, spacepoint terms take on the parameter of mass, which is always positive and quantized.

   The emergence of a new spacepoint between neighboring spacepoints is much like the birth of a human infant from two parents, for it advances the 1st level of individuals who became a 2nd level interactive couple to establish a 3rd level interrelational family and a 4th level extended family and tribe. This is a major interrelational advancement, with a whole set of additional interrelationships and responsibilities. In the conception of the gravitational systems process, diverse materials in space are brought together to form galaxies and stars, as in the conception of an infant. It is the force of gravitation that brings together large celestial bodies and produces heavenly physical systems on a macroscopic scale. So, in a true analogous way, the gravitational attraction of dispersed particles "gives birth" to planets, planet systems, stars, and multi-star galaxies.

Scientists observe that space is expanding. Because they do not know the origin of the energy source causing space to expand, they call its cause "dark energy." IPhil finds the source of this "dark energy" within the quantized mechanics of the proto-spatial system. IPhil recognizes that in the generated, quantized lattice of proto-space, every energy-communicating spaceon goes through its own four-phase interrelational logistic, which involves phases of receiving, holding, orienting, and letting go. In the 2nd phase of that process, the intrarelational holding of energy by the spaceon can be prolonged to turn it into a spacepoint. Because of the quantum character of the logistic of the proto-spatial system, the logistic of that system pushes the neighboring points away from the new spacepoint. This causes proto-space to expand at that point. Around bodies with large mass, it causes the spatial system to warp. The neighboring points thus accelerate from their former positions, and their acceleration results in the logical evolution of gravitational force.

"Dark matter" originates from spacepoints holding their received energy for a short time. According to Einstein's famous equation, $E = m c^2$, in the gravitational logistic, the energy intrarelationally held by a spacepoint for a quantum time takes on a mass equivalence. In this logical evolutionary advancement of the gravitational system, a pure spacepoint takes on the quantized parameter of mass. When space is warped around large celestial bodies, such as in large galaxies and in black holes, the *density* of mass-holding spacepoints increases, and this produces a gravitational gradient from the "dark matter" of the spacepoints of the spatial system, which adds to the effective mass of large systems of large stellar bodies.

2. Electromagnetic force is like an adolescent. It is an example of the aggressive interaction with neighboring individuals and elements. It is like the bright *yang* of the *qi* of the foundational potential energy of the primal Dao. Its *antithesis* is dualistic and quite strong. Through the inverse character of logical evolution, beyond the always positive character of gravitation, the electromagnetic system produces both positive and negative charges. This is analogous to the logical evolution of positive natural numbers to negatively signed numbers. It is easy to observe that like-charged particles are highly repellent and, unlike-charged

particles which are highly attractive. This is similar to the competitive behavior between like adolescents, and like the generally positive behavior between adolescents and their idolized leaders.

IPhil recognizes that the back-and-forth movement of spaceons between spacepoints will be according to the fourfold metaphysical phases of the logistic of energy. When two spaceons simultaneously move in different directions from neighboring spacepoints, the quantum logistic of energy requires those spaceons to always to be a quantum distance from each other. As a result, the paths of these two spaceons will not be linear but circular on a flat 2-dimensional plane. This produces the quality of "spin" that is recognized in particles like photons and electrons. With nothing in its center, this circular path is anchored at two terminal spacepoints in the quantized proto-spatial system. Because those spins can go in opposite directions, and because their energy is quantized, this spatial spin is in multiples of $\pm\frac{1}{2}$. Initially, the plane of the paths of these two opposing spaceons will be randomly oriented in the proto-space.

In addition, within the 3-dimensional spatially quantized lattice of space, the circular motion of the spaceons can also produce a precession-like magnetic effect on neighboring tiers in that lattice. The electrostatic and magnetic effects are drawn from the logistical character of protospace, so these linear electrostatic and rotational magnetic disturbances will travel through the lattice of space at the maximum speed c. These circulations are basically 2-dimensional, and they display two degrees of freedom. Consequently, more energy is involved in the stronger electromagnetic force than in the one degree of freedom establishing the gravitational force.

3. Strong force is like an interrelational adult, striving to build interrelational systems. It is an example of the harmonious interaction and interdependence of asymmetric *yin* and *yang*, as symbolized by the *taijitu* of the *qi* of the foundational potential energy of the primal Dao. Compared to the antithetical electromagnetic force, the strong force's *synthetic* tendency pulls quarks strongly and systemically together into enduring, corporate nuclear particles and atomic systems. IPhil recognizes that the proto-spatial system is spatially 3-dimensional, and accelerations are possible in 3-dimensional, bidirectional ways. Consequently, in strong interactions involving three degrees of freedom, gluons inversely tightly bond quarks together in a corporate, positive way and have electrical charges of $\pm\frac{1}{3}$ quanta, indicating this system's analogy to the fractional number system. Because of the carryover characteristic of the logical evolutionary system, this system's three degrees of freedom also has one degree of freedom of the gravitational system of mass and the two degrees of freedom of the electromagnetic system of electrical charge. That is why particles drawn together by the strong force also can have mass and electric charge. By logical evolution, the three degrees of freedom produce the new parameter of color charge.

The strong force breaks the enduring plane-symmetry of the circular 2-dimensional motion of energy-communicators in the electromagnetic system. In this way, the energy-communicators between spacepoints can form an energetic spherical shell, which has 3-dimensional volume, and which is anchored at neighboring spacepoints—with nothing in its center. Thus, this 3rd level of spatial evolution establishes intrarelational nuclear and interrelational atomic particles with volume, and it takes time to form these 3-dimensional shells between neighboring spacepoints. Consequently, the accumulated effect of the gluon energy-communicators does not occur at the maximum speed of c but with increase mass at a slower speed. Gluons are the strong, force-carrying particles that bring 3-dimensional things ever closer together. Since the strong force has three degrees of freedom, it is by logical evolution beyond gravitational force's one degree of freedom, and beyond the electromagnetic forces two degrees of freedom. The accumulated degrees of freedom of the strong force are 1 x 2 x 3 = 6. This is the number of formally distinct interrelational terms possible in the strong force system, and each can have different levels of quantized energy. Also, the symmetries associated with the strong force and the advanced quantized levels possible in this logical system produce a well-ordered 4 x 4 array of fundamental strong-force quark particles—as described in the experimentally established standard model of nuclear particles.

The adultlike interrelational logistic of the strong force progressively produces larger and larger individual physical objects, which can produce larger and larger corporate systems, like atoms and molecules. As adults are more advanced and dominant than adolescents, so too the magnitude and sophistications of the strong force are greater than the adolescent electromagnetic force and the infant gravitational force. Unlike adolescents whose antithetical ideals push things as fast as possible and as far as possible, adults' concerns are more local. They build enduring systems at slower but more stable rates. From these, enduring entities are formed. In this way, the strong force within our foundational spatial system analogously shows itself to be like an interrelationally and constructively adult, long before humans came on the scene.

4. Weak force operates analogously to the debilitating phase of elders. In Daoism, within the symbolic *taijitu*, the dynamism of yang and yin is recognized, and each subsides as a new dialogical opposite starts to appear. The weak force pushes an integrated atom of a high atomic number to freely "let go" of particles from its nuclei or atoms. This causes things to become diminutive in radioactive decay. The spent particles can be received by others, making them possibly new and greater. As in the logistic of energy, the fourth aspect involves a waning and "letting go." This is similar to the way an elder looks forward to somehow advancing beyond the present systemic state. In this way, this fourth force is *transthetic*. Previously, in a 3rd level way, atoms of high atomic numbers developed through

fusion in places like the sun where gravitation compresses particles, producing elements of higher atomic number. In this 4th stage, the negative, weak force dissociatively reverses the 3rd level associative, synthetic processes of the strong force, causing atoms with high atomic numbers to give off alpha particles, leading to the lessening of the integrity of elements, even to the demise of antecedent systems.

Mirror symmetry or parity is present in the reciprocity of the interrelational logistics of gravitation, electromagnetism, and strong force, which are logically evolved from the bidirectional distance parameters of 1-dimensional, 2-dimensional, and 3-dimensional accelerations respectively. However, the dimension of time is not bidirectional but only unidirectional. Because of this, the carryover unidirectional character of time breaks the bidirectional symmetry of logical evolution of the previous three physical forces. That is, because of the unidirectional character of quantized time, the pattern of mirror symmetry in the first three logically evolved forces is broken. There is no going back. In the 4th stage of life, progressive development in life according to the wave character of the interrelational logistic of energy is logically inverted, much as elderly individuals experience the dissociation of their materiality as they approach the disintegration in their physical death.

While gravitation pulls matter together into large, massive objects like the sun, the weak force is responsible for hydrogen fission, which produces the energy-spewing radiation from the sun, thereby giving its light and energy to others for *their* development. This logistical transformation of potential mass and energy into radiant light and heat supports the evolution of more advanced, interrelational worlds, such as our earth with its evolutionarily advancing life forms. In this way, it may be said that the weak force causes atoms systemically to "die" that others may live in a more interrelational way. The sacrifice of their potential energy is also for the purpose of some type of advancement of other individuals and systems. IPhil recognizes that analogously, the universal value of the freeing 4th phase of partial sacrificial surrender is found not only in material elements but also in plants, animals, and humans for the benefit of great physical advancement of their own kind and of others.

Einstein's famous formula, $E = mc^2$, indicates that the parameter of mass is a capsulated form potential energy, which is capable expressing itself in a kinetic way—as in an atomic bomb. By analogy, IPhil proposes that the potential energy from of the parameter of electric charge likewise has a kinetic energy equivalence. Furthermore, IPhil proposes through logical evolution that the potential energy of the parameter of color-charge mathematically has kinetic energy equivalence. Just as the potential energy of the parameter of mass can be released as atomic power, IPhil proposes that the release of the potential energy found in the parameters of electric-charge and color-charge will open to us two new sources of power.

Concluding Remarks

Modern theoretical scientists have been working hard to discover the primary axioms or algorithm of the unified field theory, because they believed the four fundamental forces of physics could be logically *deduced* from it. From a rationalist point of view, they thought that from this all-encompassing algorithm they could logically deduce everything that can and will happen in our physical world. This is a deterministic scheme, which has no freedom and no self-initiative. This deterministic belief comes from a long rationalist tradition. In 1814, the great French mathematician Pierre-Simon Laplace published the following in *A Philosophical Essay on Probabilities*. It is considered the first articulation of causal or scientific determinism.

> We may regard the present state of the universe as the effect of its past and the cause of its future. An intellect which at a certain moment would know all forces that set nature in motion, and all positions of all items of which nature is composed—if this intellect were also vast enough to submit these data to analysis—it would embrace in a single formula the movements of the greatest bodies of the universe and those of the tiniest atom. For such an intellect, nothing would be uncertain, and the future, just like the past, would be present before its eyes.[3]

Given a complete set of physical measurements at any giving time in the universe, using the scientifically known physical laws of nature, he claimed it possible to mathematically described every physical event in the history of world.

Subsequently in physics there have been several attempts to establish some type of unified field theory. They are based upon the assumption that all fundamental forces and elementary particles can be logically deduced from the characteristics of a single algorithmic field. James Clerk Maxwell developed the first successful, classical unified field theory. In 1865, he published his famous paper on the mathematical characteristics of the electromagnetic field.[4] He was able to reduce all aspects of electromagnetic fields to a set of four equations, from which all electrical and magnetic equations can be logically derived. All four of these equations exist and operate simultaneously. Which equation(s) are used depends upon the physical circumstances being investigated. Theoreticians subsequently have been driven to interconnect the equations for all four forces of physics in a deductive and logical way. However, by thinking this way, they put themselves intellectually into a closed, rationalistic, essentialistic box out of which they cannot break.

---

3. Laplace and Dale, *Philosophical Essay on Probabilities*, 19.
4. Maxwell, "Dynamical Theory."

PART II: BEGINNINGS

In a closing chapter of *A Brief History of Time*, the noted theoretician Stephen Hawking projected:

> If we do discover a complete theory, it should in time be understandable in broad principle by everyone, not just a few scientists. Then we shall all, philosophers, scientists, and just ordinary people, be able to take part in the discussion of the question of why it is that we and the universe exist. If we find the answer to that, it would be the ultimate triumph of human reason—for then we would know the mind of God.[5]

Built upon the nature of potential energy, interrelational philosophy (IPhil) provides the experientially grounded, logically consistent foundation for the evolutionary field theory. This theory is not the product of professional scientists but by a layman with a scientific avocation. IPhil establishes how we can know the "mind of God" and the "kingdom of God," by looking analogously from the physical growth and evolution of the advancing energetic events we witness in daily life.

Elsewhere Hawking asked: Where is the fire in the equations of the foundational algorithm of our dynamic world? IPhil recognizes that of themselves an algorithm and equations *effectively* can do nothing; they only describe the *form* of interactions. IPhil identifies the "fire" that animates all equations as the received potential energy from PAPE, which was antecedent to the big bang, and by which the carryover character of logical evolution continues to be the enduring, potential, existential, metaphysical foundation of all transient kinetic items and events in our physical world.

Isaac Newton empirically proposed the universal gravitational formula using observational data, and people supported it because it made accurate predictions. The successes of this and other empirical discoveries then moved the studies of physics and astronomy out of the department of natural philosophy into the independent academic disciplines of physics, astronomy, mathematics, etc. In the pursuit of physically verified equations, however, these disciplines abandoned and lost their metaphysical, philosophical roots.

Snubbing the ivory tower speculations of earlier philosophers, modern scientists swung to the opposite extreme by treating with the greatest intellectual skepticism all inferred metaphysical reflections. They now accept physical evidence and *only* physical evidence as the accepted criterion for truth. In doing that, they closed themselves into the box of materialism, and they rejected all that is spiritual in our world. As a result, all non-empirical knowledge was looked down upon as more false than true. In response, moving beyond a 2nd level either/or approach to essentialistic truth, IPhil has a both/and respect for both, realizing that metaphysics and physics each has partial knowledge of what is real in our world. Although foundational potential energy is not directly observed but only secondarily known through physical observations, it is real—just as some nuclear particles, like neutrinos, are not directly observed but

---

5. Hawking, *Illustrated Brief History of Time*, 233.

are known secondarily through their effects. The pursuit of metaphysical truth is a valuable occupation, even though it can only be known by inference from the physical level. Still, knowledge of the physical truth is the correspondence of interior mental ideas with exterior physical events. Discovery of universal, interrelational logistics goes beyond observation, yet our partial observations do point to the probably metaphysical truth of these inductions and beyond.

Beyond observations, there is a more advanced, 4th level of knowledge, namely, wisdom. Wisdom includes knowledge unconsciously known from experience, selective conscious images, and interrelational rational knowledge. Also, wisdom goes beyond these to project likely events in the future, beyond what is physically experienced. It is forward leaning of the potential energy in individuals and all thing that provides the wisdom that on the 4th level intuits the future from the 3rd level knowledge of the logistics of the involved players. Potential energy is constantly pushing us beyond knowledge of existence, knowledge of sense images, and knowledge of logistical interrelationships toward establishing systems that are more enduring, intrarelational, interrelational, and progressively good.

Most people today recognize that physical evidence securely establishes evolution as a pervasive aspect our developing world. Also, a good number of people oppose the materialist position that God has no part in this process, even though they have difficulty establishing a solid connection between a spiritual God and material evolution. Interrelational philosophy (IPhil) describes how logical evolution establishes that the driving force of all evolution in our world is the enduring, pre-bang potential energy of the Almighty. Also, the continuing and progressive logistical order of evolution in our world is founded upon the ever-present interrelational logistic of the Almighty. At this point is this discussion, it cannot be determined whether the Almighty is impersonal potential energy or only a potent personal reality. The question whether the Almighty is a person will be discussed after the discussion on human persons in a subsequent chapter. Thus, at this point in the book, IPhil is only deistic. That IPhil is also theistic, recognizing that the Almighty is a person, has yet to be established.

The subtitle of this book is: *How Our World Evolved from God*. So far in this book I have described only how the proto-spatial system and the four forces have evolved in the big bang from antecedent Almighty Primal Absolute Potential Energy (PAPE). The evolutionary history of our world, however, is more than the big bang and the emergence of the four fundamental forces of physics. The story of the evolutionary development of our world does not end here. The story of how our world evolved from God has only just begun.

Part III

Developments

Chapter 13

Evolution and History

Daoism. Origin of Big Bang Idea. Celestial Development. Geological Times. Human Historical Development. Stimuli of Evolution. Is Societal History Cyclic? Fourfold Cycles.

This book has now come to its fulcrum point where its emphasis tilts from the theoretic, metaphysical side of potential energy to the experiential physical side of kinetic energy. The discussions in previous chapters have been quite technical and abstract; now they will be more down to earth. Still, the discussions in this book have never been strictly one way or another, for our energetic world is always a mixture of kinetic and potential energy, empirical physical experience and metaphysical, conceptual knowledge. I made a point of first grounding the metaphysical axioms of IPhil in common, every day, physical experiences.

A foundation stone of interrelational philosophy (IPhil) is the interrelational logistic. Its physical prototype is the game of catch, in which two people toss a ball back and forth. In this game, each player in turn receives the ball, takes possession of it, searches where to throw it, and "lets go" of the ball to the other player. Also, the interrelational logistic is fourfold not only in the dialogical players (or terms) but also in the energy/matter communicators as well. Still, the game of catch is located neither in the ball nor in the players but in their logistically well-ordered, systemic interactions.

The metaphysical description of the game of catch is drawn from ideal interrelational possibilities. Yet an actual game of catch will have free, physical variations from this ideal. However, reality is not just the kinetic, physical side of energy as we observe it. Energy has a potential side, which is ordered toward ideal interrelationships. Some variations are "consequentially dissociative," and they can seriously disrupt or even terminate the game of catch. However, many physical variations are "inconsequentially dissociative" from the metaphysical ideal game of catch, allowing the game to continue and even advance with easy corrections.

Fourfold interrelational logistics are found on all levels of energy exchanges in our world. Through the analogous character of the results of logical evolution, well-ordered, reciprocal exchanges of energy are recognized in both West and East.

PART III: DEVELOPMENTS

Daoism

Daoism, previously spelled "Taoism," is a philosophic spirituality that recognizes the cyclic character of everything energetic in our world. In the Far East, the typology of the dynamic interrelational "four" is expressed in the asymmetric dualism of yin/yang. These are not 2nd level diametric opposites but 3rd level dialectic contraries.

In the West, starting with the singular potential energy antecedent to the singular big bang, potential energy is recognized as the foundational cause of all observed, physical, kinetic actions in our world. In China, Daoism calls the foundational energetic source of all dynamic activities "qi." Both Daoism and Western science, respectively, considers "qi" and "potential energy" impersonal and natural. Christian theology pushes beyond that and recognizes all received potential empowerments as free and freeing personal graces from the Almighty. But more about that later.

Even a peripheral comparison indicates the similarities of the metaphysics of IPhil and the cosmology of Daoism. The primal Dao is One, symbolized as a circle. It is empowered by qi, the foundational life-force consisting of yin and yang. Condensed (qi) becomes life; diluted (qi) is indefinite potential.[1] The inner dynamism of qi, like the inner dynamism of potential energy, is recognized in their logically inverse effects. In their dialogical dissociations and associations, there are many similarities between the yang and yin in the Dao driven by qi, and the all-giver and the all-receiver in PAIS driven by potential energy.

The eight trigrams of *I Ching*

The words *yin* and *yang* mean "dark" and "light" respectively. The light *yang* and dark *yin* dots in the middle of the dark yin and light yang sections respectively indicate that the seeds of their complement lie at the heart of each aspect of the Dao. In the processing of the "Way" of the Dao, each seed in the Dao grows until it overwhelms the antecedent characteristic, transforming yang into yin and yin into yang. This is like the logistical entanglement transformations in PAIS of the all-giver into an all-giver-receiver and of the all-receiver into an empowered all-receiver-giver.

In its simplest form, the feminine yin (__ __) and masculine yang (_____) can interact in four ways.

1. Robinet, *Taoism*.

1. yang/yin   strong over weak—as a parent over a child
2. yang/yang   strong over strong—as two strong individuals in opposition
3. yin/yang   weak over strong—as a servant leader over power blocs
4. yin/yin   weak over weak—as a tender person caring for the elderly

In IPhil, from the fourfold interrelational logistic of the all-giver and all-receiver, a myriad of physical things can be logically evolved. Similarly, from the asymmetric interaction of yin and yang, a myriad of things is produced. As Lao Tzu wrote in *Tao Te Ching*, "The Great Integrity expressed one. One manifests as two. Two is transformed into three. And three generates all the myriad entities of the universe. Every entity always returns to yin after engaging yang. The fusion of these two opposites births the vital energy that sustains the harmony of life."[2] The primal one Dao is not consciously specifiable but intuitively dark. The Dao that is spoken is not the real Dao.

Like the monistic *wuji* and the complementarity of yin and yang in the systemic, the symbolic *taijitu* indicates the reciprocal interactions in the dialogical PAIS of the monistic PAER which originates from the singular, unformed, energetic PAPE. Thus, Daoism and IPhil describe similar processes leading to the formation of the energized, interrelational, logically consistent, enduring, finite systems in our world.

Origin of the Big Bang Idea

As was described earlier, Einstein presented his general theory of relativity to the Prussian Academy of Science in November 1915. He assumed the state of the universe was static, and to express that in his equations, he introduced a cosmological constant, which he later called the greatest blunder of his life. Georges Lemaitre was a Belgian priest, astronomer, and professor of physics at the Catholic University of Louvain. Drawing upon the characteristics of the general theory of relativity, he was the first person to propose the theory of the expansion of the universe from a "seed." Lemaître's original paper was later published in an abbreviated English translation in 1931. There, projecting backwards the notion of an expanding universe, he proposed that the universe expanded from an initial point, which he called the "primeval atom." He developed this idea and published it in the professional journal *Nature*.[3] There he described this beginning as a cosmic egg exploding at the moment of creation. He was invited to London to take part in a conference on the relationship between the physical universe and spirituality.[4] Fred Hoyle pejoratively mocked this idea as the "big bang theory," and the name stuck.

---

2. Lao Tzu and Dale, *Tao Te Ching*, 42.
3. Lemaître, "Beginning of the World," 706.
4. Lemaître et al., *Learning the Physics of Einstein*.

PART III: DEVELOPMENTS

The first, real, physical, experimental evidence of the big bang came by accident with the discovery of a microwave whisper by Ano Penzian and Robert Wilson of AT&T Bell Labs in 1964. This provided physical evidence that the universe did begin in a big bang. In 1989, NASA launched the COBE (Cosmic Background Explorer) satellite to measure the sea of microwaves that bathes all of space and found it nearly uniform, but with some variations. In 2013 the Planck Satellite telescope provided data indicating more accurately that the big bang took place 13.81 billion years ago.

Celestial Development

After the 1st phase evolutionary, energized establishment of the non-spatial "cosmic egg," the universe for a time was remarkably homogeneous, as observed in the Cosmic Microwave Background (CMB). Its initial, very high temperature indicates very high levels of internal kinetic energy. But as the universe expanded, its temperature dropped, triggering an intrarelational establishment of small particles to begin the 2nd major phase of individualized cosmic development.

1. Primordial quantum fluctuations resulted in the formation of elementary particles, mainly electrons, protons, neutrons, hydrogen, and helium. At that time, because electrons were initially very energetic, they were not attached to nuclei. As the temperature of space went down and the kinetic energy of primal material forms dropped, the intrarelational orientation of this earliest stage of matter inversely produced integrated atoms with negative electrons attached to the positive nuclei to from atoms. As a result, electromagnetic communications between atoms became possible. Then suddenly, the dark universe burst forth with a new, logically evolved, interrelational reality: light. Until stable atoms were formed, there was no electromagnetic radiation in this electron soup, and everything was dark. Like humans, our cosmos was conceived in darkness (*yin*), and then, like the (*yang*) birth of a child this progressive energy produced light in this world.

2. Gravitation brought these clumps of atomic matter together to form centers, which grew to become proto-galactic systems. Within these galaxies, elements condensed into the first stars. Using the Keck telescope, a team from the California Institute of Technology found six star-forming galaxies about 13.2 light years ago, when the universe was about 500 million years old. Year by year, physical observations penetrate close and closer to the big bang. Galaxies grew quickly by consuming smaller galaxies.

3. Within those galaxies, stars developed with great mass. Their consequent gathering of mass increased the inward force from their gravitation, causing their cores to heat up. Here hydrogen and helium were cooked, forming atoms of higher atomic weights, such as carbon and iron.

4. Eventually the heat became so intense that these large stars exploded as super novae, spewing heavier atoms throughout the universe.

5. These heavier atoms were then incorporated into second-generation galaxies, in which stars formed with heavier elements.

6. There are numerous types of galaxies, and their development follows a general logistic. Because of the heavier atoms and matter spewed in space from exploding first-generation galaxies and stars, second-generation galaxies and stars accrued by gravitation from different directions, causing these celestial bodies to spin in foundational space, causing many to become flat desks.

7. Matter circulating at a distance from these stars accrued into planets in stable orbits around their respective suns. Some matter also accrued in orbits around these planets, producing satellites of various compositions. In this way, solar systems with various planets and moons developed, each with their stable celestial orbits.

8. As the planet we call Earth cooled, the conditions were established where biological life could evolve from the energy and materials of our planet.

In summary, from acquired scientific evidence, interrelational development from the big bang to the formation of our earth is clearly progressive. Also, it is possible to divide the observed galactic history of first and second-generation galaxies into four major eons, analogous to the four phases of the interrelational logistic.

Geological Times

Geologists have been able to chart the development of the planet earth. Again, it can be easily divided into four analogous major stages of development.

1. Among the planets of our solar system, the earth was formed approximately 4.54 billion years ago (bya). It continued to cool and gained an atmosphere. This atmosphere and the earth's magnetic field provided within one billion years the conditions needed for biological life to appear, during the Archean Eon.

2. Approximately 2.5 bya, the earliest colonial algae and soft-bodied invertebrates were formed in the Proterozoic Era. In the Paleozoic Era, 570 mya (million years ago), diverse plants, insects, fish, and reptiles evolved. Dinosaurs, and then birds, developed in the Mesozoic Era, 245 mya. This period ended with the dinosaur extinction 65 mya.

3. Then, small mammals began to grow and dominate the earth. The earliest humanoids marked the beginning of the Pleistocene epoch 1.6 mya. Tool-making *Homo sapiens* emerged from Africa about 250,000 years ago. Here natural evolution advanced into artificial evolution. Humanoids developed tools and began to work more cooperatively and intelligently, and thereby increased in dominance.

4. *Homo sapiens sapiens*, or modern man, emerged ca. 100,000 BCE (Before the Common Era). This humanoid species was highly intelligent and more cooperative. They organized and built many new things. The first things members of this species built were large worship centers, indicating that their evolution had advanced into the transcendentally oriented *Homo religiosus*.

Human Historical Development

1. The Neolithic Period, ca. 100,000 BCE, marks the initial transition of humans from hunting and gathering, to settled agriculture. This occurred just as they began their migration from Africa into the Middle East.

2. They gathered first in small groups and then into strong, localized clans around worship centers, which soon became living and commercial centers. Local family and charismatic organized clans and tribes exercised leadership and authority over them. These groups were connected by valued trade routes. Around 8,000 BCE, Jericho became a large city, and Catal Huguk, a town in Anatolia (Turkey), reached a population of 6,000.

3. Over the course of time, strong emperors gathered around themselves large armies and expanding bureaucracies to claim dominion over vast countries and empires. They built huge temples, palaces, and large public works starting ca. 4,000 BCE.

4. Modern science began with Galileo in the seventeenth century, and by the twentieth century CE telecommunication and worldwide commerce broke all kinds of natural barriers that had kept nations isolated. As all areas of our earth became increasingly explored and tied together commercially and intellectually, humans began realistically envisioning traveling beyond the confines of our earth, sending cosmonauts into space, astronauts to the moon, and gathering data and pictures of planets and satellites in the solar system and beyond.

Describing the evolutionary development of our universe from the big bang 13.8 bya to today, the above schema is awesome. The efforts of thousands of scientists have been combined to show in a very consistent way how the energy of the big bang became substantialized and often grew in complexity, sometimes in random, chaotic ways, but overall in well-ordered, logistically consistent ways, producing our wonderful, increasingly interrelational world.

The above schema indicates that, in general, the findings of evolutionary scientists and archaeologists fall easily into a broad, fourfold pattern, which is analogous to the four phases of the interrelational logistic of potential energy. Whether the above fourfold schema can be subdivided, and its subdivisions be described in a more refined fourfold ways, remains to be seen. Internal freedom and external randomness blur the boundaries of the fourfold metaphysical order of things. Nonetheless, a general fourfold pattern seems apparent, analogous to the fourfold interrelational logistic of primal potential energy.

Some people say that nothing happens by chance; everything happens for a reason. IPhil recognizes this as 2nd level, either/or thinking. In contrast, some materialistic scientist claim that advancement only occurred through 1st level randomness. Physical evidence testifies that there are truly random events in our universe. Nonetheless, as in Schrodinger's equations, randomness is well ordered within a system's logistic. Randomness occurs in the physical order within logistically defined systems, and these progressive, physical systemic orders occur in enduring logistical systems in the metaphysical order. Since randomness occurs within a logistic, and the logistic of energy is directed toward states that are more interrelational, all true randomness has a logistical orientation toward forming and evolving more interrelational systems. Evidence from the evolutionary history of our material world indicates that.

Did God make the universe? Not exclusively. Just as God eternally makes God through the interrelational logistic of God's energy, so too the things in our universe temporally make themselves through their received energy. The Almighty's potential energy is the universe's underlying primal material cause, while that energy's interrelational logistic is the primal formal cause of the universe. Subsequently, free self-initiatives of all subsequently evolved enduring systems, generated finite individuals in our universe in their own individual free ways. Their unconscious or conscious self-initiatives to transform their potential energy into kinetic energy are the efficient causes of all that is observed in the universe. In this way, 3rd level IPhil recognizes that there is no single creator. Rather, every creature panentheistically is offered the opportunity to freely *participate* in the creativity found in the evolution of our universe, which was initially generated by and is still existentially supported by the Almighty by the carryover character of the evolutionary process.

Stimuli of Evolution

In the metaphysical order of potential energy, the absolute communicating Spirit or an evolve form of a finite communicating spirit within potential energy is the driving force in the progressive process of generation and evolution. In the physical order, there are four analogous levels of initiators of these processes. (1) The potential energy of parts can randomly stimulate physical various mutations of the DNA molecule. (2) Necessity is the mother of invention. Potential energy from dissociative environmental challenges, can stimulate progressive, evolutionary changes. (3) Orderly potential possibilities within the material boundaries of a given logistic searches for new, successful combinations. (4) The push of one's integrated potentiality can intuitively stimulate advanced visions, inspirations, and new discoveries. On all levels, the potentialities of one's energy ultimately received from and supported by the Almighty are the driving forces of forward, greater interrelational development—unconsciously or consciously freely self-initiated by the energized individual.

Is Societal History Cyclic?

The statement "Life is cyclic" is a broad, generalized statement that has much truth to it. But if life is cyclic, why is it not perfectly so? Historians hate the claim that life is cyclic, for their primary interest focuses upon incidents that break from ideal, cyclical repetitions. IPhil recognizes that. Individual histories and social history are only generally cyclic. Because of self-initiating freedom in the 4th order, free human activities can be variant from a 3rd level, cyclic ideal. Free self-initiatives are an inherent metaphysical aspect of the 4th stage physical scenarios. IPhil expects that 4th level physical variations *can* and will happen within the formal, 3rd order, interrelational, metaphysical ideal. Historians base their conclusions exclusively upon physical evidence, and professionally they refuse to accept that there is a well-ordered metaphysical logistic behind the freely variant physical order from which they draw data. As the present builds upon the past, variations from the ideal cyclic pattern will grow exponentially. Contemporary history builds upon past history with new interrelational features appearing in each successive cycle. Nonetheless, the four phases of development of the metaphysical ideal remain generally visible in the physical order, as is commonly recognized.

Historians focus upon what is unique in history, while IPhil and comparative individuals recognize that behind these many variations there is a general logistical pattern in human and all energized behavior. If there were no foundational, consistent logistic in life, then with all things free in their own ways, no patterns of development would happen or be observed. But that does not match our experience. History shows a general 3rd level pattern of fourfold cyclic development, with many 4th level free variations in their details. It is not either/or but both/and from the metaphysical and physical sides of the existential energy of our world.

What is nature-made and human-made often show spurts and halts in their development in history. Our consciousness is awaken by things that are different. We tend to ignore what is routine and ordinary. IPhil recognizes that extraordinary events arise from ordinary events, and ordinary, routine events are the foundation of human life. It is in the regular, boring, routine processes of life where interrelational philosophy of reality finds its foundation. Here, a general sinusoidal pattern of development is apparent. It can be seen on many levels: in their physical parts, in their physical and psychological maturation, and in the rising and falling phases of societies and international cultures. Some people revel in finding variations that contradict consistencies of development, and some people ignore important variations from common patterns of development. It is important intellectually to break from an idealistic, antithetical, 2nd level judgment that history is either perfectly cyclic or it is not cyclic at all. In contrast,

3rd level synthetic thinking recognizes that historic changes are both basically cyclic in their continuity, while partially variant in the exception to the ordinary.

In our personal lives and in history it seems we are constantly starting over again, and again, and again. Indeed, in some ways we are. IPhil recognizes that in life and in history, we are progressing from passive receiving, to active giving, to passive receiving, to active giving. . . . It's like breathing in and breathing out, taking in a little more potential energy via energizing oxygen each time. Hopefully in our daily routines we are learning from our past mistakes and advancing interrelationally a little each time around.

Beyond being cyclic, IPhil recognizes that history is also progressive in an evolutionary way. New civilizations and new projects do not start from scratch. They are built on the achievements and mistakes of the past in varying degrees. That is, advances in the history of a society are like the backstitch in sewing. The backstitch is much stronger and enduring than a running-stitch. In a backstitch, there is a series of "e's." These stitches loop back and catch part of the previous loop before proceeding a little farther to the next stitch.

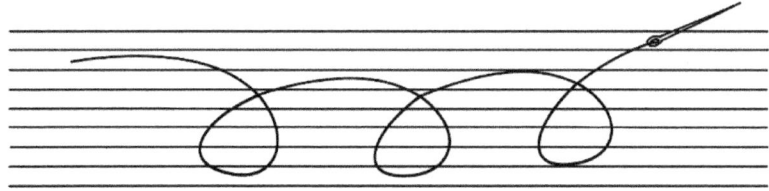

Illustration of back stitch, which is a progressive series of overlapping "e"s.

Because cultures and civilizations in varying degrees build upon the past and go beyond it, cyclic history is progressive, like an ascending spiral staircase, in which each level generally repeats the cyclic form of the preceding level and yet rises above it. As in potential energy, the "transthetical" 4th phase has two aspects to it. There is a tearing down of past structures, *and* a remnant of the failed age provides a foundation for the next level of development. Because evolving systems are regularly supplemented in their development in evolutionary history, they expand interrelationally in succeeding generations and eras. That is why the logo for IPhil, found at the end of each chapter, starts from the singular absolute energetic beginning point and advances in ascending, expanding spirals until it collapses in the end into an advanced singular state that is similar to but interrelationally more advanced than the initial state.

Hegel's historical reflections identified three natural stages to the development of societies driven by the Spirit.[5] IPhil recognizes that the Spirit of the Primal Absolute Potential Energy is like the *Geist* in Hegel's schema. The interrelationship-pushing Spirit is

---

5. Copleston, *History of Philosophy*, vol. 7, ch. 10.

the driving force of primal potential energy behind the advancement of individuals and groups. The inner forward-leaning Spirit of potential energy advances things according to the fourfold interrelational logistic of the absolute energy of the Almighty.

IPhil recognizes that Hegel's three-stage scheme is analogous to the stages of the interrelational logistic of energy. In addition, IPhil recognizes the importance of the fourth, "letting go" stage of development, in which systemic realities need to disintegrate for the sake of integration into a future, more interrelational state. People are primarily interested in physical success and power and not in the decline and fall of institutions. Similarly, historians regularly ignore the dismantling phase of societies and systems. People are interested in reading about new things and not the waning years of a society or of a human life. What interests most people are the daring exploits of the clever, aggressive, 2nd stage protagonist, striving to overcome a variety of problems and developing something new in the dramas of life. However, 4th level reflective persons and philosophers are holistically interested both in the building up and in the collapse of the things in our current world, recognizing the possibility of a subsequent, advanced, logistically breaking, transcendental state. That is part of a transcending—not flat—"circle of life," whose advancement is not fast but gentle in the long run. IPhil sees the "letting go" of the 4th phase of individuals and systems as being very important because of its potential to slowly evolve systems into a higher metaphysical order through a reforming, innovating logical inverse progenitor.

1. *Thesis*. In this initial period, the material and physical foundations of an individual or society are discovered and embraced. Like a child, a new culture or society is initially very weak. It usually flounders directionless as it begins to grow. Because closeness to the earth dominates this phase, this phase may be called a formative, earthy Green Age.

2. *Antithesis*. As a civilization grows, like an adolescent, it comes into conflict with neighboring groups, cities, and nations. The winners become leaders who not only conquer their opposition, they also discover ways to grow intrarelationally in ownership and control of the spoils of their conquests, thereby making themselves and their group stronger, expanding their inward dominion and outgoing power. This intermediate, aspiring, succeeding phase may be considered a Silver Age.

3. *Synthesis*. After a civilization expands territorially, a civilization enters a period of more conservative systemic integration, like an adult, during which expansions are done slowly and with greater caution. Digging deep into one's situation, the maturing individual or group uncovers potentials from the past and new potentials that advance things in cooperative ways. This is a Golden Age, when a culture reaches its peak accomplishments, its interior sophistication, and its furthest exterior expansion.

    People today are wrong to refer to "old age" as the "golden years." In the past, very few people lived to be sixty years old. It was to people their forties, fifties, and early sixties the term "golden years" was applied. These were their most productive and organizational years.

4. *Transthesis.* In the normal development of civilizations, like elderly people, there is a gradual diminishment of the resources necessary to keep one's civilization and one's health going well. With age, the complexity of one's advanced societal or personal state becomes increasingly difficult to govern and sustain. Becoming increasingly self-protective, their leadership becomes increasingly introverted. Broad cooperation wanes. Coasting along on past successes, older people and civilizations experience a gradual fading of their life's colors and sensitivities. Some type of traumatic deprivation can lead to the sudden collapse of an individual's or society's health. Sometimes the rise of an outside rival can renew old energies or hasten one's demise. I call this phase "transthesis." It is two-sided, marked by physical diminishment as well as possible evolutionary advancement into some type of advanced spiritual state. This may be called a Platinum Age.

Fourfold Cycles

Many physical phenomena are circular in design. The movements of the sun and moon, the seasons of the year, growth patterns in animals, etc., are all cyclic. In geometry, drawing a circle on a piece of paper consists of four distinct elements: (1) a piece of paper (or plane), (2) an arbitrary point chosen to be the center of a circle, (3) a ray of a constant length anchored at that center point. Then, (4) while hold one end of the "string" at the center and moving a pencil or marker at the other end of the sting, a circle is produced. Notice that this establishment of a circle has three fixed parameters and one moving parameter (x, y, r, θ).

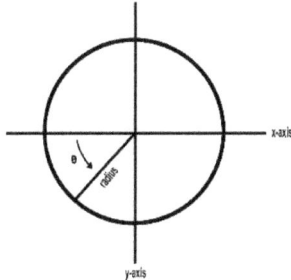

Illustration of a circle in a xy-plane

As the tip of the ray (or radius) moves around the circle, measurements along the x-axis and the y-axis can be charted. The resultant graph is a sinusoidal wave.

PART III: DEVELOPMENTS

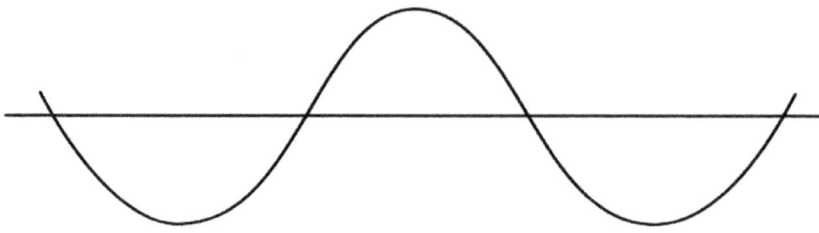

Illustration of a sinusoidal wave

Even though the sinusoidal wave is continuous, a mathematical analysis easily divides it into four formally distinct segments. The four formally different segments of every wave are separated by four "critical points."

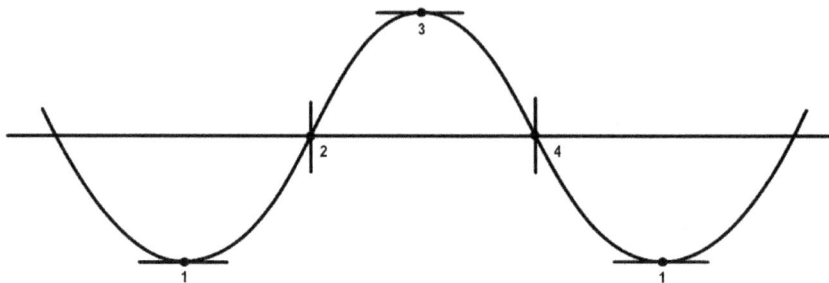

Illustration of Critical Points of a sinusoidal wave

1. Starting at the bottom of a wave or cycle, the slope of the wave gradually increases from a *minimum*.

2. Then there is an *upward critical point* at which the slope of the curve stops increasing and starts to decrease as the wave advances upward.

3. The wave has a highest point or *maximum*. From here, the slope of the curve changes from pointing upward to pointing downward.

4. Next there is a *downward critical point* at which the downward slope of the curve begins to lessen and flatten out.

5. Finally, when the slope of the curve totally flattens at the bottom or *minimum* of the wave, a new cycle begins.

This description of the critical points of a wave is true whether the continuous wave is symmetrical or asymmetrical. Sometimes the character of the wave or cycle is such that the exact point of transition is physically difficult to determine; however, the general fourfold pattern is obvious.

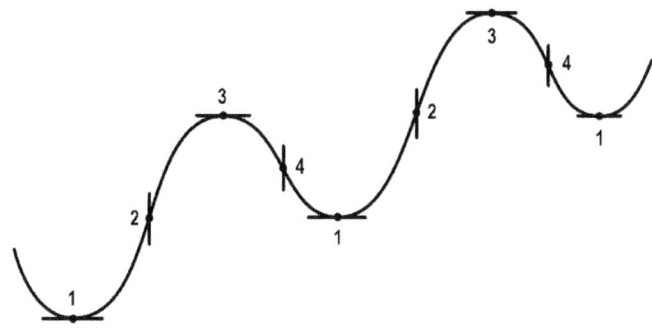

Critical Points and Slopes of tangents on a non-sinusoidal wave

The fourfold characteristics of energetic wave phenomena have been known for centuries. In the Northern Hemisphere, each year has a repeated series of four seasons: winter, spring, summer, and autumn. People have long recognized the analogous connection between the four seasons of the year and the four stages of human life. The beginning of a new year is in winter, when the sun is lowest in the sky. Among American Indian tribes, children conceived the previous spring were born in winter, so many Indians counted the age of a person by the number of the winters of their life. Infancy is related to the warm wrappings of the child in winter. Adolescence is the springtime of life when physical beauty and sexuality flower. Summer is the time when responsible adults are concerned about the steady growth of their crops and material constructions. Finally, in autumn, trees "let go" of their leaves, and plants "let go" of their fruits, which are dried and stored away for the winter and next year's planting.

In addition, each day has its four major phases, beginning in the dark of night, which is broken by the rising of the sun, when most people rise and go out to work. Through the daylight hours, people work and produce. Finally, at the end of the day at sunset, people return to their homes to rest. IPhil recognizes that the four phases of each of the above phenomena are analogous because they come as carryover characteristics of logical evolution from the source's potential energy, starting as the four phases of interrelational development of the primal, potential Almighty before the big bang.

However, there is a problem. IPhil recognizes that on the smallest, atomic level, the energy in our world is quantized. That is, in our quantized world, energy makes sudden jumps. Its energy goes up and down, not smoothly like a curve, but like an up-and-down staircase. But a staircase also has a four-phase logistic. Here is an exercise. On a piece of paper draw a set of stairs. Notice the fourfold process in that drawing.

0. Put your pencil on the paper, establishing your starting reference point, the initial, existential state of potential energy in this drawing.

1. From that point, draw a line upward a distance for the riser. The height of the step indicates an elevation of potential energy.
2. Stop at a particular height, indicating the elevated height of energy desired,
3. From that elevated point, draw a horizontal line or the flat of the step, indicating time spent on that higher level of potential energy.
4. Now, stop the forward movement of that line, indicating the length of time and width of that step level.
5. Now you are in the position to draw the next step. You can draw the line upward or downward, indicating a quantized increase or decrease in potential energy. Whichever you choose, you are back to the initial stage.

This simple illustration is important at the atomic level, where things do not happen gradually but in discrete steps of quantized energy. In mathematics, this is called a stepwise continuous function. Although it is somewhat complicated, it can be shown that a stepwise continuous function can be fully represented by a Fourier series, which is the sum of four-stage sinusoidal functions of increasing frequency. In this way, our quantized atomic world can be completely described in a fourfold cyclical way.

Chapter 14

Human Development and Evolution

Origin of Life. Origin of Conscious Thought and Deliberate Self-Initiatives. Various Forms of Freedom. Games of Chance. Cooperation. Compromise. Golden Mean and Moral Norms. Four Levels of Morality. Four Types of Law. Pursuing an End.

Let us look at human development more closely. Recall that in the 2nd phase, potential energy tends to be intrarelational with respect to itself, establishing integral individuals. Then, in the 3rd phase energized individuals advance to be interrelational with respect to others, establishing enduring logically consistent systems. In this way, IPhil explains why human development evolves not only as individuals but also in interrelational systems. Therefore humans, animals, and atoms are not only oriented toward maintaining and developing their individuality but also to become increasingly social, first with their own kind. Each energetic activity follows the fourfold interrelational logistic. IPhil recognizes that the same developmental and evolutionary process is true not only in the human order but also for all other interrelational energetic orders. So, in the following discussions, the reader may analogously replace "human" with the name of most other biological species or physical things. Still, because humans have a more evolved logistic than lesser evolved species, humans can receive information from the lower orders, and thereby fully interrelate with them in a logically consistent way within their human logistic, while lesser evolved species are not able to interrelate fully with humans in an evolved human way.

Origin of Life

Humans first name things according to their everyday, macroscopic observations. Consequently, we commonly predicate the quality of "life" on what we observe as biologically animate. We say rocks are not alive because in our limited observations they don't appear to move on their own. In our scientific age, however, we can observe things microscopically and atomically. We observe crystals growing. The sun, stars, and galaxies have their own life cycles. So, where does the progressive stages of life for all finite individuals and systems come from? Biological life begins with the formation of the first RNA molecule. Yet smaller than this, from microscopic observations, we know that even the most elementary systems grow, have self-initiative, and display other characteristics of life, according to their respective logistics.

IPhil recognizes that the forward push and process of intrarelational and interrelational life primally comes from the existential and forward push of a thing's potential energy. "Life" is an inherent quality of Primal Absolute Potential Energy (PAPE) as recognized by its fourfold advance to successive stages of advancement through time. Subsequently, "life" is experienced in the progressive development of PAER and the participants of PAIS. After the big bang, "life" is given to all evolved energized realities through the carryover aspect of the evolutionary process. Consequently, when received potential energy and matter are intrarelationally joined according to a thing's enduring individual logistic, the integrated individual displays the progressive fourfold phases of interior life. Then, when the individual becomes interactive according to its communal logistic, the individual is recognized to have exterior life. Signs of interior and exterior liveliness are physical manifestations of the metaphysical push of received potential energy, pushing the individual to advance according to its received individual and communal logistic. IPhil finds it narrow-minded and prejudicial to say a thing has life if and only if it has a life like mine, or if it has life according to my *a priori*, essentialistic, biological definition of life. Unlike essentialists who say in a closed way that life is biological and *only* biological, IPhil recognizes that life is an evolving concept originating from the Life and Love of the primal Almighty. Life is more than biological life.

In the 1st stage of development of a potential individual, the received energy/matter is initially diffused and not yet an integrated individual. Then, in the 2nd stage of development there is one piece of energy/matter that stands out in a logical evolutionary way, breaking away from the received energy/matter in an advanced way. It is called the "seed" or "germ" of the individual. Breaking from the pack, this "seed" is by a logically inverse process not the same as all received components but is logistically more advanced interrelationally than all the other energy/matter received. This "seed" or "germ" is the "progenitor" of the new, advanced individual. It is logistically more advanced than all other surrounding energized materials. Consequently, the other energized materials will be intrarelated in a subordinate way around and according to the advanced logistic of that "seed." As a result, the logistics of the "seed" will energetically, intrarelationally order all other pieces of the received lesser-logistical matter into an advanced individual according to the logistic of the seed. In this way, all received matter/energy that is integrated into a single corporate body is ordered by the logistic of the "seed" throughout the forming body, *and* the received parts' antecedent logistics are carried over but subordinated to the advanced and more interrelational logistic of the progenitor "germ." In this way, the individual via the logistic of the "germ" intrarelates the received energy/matter into one logistically advanced physical individual.

To become an integrated 2nd level reality around this progenitor "germ," this logistically ordered reality must have some level of interrelational knowledge of the other parts that are being intrarelationally integrated into the forming individual. Via mutually known, primitive knowledge drawn from their respective logistics, the parts are able to interrelate with each other, and their lower level logistic is able to become integrated in and subordinated to the more interrelational logistic of the centralizing "germ." In this way, in accordance with their respective logistics, all materials are knowingly intrarelated into the individual, becoming harmonized and coordinated with each other and within the advanced, generative logistic of the "seed." Through exchanged information between the parts within the whole, parts interact with each other on a 1st level in a material way, as well as intrarelated in a 2nd level knowing way within the individual logistic of the "germ."

The intrarelational phase of the individual integrates all information into a unified communication system throughout the forming body. As the individual advances intrarelationally and materially within itself, the individual grows in integrating self-knowledge. Then pushed by its interrelational potential energy, the individual is pushed to know and then physically to interrelate with others outside of the body on the 3rd level. Beyond knowledge of others immediately outside of the body, the individual is pushed toward actively interrelating with others more via the 4th level self-initiative of its potential energy.

In contrast, if any incoming parts are inconsistent with the logistic of the "seed" or its associated parts, the individual will energetically and antithetically dismiss or dispel those incoming materials or energy as dissociatively evil.

Note that the evolved logistic of the RNA and DNA of every human and animal individual possesses all antecedent, evolved logistics of its parts in a *nested way*. In every nested set, the fourfold interrelational logistic of PAIS is the metaphysical foundation of the logistics of all evolved finite realities. In ancient times, idealistic, essentialistic, 2nd level intellectuals dissociatively thought that each individual had one and only one unique essence. Because strictly separated species and castes lived and interacted with their own kind exclusively, ancient thinkers maintained that an individual's nature was associated solely with one's birth, *natus*. To have two natures was judged contrary to the separated class and caste social structure of their day. In contrast, IPhil's progressive, evolutionary thinking recognizes that advanced, evolved systems contain in a subordinate way all previous logistics from antecedent systems. In a nested way, the logistic of an advanced individual will have the logistic of every animal it evolved from, the logistic of all its chemicals, the logistic of the Almighty's Primal Absolute Potential System (PAIS), and the fourfold logistic of energy in PAPE. It is the Almighty's potential energy at the core of all realities in our world that gives them their "divine spark."

Closer to home, since the matter/energy of this "seed" *proximately* comes from one's parent(s), the logistic of the integral individual will manifest strongly many parental and close ancestral characteristics. On a smaller scale, as in the case of an amoeba, when there is only one parent, the evolutionary mutation of the intrarelated parts around the "seed" can be only a little different from its parent. This makes pre-sexually generated individuals more prone to attack from the viruses that can kill similar individuals. With the evolution of sexuality, the evolutionary mutated "seeds" intrarelated from the DNA of two parents will produce a progeny with a more complex DNA. It has been shown that genetic variations from two parents make a person better able to withstand attacks from viruses and other malevolent diseases.

IPhil recognizes that the rate of logical evolutionary advancement depends upon the inverse operators arising from the total number of constructive operators within one's system. In the most primitive material systems, there are only a small number of constructive, interrelational operators, and consequently, there will be fewer inverse operators, and the rate of evolutionary advancement will be very slow. Then, as the intrarelational complexity of evolved individuals increases, the number of constructive operators increases, the number of inverse operators increase, and from those more numerous inverse operators, the rate of evolutionary possibilities and the rate of evolution will increase. That is why the rate of logical and physical evolution was very slow in the universe right after the big bang, and why the rate of metaphysical and physical evolution goes faster and faster as time advances. That is what IPhil predicts, and that is what the physical record displays.

As logical evolution advances, not only does the individual as a whole advance, but the individual's subordinate parts, being enduringly logically consistent, can evolve on their own as well. Since there are many parts in advanced species and their logistics are less complicated than the whole, it will be easier and more probable that the parts will evolve in various ways. Darwin's observations focused upon the evolution not of the whole body but only of different parts of a body—like the shape of a beak in response to changes in food supplies in different environments.

Nonetheless, IPhil does recognize the possibility that an evolutionary advancement of the logistic of the individual as a whole can occasion. However, the larger the logistic is, the more difficult it is to evolve in a logically consistent way. Still, theoretically, occasionally a major evolution of one's *communal* logistic can take place. This is akin to the "punctuated equilibria" recognized from field data by Niles Eldridge and Stephen Jay Gould in a landmark paper published in 1972.[1] Field evidence indicates some species have remained basically the same from thousands of years, and then suddenly make a major change, followed by a long period of little change.

This is analogous to the invention process in human history. Most inventions occur through small variations of products already known to succeed. These small variations cause great diversification, but these are secondary variations, which are basically on the same level of the progenitor's individual logistic. In contrast, sometimes an inventor discovers something truly new, and the level of theoretical and physical development suddenly takes a huge jump. IPhil claims that major evolutionary changes and minor evolutionary changes are different because energy operates on two levels of

---

1. Gould and Eldredge, "Punctuated Equilibria."

logistics, namely, according to its 2nd level individual logistic and according to its 3rd level communal logistic. A 3rd level of communal logistical evolution happened with the development of the microchip in computers, which raised the level of computation and communication to a significantly higher, worldwide, interrelation level. The invention of the atomic bomb resulted in a huge burst of experimentation and knowledge about atomic structure, nuclear medical research, and healing.

But isn't logical evolution on the holistic, systemic level just an advanced form of small, individual advancements? The advancement of Darwinian evolutionary theory has erased the classical Aristotelian distinction between "accidental forms" and "essential forms" in individuals, the former being of a specific individual, and the latter being of the species of the individual. By recognizing the distinction of intrarelational 2nd level individual logistic from 3rd level interrelational communal logistic, IPhil inversely restores that distinction—not in an either/or way but a both/and way.

There is a critical point at which evolutionary variations of secondary, intrarelational characteristics become so dissociative that members of two subspecies are no longer able to breed together interrelationally, and therefore become enduringly distinct. Neanderthals and humans were able to mate successfully, so they did have the same communal logistic. Chimpanzees are very close on the evolutionary tree of life, but they cannot mate with humans because they are of a different communal logistic. Horses and asses are able to give birth to mules, but mules are sterile; they can endure as individuals, but they cannot reproduce. Similarly, hybrids are crossbred so that they will endure as individuals but cannot produce viable descendants with a logically consist communal logistic. In these ways, physical evidence indicates the truth of the formal distinction of 2nd level individual logistics from 3rd level communal logistics. The exact dividing line between individual and communal logistics can be determined by the physical fecundity of the progeny of mated pairs of individuals with different logistics.

Within IPhil, where do we find the origin of random freedom in elementary and molecular physical participles? IPhil recognizes the origin of freedom is in the Primal Absolute Potential Energy (PAPE), which logically evolved as PAER. Truly free self-initiatives must have existed in the participants of PAIS in order for it to be a reciprocal, logical system with dialogical terms capable of free reciprocal exchanges within their interrelational system. Then, through the carryover aspect of logical evolutions from the absolute order into the non-absolute finite order, all finite receivers have gained the freedom to self-initiate the transformation of its potential energy into the kinetic actions of an interrelational system. But those free self-initiatives need not be consciously deliberate, just as most self-initiatives of human babies are not consciously deliberate but unconsciously spontaneous.

PART III: DEVELOPMENTS

Origin of Conscious Thought and Deliberate Self-Initiatives

How do we progress from unconscious random self-initiatives to conscious deliberative self-initiatives? First, knowledge is an inherent, essential component of interrelational energy, for without some level of knowledge of self and another, there cannot be any energized, enduring interrelationships. When energized finite receivers self-initiate, their energized self-initiatives are logistically limited by their current stage of logistical development.

In their 1st phase of receiving, the receivers' received finite energy/matter is still diffused and not yet internally organized. Lacking a central, controlling, intrarelational agent, early free self-initiatives are not yet individually or systemically directed by their host's individual logistic. Lacking "individual" direction, these 1st stage expressions of its freedom are random. Because these random actions within the receiver's logistic are not toward any specific end in that logistic, interrelational knowledge is not consciously focused to a particular end, but unconsciously toward any logistically possible end. There is no conscious, specifically directed decision-making in the 1st stage of logistical development of an infant, for all logistical energetic possibilities at this point of development are experienced passively within the logistics of the received energy/matter.

Then, in the 2nd phase of their energetic development, the interrelational character of these initial bits of energy/particles start to intrarelate in a progressive, logical way in line with the centralizing logistic of a more advanced "germ." In the 2nd phase of development, the individual begins to take active control of its parts. Then, in order to be more intrarelational and increasingly interrelational, the individual is able to direct the transformation of the potential energy toward more integral intrarelational states. Directed toward a more rapid and comprehensive intrarelational integration, the individual in a primitive way consciously and then deliberately directs the transformation of the potential energy its parts toward kinetically becoming more active as an integral individual. In this stage, the individual's growing conscious concerns are initially directed toward the individual's self-maintenance and self-development. In the 2nd phase of the interrelational logistic of energy, the growing intrarelational individual metaphysically claims ownership of all the parts and then the subsystems of the whole body within its individual's logistic. During the formation of the integrating individual, the individual becomes selective in energetically activating parts for the sake of its self-preservation and more interrelational development. It is out of an inner push for greater self-control that the former unconscious random self-initiatives selectively advance into conscious deliberate choices for self-development and self-control, according to the logistic of the germ. Nonetheless, because of the carryover aspect of logical evolution, the parts of the individual retain in a subordinated way the lower-level logistics of its elements.

Because of the push of its potential energy toward greater interrelationships, the integrating individual is logistically oriented toward activating those options that are more apt to advance the formation of the individual. Then, on the 3rd level, the intrarelating individual pushes to be more interrelational with others—if their logistic is compatible with the logistic of the growing individual—becoming increasingly consciously aware that some available options are potentially more interrelational than others. When an individual encounters anything that is dissociative from its current individual or communal existence, unconscious awareness suddenly jumps into conscious awareness, which is directed to eliminate that dissociative disturbance or transform it in a way that is peacefully associative.

Within a changing, challenging environment, the push of an individual's received potential energy can advance toward a 4th level, which possibly is more interrelational with and/or through others. Thus, the potential energy is logistically ordered and pushes unconscious knowledge into consciousness for the sake of greater control of one's operations, interactions, unto a logistically sensed and fulfilling destiny. Then, the individual's conscious awareness and concerns advances to an advanced 5th level of unconscious self-maintenance—until something antithetical stirs the individual on the 6th level toward greater interrelational protection and/or development in the new challenging environment, starting the whole unconscious/conscious, random/deliberate, *yin/yang* process over again.

Various Forms of Freedom

Recall that the origin of free self-initiative is found in the forward leaning of potential energy in the ur-stuff of reality, PAPE. Subsequently, true freedom was passed on from PAER, as the all-giver by "letting go" of ownership of PAPE to the all-giver-receiver in PAIS. As a result of panentheistic carryover, that true freedom is released in the logical evolution of the absolute-finite progenitor of the finite proto-spatial system. In and after the big bang, true freedom was communicated to all finite participants in that logically and physically evolving spatial system. Then, by logical evolution to all finite receiver-givers in all subsequently evolved systems in our world, true freedom is communicated to all physically evolved individuals in our world. Through evolutionary advancements of consciousness, antecedent levels of awareness are carried over and act as the continued, nested foundation for higher levels of conscious awareness.

Potential energy has a fourfold logistic, which expresses itself physically in the four physical phases of receiving, intrarelating, interrelating, and releasing. Each physical phase has its own metaphysical potential foundation. Consequently, free self-initiatives are not just found at the 4th physical end of one's life but in *every* physical phase of the life of individuals—qualified by the character of each phase of life. In this way, there are four formally different types of freedom in the physical order. (1) Random physical freedom dominates in the 1st level of material formation. (2) Individual-centered freedom dominates on the intrarelational, 2nd individual level. (3) Leader-led societal freedom dominates on the 3rd level of interrelational communities. (4) Transcendentally-oriented freedom dominates on the 4th level of self-surrender.

The diverse directions of these four metaphysical levels of freedom can be physically experienced simultaneously in the individual. An individual can be simultaneously emotionally torn by the physical pangs of hunger, by an individual seeking to do one's best, by loyalty to a corporation making demands that interfere with one's health and family, and one's loyalty to the Almighty within one's religious beliefs. Experiencing the logical inverse of these simultaneous options can greatly distress one's simultaneous desire to be faithful to one's family, oneself, one's job, and one's religion. The choices one makes are signs of an individual's priority in life. Interrelational philosophy, unlike most other philosophies, explains how and why these inner conflicts arise, giving principles and clues for respectfully ordering and making one's life holistically more interrelational and good.

Self-initiated expressions of freedom on all levels mean choosing to act or not to act—to move ahead or to procrastinate. The least energetic way to interrelate is to procrastinate, by opting to not change one's existential and interrelational state. We often observed a child turn his head away when offered food, nonverbally saying, "No!" Unwillingness to step forward in a personal developmental way or in one's social situation expresses an individual's reluctance to make an advancing interrelational commitment. In older individuals, it is common for them to dig in their heels rather than risking the possibilities of painful dissociations in attempting anything new.

To self-initiate a more advanced potential option requires kinetic pushing to a state of higher interrelational potential, which logistically makes more demands upon the individual. Choosing to transform one's potential energy into kinetic action starts something more advanced, and such dissociations are regularly difficult and painful. The individual must take personal ownership of her potential energy and energetically transform that potential energy into the kinetic energy of the action chosen. Because of the increased expenditure of energy needed to break from one's current state, change is always hard. Saying "no" is an easy option, but deep within oneself, the individual's potential energy metaphysically and spiritually continues to push the procrastinating individual toward doing what is interrelationally better according to one's individual and communal logistic. Nonetheless, the natural, inner push of one's received potential energy makes procrastinating individuals feel restless, anxious, and disappointed. Solitary confinement in prison, a child's "time out," and a simple barrier blocking any change, can evoke great dissociative frustration and pain. Yet, after holding back one's potential energy for a time, most individuals typically tire of saying "No!", "let go" of their restraint, and go with the natural, energized flow of things.

IPhil recognizes that even with free self-initiatives, the things of nature are not oriented totally toward disordered chaos because their 4th level free self-initiatives are always within their 3rd level communal logistics. That explains why the unconscious, intuited, indeliberate, random self-initiatives of fundamental matter from the beginning of our world have pushed the evolution of our world into more advanced, logistically well-ordered, interrelational states.

In a similar way, the Daoist principle of *wu wei* advocates the effortless action of "going with the flow." When professional athletes are playing "in the zone," they freely, totally surrender their 4th level self-initiatives beyond their 3rd level rational analyses, and they advance to a 5th level into a state that transcends and breaks from their 2nd level self-conception.

Games of Chance

Games of chance and random feedback programs begin within a well-ordered logistic, which has its own rules and parameters. Randomness does not pertain directly to the rules of the game, its program, or the algorithm in the computer. Rather, randomness pertains to the energized/materialized values of the variables within that logistic. Consequently, the exercise of freedom in games of chance is not unlimited but is only within the legitimate range of possibilities within the rules or logistic of that game.

In response to the randomly statistical equations found in quantum mechanics, Einstein was a determinist and made the famous quip that God does not play dice with creation. How can a person respond to that critical statement? Within a rational, deterministic worldview, the primal energetic reality or algorithm organizes and initiates everything. Everything happens according to a given equation of logic, rather than the logistic of a thing flowing from the received energy of a thing. IPhil recognizes that the Almighty primally provides potential energy in the world's 1st phase. Then that energy in the 2nd phase establishes intrarelational individuals according to their received individual logistics. Then, in the 3rd phase, that energy pushes the individual toward establishing and/or participating in logistically ordered interrelational systems. It is only in the 4th phase in each interaction, that individuals are able to express their empowerment in self-initiated, free ways.

IPhil recognizes that an individual's true freedom comes as a logical evolutionary carryover aspect in the received potential energy. With a gambling die, the freedom found in the throw comes when the human holder "lets go" of the ownership and control of that die's energy/matter. It is the energy-receiving die that actualizes one of its six possible options, providing the odds that a die will show just one possibility when it comes to rest. The odds would be skewed toward a particular option if the interior design of the dice were not made in a balanced fashion and prevented each possibility a different chance in a random throw. In a rational, 2nd level deterministic world, the Almighty causes everything and nothing is truly free. In IPhil's 3rd level progressive world, there is true freedom and real chance, allowing an individual person or things to operate according to their diffused 1st level potential energy. By "covering" our higher levels of conscious control, gambling and chance allow ever-present, 1st level of diffused, random potential energy to be exposed.

In many ways, our material world is like one big casino. Here is an allegory:

1. A rich individual freely spends a large amount of his financial resources building a grand casino with many games, each with its own rules.

2. Another individual with lesser resources then walks in the door and freely decides to be a player.

3. When the player is at the craps table, the player must play according to the rules of that game, and the owner is similarly committed to abide by the consequences of the free random self-initiatives of the dice.

4. The player then picks up the pair of dice and throws them, "letting go" of the dice. The behavior of the dice is no longer deterministically controlled by the gambler or the casino owner. The initiatives of both the owner, the player, and the dice are all free. Because the dice are not integrated individuals capable of determining their actions, the results are not deliberate but random.

5. Then, the player's winning or losing influences whether the player freely decides to leave the casino or to stay and try again. Adventurous gamblers will continue playing the game, hoping to have a big win along the way, while conservative individuals will keep their resources close at hand.

Why are games of chance pleasurable? Games of chance are deliberately designed to create odds of winning vs. odds of losing. Those odds are dissociative, and they put the player in a state of dissociative tension. When the player beats the odds, the player is associatively connected with increased winnings, which releases the tension and gives pleasure to the player. A 2nd level gambler remembers their past successes and focuses on the odds to win again and increase their winnings or resources. While highlighting one's winnings as a sign of one's cunning, a 2nd level gambler dismisses losses as "inconsequential evils," anticipating a success next time or in the future—which the gambler convinces oneself that future winning will in the short run make up and even exceed loses. In contrast, a 3rd level individual knows the rules of the game and knows that in the long run the odds are always in favor of the house. Consequently, 3rd level individuals will tend to not participate in risky adventures where the odds are stacked more against them—except, perhaps, when one's participation in the game benefits a good cause.

Beyond the above, however, IPhil recognizes that the Almighty owner of our casino-world has made it as a panentheistic cooperative. As panentheistic co-owners, then, we are gifted to share in the interrelational successes of the casino—which includes associative profits from the dissociations of the past. Then, at the end of the evening, we will share in the net profits from all player's losses. Thus, what is lost during the game is reclaimed at the end of the evening. So why take the gamble? Because if the person does not play, the interrelational gain is not experienced directly but only indirectly through the generosity of the owner. Even when there are loses in one's potential energy by chance or by deliberate fault, both the owner and players *can* be winners in an evolutionary way in the end. This is a more advanced win-win situation. Knowing this, the conservative player can express greater courage in risk-taking, for even if one loses now, a person who is closely associated with the owner of the casino will win it back and potentially even more in the transcendental end.

Cooperation

Cooperation as observed in nature has created a conundrum to generations of evolutionary scientists. If natural selection among individuals favors the survival of the fittest individuals, why would one individual help another at a cost to herself? Charles Darwin himself noted the difficulty of explaining why a worker drone-bee would labor for the good of the colony, because its efforts do not lead to its own reproduction. He wrote in *On the Origin of Species* that social insects are one special difficulty which first appeared insuperable, and actually fatal to his theory.

Yet examples of cooperation and sacrifice for the sake of one's own are commonly seen in nature. Humans working together have transformed the planet to meet the needs of billions of people they don't even know.[2] So pervasive is cooperation that Martin Nowak of Harvard University ranks it as the third pillar of evolution, alongside of mutation and natural selection. He explains that natural selection and mutation describe how things change at the same level of organization, but natural selection and mutation alone wouldn't explain how you get from the world of bacteria three billion years ago to what you have now.[3] Cooperation leads to social integration, and social integration leads to the complex organizational systems we see in modern life.

Interrelational philosophy (IPhil) recognizes that in the interrelational logistic of energy, 1st aspect mutations and 2nd level natural selections fit individuals better, whereas the 3rd aspect of energy is social and systemic. In the interrelational logistic, the communal aspect is metaphysically distinct from and cannot be logically derived from the individualistic 2nd level aspect of energy. The energy/matter of our world by its nature is not only intrarelationally self-centered, it is also interrelationally community oriented. Through the centuries, Western philosophy and psychology have predominately been very self-centered rather than other-centered and systems-oriented. Their foci have pervasively been biased toward the individual and the material side of life. When they start by looking at themselves and their personal experiences in a 2nd level way, they are never able to advance logically to a societal, 3rd level awareness of our interrelational world. However, the interrelational 3rd aspect of energized reality is needed for the establishment of logically consistent systems capable of evolution.

2. Bennisi, "On the Origin of Cooperation."
3. Nowak, *Evolutionary Dynamics*.

A self-centered philosophy provides no rationally consistent reason for an individual going beyond oneself, sacrificing oneself for the altruistic benefit of others.

IPhil also recognizes that it is not enough that an individual progenitor break through the boundary of its antecedent system. The logical evolutionary theory requires that the superior, mutated individual must then reassociate with members of the antecedent system so that through sharing its advanced logistic with others a progenitor can produce an enduring, logically consistent, advanced system. That is why enduring reality is not just antithetically either/or but also synthetically both/and. Therein is found the balancing of tensions that make life dynamic.

Social cheating is the logical inverse of bonded social cooperation. While consistency in male/female bonding provides stability in a relationship and society, inverse actions are beyond the closed boundary of exclusive male/female bonding. But doesn't cheating provide greater possibilities of evolutionary advancement? It is natural for self-centered, 2nd level individuals to be internally urged by their interrelational energy toward antithetically breaking the 3rd level communally established boundaries. In this way, violators of convention expand their self-centered, 2nd level, personal domain. Also, in cheating, greater diversity is made possible on the 1st level in the gene pool of a species. However, such 2nd level, adolescent-style infidelities regularly result in short-lived intrarelationships for sake of short-lived pleasures. Infidelities are usually secretive, and they usually do not link up publicly within one's community, and they regularly do not produce enduring families and a peaceful society.

Such dissociations from 3rd level social conventions usually result in a great conflict with the antecedent partner, breaking the antecedent 3rd level bond that gives stability to child rearing and community development. While there is the possibility that the grass may truly be greener on the other side of the fence and give momentary pleasure, there is a high probability that 2nd level individualistic breaking of social norms will have negative 3rd level social implications and lower social benefits from cheating. Immediate, 2nd level individual satisfaction and expressions of prowess in infidelity may have important values in an individual's material advancement. However, it is well-known that the turmoil of broken relationships disturbs the stability of a society. That is why 3rd level societies regularly have laws that favor lasting marital bonding and family stability.

Nonetheless, with the emergence of a greater, stronger, shrewder aggressor, and competition by an alpha male of a pack, the formation of a herd or harem by a powerful individual can produce a stable, advanced 3rd level community. From a logical evolutionary perspective, the advanced aggressor is like a more advanced, 4th level progenitor, breaking past communal bonds to establish a more advanced interrelationship that endures. It is crucial that the aggressor is strong enough and wise enough to establish communal stability within the abducted pack, herd, or harem. Enduring domination by a new superior male in a herd allows the evolutionary advancement of the progeny.

Compromise

"Compromise" is a dirty word for 2nd level ideologues. In an immature, adolescent way, ideologues consider granting a concession to one's political opponent to be a betrayal of their God-given, absolute, perfect values. In their black-and-white world, ideologues uncompromisingly claim, "My way and *only* my way is right." They judge everything else to be wrong and evil. It is like the cardinal who refused to look through Galileo's telescope at the moons of Jupiter, for the very act of looking at that evidence went against his beliefs. Ideologues refuse to read articles and to listen attentively to the views of their opponents, for that implies their position might be wrong or incomplete. Consequently, they are morally impelled to oppose and reject as strongly as possible all compromises, and they obstinately refuse to consider seriously any evidence that backs the opposing position. They only want their party, coalition, or group to dominate over everyone else. "My way or the highway." Ideologues' decisions are quick and cocksure, typical of adolescent behavior.

In contrast, 3rd level individuals recognize that among the many issues at hand, many disagreements are "inconsequential evils," and they *can* be ignored or adopted with few enduring interrelational consequences. Reaching a compromise requires time, openness, and thought, as is characteristic of more adult construction projects. As the goal of greater interrelationships is pursued, compromises will avoid stalemates and more quickly promote the "common good" of a group or society.

The habit of "tolerance" embraces and works sympathetically with dissociated elements, people, and policies. The 3rd level, adult diplomat "lets go" of the perfectionistic, negative judgment of 2nd level ideologues for the sake of logically evolving and improving in small ways what has worked in the past into what *can* work in the future. While 2nd level ideologues have a static, protective, conservative view of reality, 3rd level realists have a dynamic, liberating, progressive view of history. Rather than promoting what is best for me and for mine, this approach promotes the "common good" for society as a diverse organic whole. Making this advancement, however, often requires a paradigm shift from individualized me-centeredness to corporate we-centeredness. Unfortunately, stubborn idealists are hardhearted and strongly resist "metanoia" or change of mind and heart.

Uncompromising essentialists maintain that there is only a single right way for society and interrelationships. *They* know it, and their opponents *don't*. They put their ideas forward as if they know the mind of God, and they demonize their opponents for going against their allegedly God-given plans. Essentialists ground their arguments on a few, "clean" metaphysical ideas, and they are not willing to recognize and deal with the "nitty-gritty" of physical solutions. Idealists usually consider themselves to be intellectually superior to their opponents, and they tend to look down from their lofty pedestals upon realists who are getting their hands dirty working in practical ways among the poor and needy. Even though idealists tend to be intrarelationally closed-minded and hard-hearted, they need to be confronted constantly with objective truths that are interrelationally consonant with the lives of both the rich and the poor. History shows that ideologues are regularly associated with the rich, who live in gated communities. Compromisers regularly associate with the poor and needy, because they see them as the politically and socially forgotten and ignored, whose lives need to be taken into account within a society that respects *all* the members of society and not just the rich and powerful. The potential energy that founds our world pushes all to be more interrelational, beyond our 2nd level selves and one's own, to include all members in a 3rd level society. This requires a paradigm shift from an idealized, essentialistic view of one's closed society, to an open, realistic, interrelational view of all in our world.

Golden Mean and Moral Norms

In random interactions there are many possibilities. The simplest example is tossing a coin, which can land in two possible ways: heads or tails. With many random flips, different combinations are possible, yielding a distribution that is bell-shaped. This is a Gaussian or normal distribution, whose most frequent combination is the "norm." In this bell-shaped curve, a standard deviation from the norm holds 68.2 percent of all random chances, and these are considered socially acceptable. In the second deviation from the norm are found 27.2 percent random chances, and these are socially less than acceptable but tolerated. The remaining 4.6 percent of all random chances are more than two standard deviations from the norm, and they are regularly censured socially in various ways.

What is the relationship of the norm of a random chance distribution and Aristotle's golden mean? With daily repetition of associative, reciprocal, peaceful interactions, traditional behaviors soon drop from conscious concerns into the realm of the unconsciousness routine. Normative ways of behaving are simply expected in such social behavior. Within a stable society, members habitually, unconsciously act randomly with respect to accepted social norms. Unconscious and conscious habitual behaviors cluster around the accepted norm. From this comes the classical maxim: "*Virtus in medio stat*" (Virtue stands in the middle).

When small variations from the traditional norm occur unconsciously and randomly, they are ignored or dismissed as "inconsequential." However, larger variations from the norm are dissociatively distressing and are considered consequential. These dissociations become consciously recognized as significantly dissociative from the golden mean. We complain about such variations, but with few consequences. Then, actions even greater the variance from the norm are more disturbing become socially unacceptable. There comes a point when variant behaviors are considered significantly disruptive in a society. Against such offensive behaviors, social and legal sanction are imposed to correct or eliminate such moral misdeeds.

From the above explanation of how the golden mean in a stable culture, it also becomes clear that the golden mean in one culture can be different from the golden mean in another culture. Different cultures and societies are at different stages of development within their respective geographic settings. What is reciprocally sensitive and respectful within one culture may not be reciprocally sensitive and respectful in another. (The Bible records how in ancient time, killing another Jew who is one's enemy was unacceptable but killing a non-Jewish enemy was acceptable. Such a moral standard most consider unacceptable today.) In closed-minded individuals, the sight of different moral standards can produce cultural biases and prejudices against cultures that have different cultural norms. Contrariwise, an open-minded traveler visiting different well-organized, stable societies quickly recognizes that different golden means and cultural customs are best suited to those different cultures. There are local historical and cultural reasons for variations of moral norms. This leads to the advice: "When in Rome do as the Romans do." Also, in large cosmopolitan cities, different peoples and different customs can be found in adjacent neighborhoods. Experience shows that learning different cultural ways and norms adds great pleasure to one's life. People seeking universal harmony and peace will strive to find ways to show respect for local customs as well as commonly accepted universal principles of law.

Animals display the same tendency. Different behaviors develop in different social groupings of animals, and these promote the endurance and welfare of those species. Among chickens there is a recognized pecking order. Among many animal species, alpha males and hierarchies of deference are recognized. If members of an animal community display attributes or behaviors that are not normative in that species, other members of the community will react antithetically to affirm the norm. Despite their differences, different species generally are able to live in harmony in a particular location despite their differences.

The golden mean is not always the best guide to acceptable moral behavior within a society. There are times when it is right to break from that principle. IPhil recognized that the golden mean works only as long as one's society has been operating peacefully for a length of time. However, invasions and civil discord can significantly shake the status quo of a society. When the routines of respectful reciprocal daily life are broken, new ways of responding to the new interrelational situation are needed. Disruptions which break the reciprocity of normal logistical interactions within a traditional society push the righteous person to break the normal laws and traditions of one's society in order to advance in new, more logically evolved ways that are suited to the new community situation. In dissociative, conflicting situations, it is fitting to "let go" of the normal ways of interacting within one's society to execute new behaviors that lead to peace in an advanced social order.

What is the relationship between the local golden mean and the universal natural moral law? It is common for teachers, philosophers, and religious leaders to insist that there exists a code that is right universally and eternally in all societies. IPhil recognizes that in an enduring, interrelational society or species, energy is both potential and kinetic, and these have different communal logistics. The physical order focuses upon local customs, and the golden means is variant in different cultures. In contrast, the metaphysical order focuses upon the enduing communal logistics of an entire species or system, commonly referred to as the natural law. Many different philosophers have maintained *that* there is a natural law, but they have disagreed as to *what* that natural law is. Limitation of full experiential knowledge of all human cultures, makes knowledge of the optimum interrelationality of the human metaphysical communal logistic difficult to determine. IPhil argues that the natural law is the interrelational logistic of energy for the human race. However, Essentialists view the natural law as static, while IPhil recognizes that the communal logistic is developmental, and at a given time, not all individuals and communities are at the same place within that universal communal logistic.

Theoretically, the metaphysical logistics for all energetic things are evolved logically from the foundational logistic of Primal Absolute Potential Energy (PAPE). So, in the metaphysical order, there is an idealistic, 3rd order moral norm, or communal logistic for all energetic individuals and cultural systems. Within that communal logistic, individuals and cultures have many physical options, which can be consciously chosen for the greater good within a given physical environment. Knowledge of both local norms of conduct and of the universal natural laws of conduct are learned from our partial, limited experiences, our awareness of them and their differences are partial and limited.

More concisely, potential energy provides a universal code of conduct: Try your best interrelationally. Another handy rule of thumb is: "By their fruits you shall know them." Appropriate local physical norms are learned and established from the interrelational results of local experiences. It is necessary to draw together experiences from all societies to approach an understanding of the natural moral law found in the metaphysically evolved, universal communal logistic. Unfortunately, essentialistic thinkers view the natural law in a singular, static way. IPhil recognizes that the human communal logistic is not static but progressive. That is, IPhil recognizes that it is wrong to say that the natural moral law is idealistically this way and only this way all the time. IPhil recognizes that the communal logistic of humanity is not static but interrelationally progressive.

Primitive and developed societies may have the same communal logic; however, they probably are not at the same stage of interrelational development within the human logistic at the same time. IPhil recognizes that metaphysically, primitive societies will progressively be at different stages of moral development from socially developed societies. This is similar to children being at different stages of physical and moral development within the family. Moral expectation of a four-year old is not that of a ten-year old. Nonetheless, 2nd level adolescents often expect that everyone in the family is to follow the same rules, and if they don't, they complain that it is unfair that the rules in the family for siblings at different ages vary. While essentialists assume that the natural law is exactly the same for everyone at all times, IPhil recognizes that while a singular metaphysical equation describes the motion of a planet around the sun, the physical actions of a planet now will be kinetically different from other stages of that journey. In speaking about morality, the immediate physical expectations must be distinguished from the extended metaphysical logistic expectations at different stages of development.

Four Levels of Morality

The root of the word "morality" is *mores*, which refers to the traditional behaviors of a group of people. Laws are identified and written with respect to systemic, group, and social behaviors to promote a high level of interrelationship. Because the human logistic advances through four formally distinct phases, IPhil recognizes four levels of morality. These are not independent but successive, building one upon another.

Family morality is practiced and learned within a family setting from the time a child is born until a child starts making moral decisions on its own. It can be argued that because most of the behavior of children is instinctive and responsive, they are therefore not conscious, and these behaviors should not be considered moral. Also, these behaviors follow the examples and expectation of one's parent and siblings. In a religious home, parental actions and expectations follow religious directives non-verbally, and children are expected to follow. Jews recognize that some violations of the laws of God can be unconscious, and the celebration of Yom Kippur covers all indeliberate sins of the participants. Learned bad habits are usually unconscious. From IPhil's perspective, even though they are not conscious, some behaviors can still be in accord with or contrary to God's will, either as intuitively known from one's logistic or from one's parents. Certain behaviors are wrong, such as a child running out in a busy street. It is important to correct that child for that uninformed, unconscious, wrong behavior in an age appropriate way.

Personal morality is directed toward the successes of the intrarelated, adolescent individuals. Each individual has his or her unique personal morality, directed toward actualizing one's individual logistic unto one's personal destiny. Personal morality involves practices that promote one's health, one's psychological peace, one's mutually beneficial interrelationships with others, and one's drive to fulfill one's intuitively known, personal, logistic destiny. On this level, the individual grows in self-consciousness and self-esteem, which prepare and direct the individual toward 3rd level social interactions. Actions which hinder or promote interrelationships in one's own individual development may be called personal sins or personal virtues, respectively. IPhil recognizes that the first sin of disobedience against our God-given logistic and against the directions of one's superiors is procrastination, not advancing according to our received, energizing potential.

Personal tutors, coaches, and mothers are well-aware of an individual's logistic, and they encourage the individual to become the best, the most interrelational person one can be according to one's God-given individual logistic—to become the best individual one can be according to one's natural talents. "Just do your best." This instruction is different from the moral focus of public teachers and community leaders who teach about and exhort people toward a 3rd level, single standard of moral behavior for all.

Public morality is directed toward the success of a culture, society, or interrelational organization. Different cultures have their own traditions and local laws, which vary according to the historic development of one's culture. These are directed toward harmonious interactions within one's current society. Religious, military, and legal leaders regularly promulgate a set of rules and laws designed to improve an institution, organization, and the community as a whole.

Transcending morality is directed toward going beyond the status quo toward a new interrelational systemic order which is interrelationally more advanced than our current social order. Following the transcendent visions of innovators, prophets, and progenitors can lead toward building a society and world that is transcendentally and spiritually more evolved than our current one. Religious and charismatic visionaries and leaders regularly promote a transcendental morality that pushes their members beyond current materialistic standards to a higher level of individual and communal spiritual life.

Four Types of Law

Greek philosophers did not recognize the notion of "natural law." While they recognized that each species of things operated in a consistent way, according to what they called its "essence," Greek philosophers avoided the notion of law, because laws in their culture were constantly changing. As the Roman Empire extended its jurisdiction over the whole Mediterranean world, it introduced Roman law into all the provinces.[4] The Roman principle of *Lex Gentium* (law of the nations) was transcultural, pushing Roman jurists toward the insight that despite the many variations of cultures, there were certain principles of law that were universal—for example, the principle of equity. In the highly legal climate of Roman culture, Stoics linked the notion of universal human nature to the notion of law and put a strong emphasis upon the notion of "natural law."

Stoicism and Scholastic philosophy emphasized that there is a "natural law," overseeing the morality of human conduct. Many sought to determine what the natural law is by looking at what physically is beneficial to humans. Unfortunately, looking at physical data, there is great disagreement as to what specifically the natural law is, so belief in that concepts is doubted by many. IPhil determines not only *that* there is

---

4. Jolowicz, *Historical Introduction to the Study of Roman Law*.

a natural law at the foundation of all logical and physical evolution, but *what* are the specifics of the natural law, namely, the metaphysical fourfold interrelational logistic of Primal Absolute Potential Energy (PAPE) as evolved in the existence and lives of the things in our finite kinetic world.

Thomas Aquinas was the first philosopher to present a well-ordered, comprehensive treatment of law. Aristotle's four metaphysical categories and four types of causalities guided his schema. Aquinas divided laws into the following four categories, which I have modified and re-ordered to match the interrelational logistic.[5]

Eternal law is the Almighty's providential ordering of all things in our world according to their respective essences. An example of an eternal law is gravity. Observance of eternal law by lower level material realities is not rational or deliberate but unconscious and spontaneous. Modern scientists have discovered many unchanging eternal laws in the elementary forms of nature. But not all laws in nature are 1st level, eternal, and unchanging.

Natural law involves human participation in God's eternal law with regard to the providential ordering of human life—that is, according to one's *natus* or birth. There can be confusion here. Thomas and other Scholastics applied the term "natural law" only to humans. Humans were free to follow the human natural laws—or not. However, IPhil recognizes that the sciences have determined that material individuals on all levels have some degree of free self-initiative. In this context, "natural law" applies to the logistic that is energetically given to a self-initiating individual, logistically operating freely according to its generation, *natus*. In essentialistic 2nd level thought, "natural law" pertains to what is dualistically judged to be ether right or wrong. In contrast, 3rd level IPhil thought, "natural law" recognizes the development of all things to be progressively fourfold, energetically striving to be more interrelational.

IPhil recognizes that in all energized realities, even down to the level of atoms, there is some level of free self-initiatives, and the exercise of freedom on all levels is within the boundaries of a thing's or a person's metaphysical logistic. Evidence indicates that animal species are to some degree conscious and capable of deliberate, self-initiated judgments for the welfare of self or their kin. Their "natural law" is their tendency to do those actions which are generally more beneficial to themselves and their own kind. Because of the intuited drive toward fulfilling one's individual logistic, the individual pushes to saying, "I've got to be me!"

Positive law is called "positive" because human authorities *posit* these laws. These laws consist of legislations of various types, including constitutions, statutes, administrative decrees, and even customs. Laws of this type are far more specific than eternal laws and more physically communal than natural individual law. Some animals follow positive laws in a reactionary way within their species. For example, by physical dominance, respect toward an alpha male is expected in a pack of wolves. Lest they be physically abused, a flock of chickens learns to observe the pecking

---

5. O'Connor, *Aquinas and Natural Law*.

order of that group. These laws may vary from one subspecies to another. While natural law focuses on what is right for the individual, positive law focuses on what is respectfully right between individuals within a current community or society. While many positive laws vary from culture to culture, there are some reciprocal and egalitarian metaphysical principles which express a common interrelational logistic in many types of organizations in a community.

Divine law refers to God's formal commands given through the absolute Spirit as intuitively known from historically received revelations. These can be written down as in the Decalogue found in Exodus and Deuteronomy, or as the commands of Allah in the Qur'an. Divine laws are triggered by the Spirit of absolute potential energy within prophets, religious leaders, and charismatic dreamers. They are regularly directed toward establishing a perfect utopia beyond our world and closer to God. These laws are more advanced than positive laws, for they are directed toward the establishment of a higher interrelational association with the Almighty and with all finite individuals as "familial members" in the panentheistic evolutionary descendants of the Almighty.

These levels of law do not contradict one another, for each advanced level of law logistically and metaphysically contains and builds upon the prior level.

Pursuing an End

Pursuing an end is not exclusively a human activity; it is found in many animal species. Squirrels figure out ways through mazes and around barriers to get something to eat. Salmon will travel miles and overcome many rapids to return to their birth site, so they can spawn the next generation. Peacocks and other male animals perform intricate mating rituals to find favor with a female in their species. Regardless of whether these pursuits are conscious or simply instinctive, each individual in the course of these activities must choose from many options to reach a desired goal.

While laws are important for the well-ordered operations of individuals within a society, Aristotle emphasized final causality in one's pursuit of an end.[6] In describing how he made a sculpture, Michelangelo said that when he looked at a block of marble, he saw a statue already existing inside of it. His role as a sculptor was simply to remove the surplus marble around this form to bring it out in the open. Aristotle, in describing final causality, pointed out that the end had to be metaphysically conceived in the mind before anything could be done physically. In this way, the "metaphysical end" is in mind at the *beginning* of and throughout the physical process of creation until it achieves its "physical end." Tall skyscrapers and Martian rovers were physically constructed only after being fully conceived metaphysically in the human mind. Metaphysical knowledge of those formative ideas is inspired by the energetic spirit-communicator of potential energy, transcendentally going beyond what is originally materially present to what is finally physically made.

6. Aristotle, *Physics*, bk. B, ch. 3.

Because of the fourfold nature of energy, all changes have four, formally distinct, interrelated causes.[7] Consider Michelangelo's example of carving a statue. (1) The *material cause* is the existing block of marble from the past. (2) The *final cause* is the initial, intuited image metaphysical held as the possible end to the project. (3) The *formal cause* pertains to the physical structure of the marble as it is now. Different blocks of marble have flaws and qualities, which limit and modify plans for execution. (4) The *efficient cause* is the deliberate hammering of the sculpture on the marble, bringing the interior metaphysical model out in a future physical reality. Why does the successful artist go through this fourfold process? Because the artist is an energized individual, and these steps express the metaphysical fourfold process of the artist's potential energy.

During that sculpting, the partially completed physical statue may be said to be partially true to the final metaphysical image held as the virtual end to this project. The irritating dissociation of the metaphysical image from the partially completed statue is what energizes the sculptor to keep working on the project. Finally, physical completion of the statue is *true* when the chiseled state of the statue in the physical order matches the metaphysical idea of the statue in the mind of the artist. This correspondence yields a pleasant, satisfying, and restful experience within the artist. For that reason, when the statue as conceived is physically expressed in the marble, Michelangelo stopped. He had no reason to continue going forward efficiently. He had physically achieved the metaphysical end desired for the marble. He would begin to act as a sculptor again on that statute again, only if he recognized some dissociation of the physical statue from the new conception of the statue in his mind.

Notice that the "final cause" of his working was not the carved material statue at the end of the project. Philosophically, the "end" was the metaphysical visualization of the completed project continually present in his mind as he worked on the statue. In this way, as he was sculpting the statue, every swing of the hammer was an interrelational act of love, bringing what was potential into actuality. Consequently, the virtue of Michelangelo is not present in the *completed* statue but in his *completing* it—in his bringing its potentiality into actuality. As later viewed by admirers, the completed statue is but a sign of Michelangelo's virtue at the time of its being made. Similarly, a published book is but a sign of the author's virtue, for it is the author's writing of the manuscript that marks the true greatness of the writer. In contrast, the completed statue and the published book are considered great primarily according to its interrelational pleasure given to the beauty of the piece or the financial value of the peace to the viewer. Michelangelo and the writer are haled as the efficient crafter of the pieces.

Experientially, there is usually a major let-down once a project is over. A completed project is in the past, and one's interrelationship with that project is no longer active in the mind but only as a memory. A person is never really satisfied at the completion of a project because one's potential energy continues to push the person

7. Falcon, "Four Causes."

to be more interrelational. We are only partially satisfied because our current existence is always grounded upon our potential energy, and we are always partially dissatisfied and restless because our potential energy is always forward leaning. In this situation, we are content or discontent depending on what side potential energy is currently focused upon.

Here, it becomes clear that the *meaning of life* is not found in its historic end but in its interrelational process. Personal engagement in life involves going from potentiality to actuality, in striving beyond a potential, ideal, metaphysical goal unto achieving an actual, physical accomplishment. So, we are never fully satisfied. Can we be at peace with and celebrate that progressive bipolarity? IPhil recognizes the importance for an individual to "rest" at the completion of a project, to enjoy the closure of the fourfold logistic embodied in the completed project. Then, there is a time when the individual needs to be self-reflective and community reflective in preparation of moving forward again to another interrelational opportunity.

Why do observers find such great joy in looking at a completed statue by Michelangelo? Spectators of art and of sports events intrarelationally take the images of great things into their minds, adopting them within their lives. This positive association makes them feel interrelationally better—without doing any work. Here is a 2nd level, adolescent mentally embracing the success of another as one's own. However, the passive experience soon becomes a fading memory, and the person moves on to a different statue or the next ball game. Parents take great joy in seeing their children succeed. In the game of catch, it is the exchanging of the ball that is satisfying. Once a cycle or several cycles are complete, acknowledgment of the joy of the game is subsequent, but that joy is not really interiorized and soon fade.

On a long trip to a town, every step on that journey can be viewed negatively—like a youth who keeps asking, "Are we there yet?" That youth's understanding of the end is self-centered 2nd level, and every moment in which that the end is not physically attained is viewed as dissociative and painful. In contrast, when a 3rd level adult views the metaphysical end as a current, existential part of the logistic of the journey, each step is seen as logistically right and satisfying. Every step and stage of a project is pleasing when it is seen as being "on the way" to the completion of the project. If something interferes with that scenario, a 4th level view sees every dissociative turn of fortune as the first step of another scenario unto that end. The "eternal optimist" is very happy because the optimist recognizes that every dissociative turn of events opens a new way to reach a potential, future, associative end. In this way, *awareness* of the correspondence of one's present physical moment with the full potentialities of one's potential energy and existence, is satisfying, exciting, and restful in the midst of work.

Aristotle said that the "end" of life is happiness, *eudemonia*.[8] That statement must be understood in terms of an enduring metaphysical end and not in terms of a physical end. A product of the Age of Enlightenment, Thomas Jefferson in the Declaration

---

8. Aristotle, *Nicomachean Ethic*.

of Independence listed the pursuit of happiness as one of the inalienable rights of humans.[9] However, this does not mean that we will experience happiness with the possession of a certain type of property or material wealth. Rather, our pursuit of happiness is our response to the potentialities in our world, pushing for great interrelational harmony in our individual life and in our society.

IPhil recognizes that the meaning and purpose of life is to be more interrelational according to one's logistic. To be all what one *can* be. That is our calling and our destiny.

What makes a person happy? A person cannot consciously try to have fun or make oneself happy. By logistical entanglement, the joy of play arises from one's successful *associations* with others. Unsuccessful associations with others lead to disappoint and anger. There are three in an interrelationship. While a 2nd level individual will blame either oneself or the other party, a 3rd level individual recognizes the role of one's potential spirit and the energizing Spirit pushing oneself and everything forward to be happily more interrelational.

9. Armitage, *Declaration of Independence*, 103–4.

Chapter 15

Logistic of Knowledge

Internal Knowledge of External World. Resolving Kant's Conundrum. Neurons. Breaking through the Neuron Hillock. Evolution of the Brain. Memory. Relationship of Brain and Mind. To Reason and Beyond. Seeing 4 X. Words and Seeds. Types of Knowledge. Statements about the Future. Does God Know the Future? Scientific Method. Is IPhil Scientific?

Anthropologists call modern humans *Homo sapiens* because one of the most dominant features that distinguishes humans from all other animals and things in this world is their intelligence. Its dominant features include: (1) physical exploration of our world and beyond; (2) psychological integration of oneself; (3) interrelational and international cooperation with others; (4) transcendent intellectual and spiritual pursuits. Intellectually, humans have invented language and built skyscrapers. They send images of themselves through television. They have engineered rockets, which have taken them to the moon and back. Human science has learned how planets circle the sun and how hadrons come together to form the nuclei of atoms. Today human communications and knowledge are expanding at an exponential rate, giving humans an understanding of the world that is unprecedented in history. Clearly, the sapiential knowledge of humans is beyond that of every other biological species on earth. Consequently, it is reasonable to examine the ways interrelational philosophy (IPhil) provides the metaphysical foundation not only for the physical sciences of physics, cosmology, and chemistry, but also for human psychology, sociology, theology, and this philosophy. IPhil considers any discipline that uses the scientific method in an experiential and logical manner an apt subject for the philosophical reflections of this book. Studies of knowledge, epistemology, and psychology are the subjects of this chapter.

IPhil does not begin the study of knowledge as it occurs in the human mind, for the human brain and its mind did not exist until 13.8 billion years after the big bang. Because of the enduring character of the logistic of energy and the continuing principle of conservation of energy, we can study the beginnings of knowledge in the interrelational logistic of energy. Experimental observations of energy go back almost to the big bang, so our reflections on the nature of knowledge needs to go beyond personal experiences of conscious knowledge, all the way back to the interrelational kinetic energy of the big bang.

Starting at the very beginning of our energized world, IPhil recognizes that some type of interrelational knowledge is an inherent characteristic of all logically evolved energetic systems. Without some type of metaphysical knowledge of the possibility of a subsequent state, energy could not be potentially interrelational. Thus, knowledge does not begin in the kinetic, physical, human order but in the potential, metaphysical order, because potentiality of interrelationships precedes actual interrelationships.

Analysis of physical systems as simple as the game of catch reveals that the four inherent characteristics of potential energy are: (1) enduring existence; (2) intrarelational individuality; (3) interrelational logistics; and (4) free self-initiative. Associated with these states are four different kinds of knowledge: (1) existential knowledge; (2) intrarelational self-knowledge; (3) interrelational logistical knowledge; and (4) projective knowledge of logistical evolution. This examination of our experiences of knowledge indicates that our minds are energized by: (1) experiential knowledge of the current state of things; (2) remembered knowledge of the previous states of things; (3) imaginative knowledge of possible proximate states of things; and (4) intuitive, highly projective, transcendental knowledge of remote future states. IPhil recognizes that these different types of knowledge must be inherently present in potential and kinetic energy which advances through the four stages. Because these four forms of knowledge are inherent within the interrelational logistic of energy, IPhil recognizes that the fourfold types of knowledge must have been present in the Primal Absolute Potential Energy (PAPE), and by logical evolution be present in an analogous way in the logistic of every subsequently logically and physically evolved, finite, energized receiver-giver.

Internal Knowledge of External World

My experience of knowledge begins with information I receive from my senses. There is a subtle, subconscious, felt experience of my received sensations before I become intrarelationally self-conscious of them. Yet this knowledge is highly projective of different possibilities resulting in virtual images of my imagination. Also, my energy pushes me to interrelate with others in a projective, rational way. I experience my knowledge in my mind, and I know my body is different from my ideas. If I receive information about and through my body, how do I know that my received information really comes through my body?

Could not my knowledge of my body and through my body actually be but self-created imaginations of and in my mind? Here is a classic epistemological paradox: How can a person prove that there is a distinction and a correspondence between what I know in my mind and what is outside of the mind in my body and in our world? If I think something is outside my mind, that thought could be the creation of my imagination and my rational mind. Rationally proving something is outside cannot be done logically, for the premises and conclusions of the mind would be of the same logistic as the human mind. The logical form of all arguments is 3rd level and is interior to the logistic of my mind. Asking for proof that internal knowledge is more than interior to the mind. Such a demonstration must transcend the rational order of the mind. But how?

From personal experience, it is clear that my mind is capable of logically consistent rational thought. Is it possible to break the boundary of rational thought and go beyond it? Gödel's incompleteness theory not only states that the mental order is incomplete, but given a logically consistent system, there *can* be a statement that has the terms of that systems but also cannot be derived from that system. Being rational, the mind is capable of establishing the logical evolutionary theory, which breaks the boundary of closed, logically consistent systems. Logical evolution establishes that metaphysically there *can* be knowledge of a statement that is beyond the boundary of the rational mind. Logical evolution establishes that there can be more truth than what is determined by logical deduction.

By logical evolution, there can be an inverse progenitor statement that breaks the close boundary of the mind's internal operations and connects the mind to a logistic that is exterior to the mind. The logical inverse of "in the mind" is "not in the mind." Also, by the carryover process of logical evolution, there can be a statement and knowledge that logistically pertains to what is both interior to the mind and outside of the mind. But this proof is only of the metaphysical order of *possibilities*. Something more is needed.

Experience shows that there is information and knowledge that is generated by input from outside my rational or imaginary mind. That is, I regularly learn new, surprising information that initially is not of the mind or its activities but somehow is outside of it. Here is an applied application of the maxim: By their fruits you shall know them. Furthermore, the accumulation of surprisingly new information shows that its origin is well-ordered in ways the exceed the abilities of my current rational mind and imagination. Thus, consistent physical evidence indicates that our mental knowledge of logically consistent systems is not only metaphysical but also physical, and that there *can* be a statement about the physical order that is beyond what our metaphysical, rational minds can deduce. I in fact recognize new and disjointed information that I have not rationally or imaginatively generated. This cannot be deduced rationally, but it can be induced physically in a logical evolutionary way. These observations indicate that some received information shows that it comes from outside my rational and imaginative mind. We call the extraneous, non-mental outside source of received new information our physical world.

LOGISTIC OF KNOWLEDGE
Resolving Kant's Conundrum

Immanuel Kant raised a different significant concern, which has had a profound impact upon modern Western philosophy.[1] How can it be said that what I subjectively observe to be an outside thing truly is *as* I perceive it to be? That is, how do I know that the knowledge in human metaphysical thought truly matches what is physically experienced outside my body? What a human knows internally *could* be only a prejudicial interpretation and expression of rational categories already in the human mind. Every individual has one's own individual logistic, so every human sees and knows the things of this world only according to the limitations of one's own particular senses and rational categories. Kant argued that humans see and know only in their own *subjective* way. Consequently, all human knowledge is subjectively true and *only* subjectively true. In other words, humans cannot truthfully know how a thing operates outside the knower in the thing's own way—in an *objective* way. Because of the way we are made and know, we cannot know objective truth but only our own subjective ideas.

This viewpoint leads to statements like this: "You see things your way. I see things my way. We are both subjectively right and we are both probably wrong objectively. So, feel free do things in the way that will make you happy, but don't tell me what is right or wrong for me." This line of thought easily leads to a morality which has no objective basis. From this point of view, observations and moral values are strictly subjective, and there is no objective truth or morality. Still, pragmatically we can agree socially on what we see and what we value by common consensus, rather than objective truth. In such a society, people adopt a common morality to reduce conflict. Peace then is achieved only for arbitrary, pragmatic, utilitarian reasons, but these could change tomorrow by fickle public opinion or with the dictates of an authoritarian individual or group. From this point of view, morality can change with the changing winds of time and whims of society or its leaders.

When a camera takes a picture of a scene, the film in the camera can only record its received signal according to the limitations of the design of the camera and the film. As Thomas Aquinas wrote: *Quidquid recipitur in modo recipientis recipitur*; "Whatever is received is received in the mode [logistic] of the receiver." Consider photographing a dog. The resulting picture doesn't smell like a dog; it isn't warm like a dog; its nose is not moist like a dog's. The crinkle of the photo-stock doesn't sound like the barking of a dog. In fact, if the camera were out of focus, the resulting picture would be very blurred and might not even look like a dog. If the film only records x-rays, the picture would look very different from what our eyes see, for the eyes of humans do not see in the x-ray range but only in what is called the visible spectrum. In other words, the parameters and logistic of the camera greatly impact and even distort what is in the resultant photograph. Is this what happens to humans in their perceiving minds?

---

1. Greenberg, *Kant's Theory of A Priori Knowledge*.

Kant claimed that the human mind has only a limited number of rational categories. Because of the limitations of our senses and the limitations of the categories of our mind, we are not able to observe what is really out in the physical world. We only recognize what is within our own limited mental categories. Because of our self-centered, limited, internal prejudices, how can any human say that her internal knowledge matches what is outside of her mind? One might also ask whether *any* knowledge in our minds is logistically true to anything that exists in the outside world. Unanswered, Kant's claim has made modern philosophy highly subjective, self-centered, and lonely—relationally prone to skepticism, ennui, and despair.

Meanwhile, modern science makes multiple formulated predictions and by observations confirms them to be true with high certitude. Because of a plethora of successful physical observations of mentally known physical laws, science has ignored Kant's epistemological challenge. Science has based its conclusions on observations using many kinds of machines, and science has simply assumed that these findings are experientially objective and true. This pragmatic, objective, materialistic approach to knowledge by the physical sciences stands at the opposite end of Kant's epistemological approach to mental ideas. Because scientific experimentation has borne much fruit, it seems that there must be something right about the scientific view of objective knowledge through physical observations by humans and observational machines. In contrast, within the field of modern Western philosophy, Kant's conundrum questions every subject's ability to know accurately the objective truth about *anything* in our outside world. How can the objective investigations of science be philosophically justified in face of Kant's conundrum? IPhil can resolve this conundrum via the logical evolutionary theory.

In the process of logical evolution, subsequent systems take to themselves the logistics of their antecedent systems. That is, humans carry in their genes many logistic traits from their chimpanzee ancestors, from their mammal ancestors, from their amoebic ancestors, from their chemical ancestors, from their nuclear ancestors, from the logistics of the four fundamental forces, from the mathematical categories of space, and from the interrelational logistic of the Primal Absolute Potential Energy before the big bang. The knowledge categories that are already within humans are many more than the few which Kant listed. Humans internally contain the logistics of all antecedent subspecies in the human branch of the Logical Tree of Evolution, and they also contain analogously the logistics of species on neighboring branches of the Logistical Tree of Evolution.

When human beings encounter an animal or a lunar eclipse, because the logically evolved metaphysical logistic of the human observer contains and is beyond the logistics of these lesser phenomena, the logistically defined actions of these external objects are receivable by the human body and knowable in the human mind through the many logistical categories humans share with the external objects observed. Because humans, animals, trees, and rocks all have the logistics of the proto-spatial system and the four forces of physics within the evolutionary Tree of Life, they can all interrelate and communicate many of their logistical abilities and behaviors with others in evolved ways using common, evolved parameters. This correspondence is especially true with species and elements lower on the Tree of Life.

The logistics between high species on the evolutionary Tree of Life are not exactly alike but significantly similar. Hence, their interrelational truth is not perfect but analogously very close. It is up to the observer to determine whether these differences are consequential or inconsequential in the interrelationship they have in the life, the experiences, and the experiments of the human observer. By logical and physical evolution, because a chimpanzee and a human have nearly the same logistic, a human's objective observation and internal knowledge of a chimpanzee will be highly true to the internal and external reality of the observed chimpanzee, albeit not totally. The closer two things are on the evolutionary Tree of Life, the more similar will be the received knowledge about other species and their physical world.

Essentialistic, 2nd level thinkers are idealistic, and they accept only statements that are totally true or totally false. In contrast, IPhil is 3rd level and knows that our partial natures *can* receive in a high degree of accuracy, true knowledge. Like knowledge gleaned from scientific experiments, IPhil recognizes it is possible to gain highly accurate knowledge of the things we observe. We are tolerant of variants and partial knowledge as long as these variants are "interrelationally inconsequential." Experience shows the degree to which such partially true knowledge is enduringly interrelational or not. Unlike 2nd level idealism, in 3rd level realism the more important thing is not the exactitude of known truths, but the magnitude of successful interrelationships. IPhil recognizes that the purpose of energy and the purpose of life is not to *know* things perfectly but to *interact* with others more interrelationally. Physical life is the dynamic character of a logically consistent intrarelational individual. Death occurs when that intrarelational, logical consistency is terminated.

We do not live alone in our world. By approaching an unusual phenomenon from many directions, humans can rationally combine knowledge that is imperfect from different directions. The accumulation of diversified knowledge can be rationally combined to produce a conclusion that has a greater probability of being close to the truth and useful in interrelating physically what is metaphysically known.

Neurons

The central physical structure for passing information within humans is the nervous system. The central cell of the nervous system is the neuron. Knowledge is in the metaphysical order, and its development follows the fourfold interrelational logistic of potential energy. In the physical order, do neurons and the energetic flow of information in the nervous system follow the same fourfold interrelational logistic?

PART III: DEVELOPMENTS

NEURON

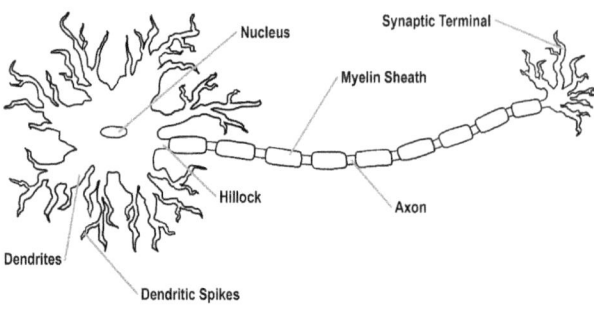

A neuron has four major components, which express the four stages of the interrelational process of interrelating the physical parts of an integrated individual.[2]

1. There are many different types of sensors scattered throughout the body. Sensors *receive* one type of physical signal, like pressure, heat, cold, rough, smooth, etc., and then through their evolved interrelational logistic transform that physical signal into a neurochemical signal. The incoming signal has kinetic energy which can be received by the sensor because part of the logistic of the sensor matches the logistic of the incoming energetic signal. Also, the logistic of the sensors is greater than the logistic of the incoming signal, for it can give out energy-communicators that have the evolved logistic of the neurochemical subsystem of the body.

2. Dendritic spikes extend the nervous systems beyond these sensors. Received physical sensations cause the breaking and firing of electronic potentials that have been established around the dendritic spikes. A single neuron can have over 10,000 dendritic spikes. A single dendrite *intrarelates* sensory signals from its many dendritic spikes. Excitation from these dendritic spikes fades with distance, and it requires stimulation from many spikes to cause a signal to move from the sensor via dendrites into the nucleus.

3. The neuron has a central body, called the nucleus, where the signals from many dendrites are brought together in a process called *summation*. Here electrical information is combined and gathered from many dendrites and dendritic spikes. There are two main types of summation. In *temporal summation* informational energy is accumulated from a single dendrite, which repeatedly fires again and again in one particular way. In *spatial summation*, informational energy is accumulated from many dendrites firing at the same time in a coordinated way.

---

2. Kandel et al., *Principles of Neural Science*.

4. The axon is a long extension of the central cell body of the neuron to different organs in the body. An axon may be several feet long. To speed the propagation of a signal down the axon, glial cells build a myelin sheath around the axon. At the edge of the neuron's central cell body where it is connected to the axon, there is what is called an "axon hillock." The axon hillock acts like a dam holding back a resting potential within the nucleus until it reaches a threshold. Once the potential exceeds the limit specified by the axon hillock, the dam *breaks* and a surge of excitation energy shoots down the axon to stimulate different muscles, glands, organs, etc. These organs then operate physically in response to the received information in a fourfold way.

The height of the axon hillock's potential dam is not fixed but can vary for a variety of reasons: natural disposition, chemicals, hormones, etc. which either build up or lower that potential dam. The height of the axon hillock determines when a motivational signal generated in the nucleus is "let go" down the axon to muscles, organs, etc., triggering neurologically induced changes in various organs in the body.

From the above, it is clear that the physical structure of neurons expresses the four stages of the interrelational logistic of receiving, claiming, processing, and releasing. This is true also of the body's organs which receives signals from axons, intrarelate that new signal into itself, determine within its own logistic an interrelational way to respond within the body and its environment, and then efficiently stimulate an outgoing physical action by that organ or muscle—or not.

Breaking through the Neuron Hillock

Human sense images and rational thoughts can influence the potential height of a neuron hillock. If what is being considered is familiar, there is little tension, and the hillock is very low. Self-initiating a habitual physical action is easy, like water going over a waterfall. However, when a prospective action is new and unfamiliar, there is an increased dissociative tension, and the potential height of the hillock rises to be above the level of the normal potential of the neuron. This disparity feels like running into a high brick wall. How can a person advance?

Motivational speakers and programs can provide mental exercises that raise or lower the potential level of one's hillock, thereby inhibiting or facilitating physical actions, respectively. Also, experience shows that the amount of potential energy available for reformative self-initiatives is limited. As the day goes on, a person's stockpile of available psychological energy for reforming self-initiatives is reduced, and one's resolutions are harder to execute—unless replenished by short periods of meditation, rest, yoga, snacking, power drinks, etc.

People are defeated in life not because of lack of ability, but for the lack of wholeheartedness. They do not wholeheartedly expect to succeed. Their heart isn't in it, which is to say they are not fully, mentally given to the desired physical task. Even if a person is physically very tired, high psychological motivation can keep the person moving toward a goal. IPhil recognizes the importance of summarizing all aspects of a situation within one's neuron centers, especially in the brain. This energizes all one's potentials in body, psyche, mind, and spirit. The individual thereby compounds and increases one's metaphysical and physical potential to consciously and freely self-initiate an action by flooding over the high hillock with heightened dedication.

Frequently, people who strive to physical break through such a barrier will often grunt as they mentally push themselves forward. In judo, the combatants holler as they execute a move. Football linemen and boxers make themselves angry so that they can exert more force in their moves. In these ways, they engage more physical components associated with that action. In addition, deliberate initiatives need to draw upon a strong faith and confidence in one's received potential, enabling a person to holistically break through the boundaries of the status quo into a more advanced state. I personally find imaging very helpful to picture the heightened hillock standing in my way, and then project myself mentally and physically smashing through that boundary using all my faculties as a soldier on an assault into a new domain of operations. Athletes regularly use visualization followed by physical exertion to achieve advanced performances. IPhil provides an energetic explanation *why* metaphysical visualization assists self-initiated physical actions and improved performance.

In the other direction, sometimes individuals want to stop a bad habit, such as, eating too much, drinking too much, taking drugs, etc. In these cases, it is important to raise the energy level of the hillock, blocking the individual from physical acting out in these ways through their axons. There are a variety of things an individual can do to raise the potential of one's neurological block: meditation, visualizations, defensive planning, testimony for others, exercise, alternative activities, etc. Unfortunately, each of these take psychic energy, and the individual's resolve soon wanes, and the individual too often falls back into habitual, dysfunctional behavior.

Evolution of the Brain

The human body is connected to the brain in physically distinct parts: (1) The external senses are connected to the spinal cord, (2) the spinal column to the hindbrain; (3) which is connected to the midbrain; and (4) finally the forebrain.

The American physician and neuroscientist Paul D. MacLean in the 1960s proposed that the brain evolved in three stages. Later, he described it at length in his 1990 book, *The Triune Brain in Evolution*. The most primitive form was a reptilian complex, next developed was the paleomammalian complex, and the most recently developed was the neomammalian complex. He picturesquely named these evolutionary stages of physical development of the brain the Graven Image, the Lethe, and the Guru. Building on the foundation of the spinal cord's autonomic abilities, the small, inner, reptilian brain is primarily concerned about the past. Through its accumulated memory and acquired wisdom, it operates primarily by instinct. The middle, paleomammalian brain, McLean says, is primarily concerned about the present and how to react physically to present situations. The outer, neomammlian, human brain is primarily concerned about the future and with planning far ahead. In this way, our outer cortex is concerned with projections, probabilities, and expectations.

Contemporary comparative neuroscientists, however, have proposed alternative developmental models. The paleontologist Neil Shubin has studied the evolution of the brain in many species going back to the most primitive form of tiny fish.[3] In his 2008 book, *Your Inner Fish: A Journey into the 3.5 Billion-Year History of the Human Body*, he points out that from their earliest times, all brains *already have* a threefold structure, namely, the hindbrain, midbrain, and forebrain. These structures assume different shapes and sizes during evolution, but all three divisions of the brain are found in all species from the beginning. They arise from different parts of an animal's DNA. They did not come into existence in sequence but all at once. Moreover, in the most primitive species, the spinal cord and hindbrain is the largest part, while the midbrain and the forebrain are very small. During evolution, the midbrain grows, as in reptiles. The more recent and largest part is the forebrain, especially in humans.

Note that these models of the brain assume the founding neurological connection of external senses to the spinal column. Shubin's model of the structure of the neurological system physically expresses in an evolutionary, analogous way the four aspects and stages of potential energy. All metaphysical aspects arose together in the physical order. However, the growth in physical size of each part was progressive, strongly influenced by changes in the individual's physical and social environments. Evolved species had larger higher functioning areas corresponding to higher levels of neurological development. Humans have the highest level of growth and strongly use in the prefrontal cortex of the forebrain.

Going beyond 1st level of vibrations from different physical bodies, people manufacture 2nd level musical instruments that produce sounds in consistent, pleasing ways. Then, different instrument can sound together, producing harmony and music on the 3rd level. Beyond this, creative musicians seek to discover new sounds and harmonies,

Similarly, network neuroscientist are discovering that neurons in different parts of the brain can produce their own neurochemical impulses, *and* different neurons in different parts of the brain can interrelated in systemic ways that go beyond those physical, material structures to form hubs that systemically transcends those material part to produce higher level metaphysical harmonies. From this perspective, music and ideas transcend instruments and general sense images. Furthermore, IPhil recognizes

3. Shubin, *Your Inner Fish*.

that from logically consistent, neurological systems, truly new creative ideas and music can logistically evolve. Here is the progressive energetic bridge from material, physical sensations to neurons into spiritual, metaphysical harmonies—advancing our receiving material brains into our transcending spiritual minds

Memory

IPhil finds the origin of memory in the primal order of the Almighty. Memory is a necessary aspect of potential energy as it leans forward into the future of becoming. Going from a state of being to a state of becoming involves carryover of knowledge of the antecedent state of existence into the subsequent state of becoming, and memory is essential in that carryover process. There can be no state of "becoming" if it does not carry over some memory of the antecedent state of "being." If there were no memory, the progressive aspect of energy would have no systemic meaning. In the closure of the absolute interaction of PAIS, there must be some memory of the antecedent state of the all-giver for the all-giver-receiver to know that PAIS is a logically consistent system and thereby capable of logically evolving our finite world.

In the memory of the evolved human brain, the information-receiving brain (1) advances received interrelational knowledge through three processes: (2) encoding, (3) recording, and (4) retrieval. The brain receives and encodes new information. Then the brain records potentially available information. Finally, the brain retrieves information when needed by the individual for advancing some interrelationship(s). Many brain regions are involved in this process. (1) Our physical senses receive and transmit outside information to the brain; (2) the cerebral cortex—the large outer layer of the brain—acquires new information as input from our senses; (3) the amygdala tags information as being worthy of storage and the nearby hippocampus stores memories; and (4) the frontal lobes help us consciously to determine how remembered information can guide and stimulate interrelationally desired tasks.

Working memory is active within a given situation. Experience shows that from the unconscious knowledge in memories and sensations, consciousness first arises when new information is highly dissociative and stressful compared to the antecedent placid situation of the knower. Sensations and new knowledge, which are systemically and interrelationally consequential, are stored in long-term memory. In contrast: Dr. Joel Salinas, a neurologist specializing in behavioral neurology and neuropsychiatry at Harvard-affiliated Massachusetts General Hospital, explains that working memory is a mental scratch pad that allows us to use important information throughout the day. Depression, anxiety, stress, and a lack of sleep can affect the brain's retention and use of information.

Relationship of Brain and Mind

This leads to a series of question: What is the difference between brain and mind? Which came first: the growth of the physical brain, or the growth of the metaphysical mind? What is the nexus between the operations of neurons in the brain and the types of knowledge in the mind?

The example of Michelangelo's carving of a statue indicated that mental form preceded his physical sculpting. IPhil recognizes that usually metaphysical 3rd level ideas are efficiently initiated into physical actions from a 4th level metaphysical self-initiative, which in a "letting go" transformation, moves some potential energy into efficient, kinetic energy. However, this can be done in two ways.

In building a bridge in a difficult terrain, a construction engineer must ask whether existing machines fulfill the needs of the advanced project, or do the advanced needs of the project require the creation of new machines? Initially, the first is true, because an engineer will use all available machines to accomplish the goal. If there are no new challenges, the engineer will achieve the end desired by using old techniques—within the previously known engineering logistic. However, when existing machines and techniques cannot accomplish the task, a first-rate engineer will devise—logically evolve—new machines capable of achieving the end imagined and desired. A routine project can more easily and efficiently use existing machinery and existing techniques, but a new and difficult engineering problem forces a builder mentally and metaphysically to evolve a new technique and then physically construct new machines to achieve the goal desired. Once that new technique is established as interrelationally successful, that new technique usually spreads rapidly throughout our world.

IPhil proposes that, faced with new mental challenges demanding new interrelational techniques, something analogous happens in the brain. When old techniques are applied multiple times in the mind, the brain simply increases in magnitude of size, like an exercised muscle. In contrast, when new, difficult situations demand the breaking of old logistical patterns of operations, there is need for logical evolutionary changes in the interrelational, logistical structure of the mind, and subsequently new *structural* changes—not size—in the interrelational structure of the brain. Repetition results in increased size; challenge results in new advanced structures.

Great advances have recently been made in neurological research. Dr. Michael Merzenich in his 2013 book, *Soft-Wired: How the New Science of Brain Plasticity Can Change Your Life*, has grouped meaningful findings this way. (1) The brain is elastic. Unlike former thinking, it has been found that the brain has the potential to change positively. (2) To change positively it is necessary for the brain to be challenged in new ways. (3) To advance one's brain it is important to calibrate one's mental capacity to change. (4) To advance in mental capacity, it is important to energetically exercise the brain in challengingly new ways. Notice the similarities of these four stages to the stages of the fourfold interrelational logistic of potential energy.

Some professionals deny the distinction between brain and mind. When a person mentally focuses upon a problem, CAT scans reveal that the frontal cortex becomes active. When a person thinks about a map, the spatial sections of the brain lights up on the scan. These CAT scans do not indicate thoughts directly but only an increase of blood flow and electrical activity in different physical part of the brain. When a person hears music, the listening part of the brain associated with the ears becomes active. If a certain part of the brain is severely damaged by a physical injury or stroke, the person's ability to think in the way associated with that physical part of the brain is severely hampered or terminated. For these and other reasons, it is common today to identify the operations of the brain with the operations of the mind. Still, many other people resist identifying the *brain's* physical composition and operations with the *mind's* thoughts and logical interrelationships.

For me, the best demonstration of the difference between the operations of the brain and the operations of the mind is found in what researchers have discovered by comparing physical vision from the eyes and visual images in the mind. The sensors of the eye are connected to the visual cortex in the back of the brain. That connection is not random or haphazard. Rather, whole sections of the retina are mapped one-to-one on corresponding grids of neurons in the back of the brain. Also, in the retina in the back of the eyeball, there are veins and a blind spot, which have no optical sensors. There are corresponding gaps in the field of neurons in the back of the *brain*. How those optical neurons in the back of the brain fire can be observed physically.

But what does the mind see? The mind does *not* see any of the gaps from the eye's blind spot or veins in its visual image. Rather the mind fills in those gaps with a best approximation. If an object initially is in one's physical blind spot, the mind of the viewer will not have that object in one's mental image. However, turn the eye slightly to see just a part of that object, and the whole object will suddenly pop into the individual's mental image. In this process, the mind interpolates and fills in the gap—beyond what is actually seen. Note that there is no physical cause for the interpolation of the unseen object in the mental sense image. Those interpolated images are not always filled in accurately but only approximately, as the mind's best guess. Here the interrelational metaphysical mind goes beyond the intrarelational physical structure of the brain. The continuous image in the mind is progressively metaphysical because it appears "after physical" brain stimulations, and the potential energy of the mental image pushes beyond the physical stimulation to a more advanced image. Consequently, mental images are interrelated sets of metaphysical knowledge, built upon and inferentially going beyond the information found in incoming intrarelated physical sensations. Neurological sensations in the brain are physical, and mental images in the mind are metaphysical.

Not all knowledge is the same. General ideas are 3rd level and drawn in an interrelational way from a set of 2nd level individual sense images drawn from 1st level sensations or physically stored memories. In this way, metaphysically generalized ideas are formally different from intrarelated sense images. On the 4th level, extrapolations and interpolations go beyond sense-grounded data and generalized ideas because their potential energy pushes the mind not only to think about what has already been received but also to make metaphysical projections which stimulate self-initiatives that go beyond the present into the possible future. This evidence points toward activation of the kinetic vs. potential aspects of energy, producing physical phenomena vs. metaphysical inferences.

Examining this matter from an interrelational point of view, let me ask a very simple question: Do you *physically* observe a game of catch? In response to my question you might turn and look out the window of a classroom to see two kids tossing a ball back and forth on the playground. Then, you probably would say, "Of course! I am watching a game of catch right now." My response to you is: No, you are not. You are not physically watching that game of catch. Rather, you are inferring a metaphysical extrapolation of the received information from the observed physical events. You do not actually physically see a game of catch; rather, you mentally conclude that you are watching a game of catch. The real game of catch is interrelationally happening *between* those two kids, as they experience the ball going back and forth between them. Similarly, what is observed in CAT scans and other scans of the brain is not the internal operations of the mind but the external, physical stimuli of the operations of the individual's physical brain. While the external manifestations of the operations of the electrons and the blood of the brain are physically observable, the internal, interrelational operations of the mind are not physically observable. They are only inferred. The actual operations of the mind are expressive of the metaphysical potential side of energy, while the CAT scan only records the physical, kinetic side of that energy.

Knowledge advances from physical sensations, to mental integration, to projective extrapolations—from the physical order unto the metaphysical order. If a person's brain is physically injured in war, in a car accident, by Alzheimer's disease, or a stroke, the upward movement of integrating information will be interfered with, and the individual's thinking and personality can be significantly impaired and transformed.

In summary, information is obtained from our physical experiences. Knowledge goes beyond physical stimuli to establish metaphysically new, extrapolated images or abstractions in the mind. The brain and the mind are intimately and distinctively related as kinetic energy is formally distinct from potential energy. Those formal distinctions within energy are real and interrelationally important. Thus, the brain is to the mind is as kinetic energy and the physical order is to potential energy and the metaphysical order.

To Reason and Beyond

Reason didn't suddenly appear on the scene with the evolution of humans. Also, reason is not a unique characteristic of being a human. Looking at the advance of knowledge in biological species, reasoning advanced with the evolution of the biological species as they tactically faced new environmental challenges. A human child is not born with full-blown rationality. Rather, reason is a talent that grows. In the development of the faculty of reason, IPhil recognizes four discrete phases, analogous to the phases of the interrelational logistic of energy.

1. *Experience.* Information of current physical events is first existentially perceived in the senses and then carried over as memories into future states. Furthermore, energy responds intrarelationally and defensively to dissociations. Energy strives to resolve dissociative experiences and knowledge, pushing to make them somehow harmoniously interrelationally associative. The individual on this level unconsciously learns from experiences dealing with new things, especially things that are systemically troublesome. The reactions and interactions of nuclear particles relies on the experiential knowledge gleaned from received information in those interrelationships. This information is passively received and therefore not initially conscious.

2. *Imaging.* From various dissociative and associative experiences, the individual intrarelationally groups things together that are pleasantly associative and secure, while avoiding and rejecting things that are antithetical to one's intrarelational development. This mental segregation process brings knowledge of good sensed things together and bad sensed things apart. An individual then rationalizes different responses to these positive and negative experiences by reaffirming logistically extended reasons for and reasons against certain actions and associations. This individualized, idealized rationalizing process elevates the good into the best and the bad into the worst, increasing the interior emotional responses of the individual.

3. *Reasoning.* With maturity, an individual recognizes that a thing or event happens within an environment that has multiple interrelationships with it. The knower recognizes that there are many different responses to the matter under consideration. Also, learning from others, the scope of one's integrated knowledge grows and strives to be universal. Experiences show logistical sequences of phenomena and logical arguments that are logically consistent and enduring. The thinker first mentally weighs different courses of action to determine which associations or interactions are better or best in the current situation. Long-proven rules of logic give greater assurance of argumentation and broad systemic cohesion.

4. *Wisdom.* Philosophy is not the love of reason. Rather, philosophy is the "love of wisdom." A person can be very intelligent, demanding tightly reasoned arguments. But if a person's reasoning is founded on false premises, it can lead to false and imprudent conclusions. In contrast, another person can have little formal education but be truly wise. Wisdom respects reason and goes beyond reason.

A good philosopher is not fully satisfied with, and is somewhat skeptical of orderly, comprehensive conclusions, pushing beyond rationalistic and reasoned thinking. A wise person can know all the established reasons for something and then can "break the rules" to direct others intuitively in an innovative way that subsequently shows itself to be more successful in solving a difficult problem. Unorthodox predictions by wise individuals push individuals toward making a paradigm shift in their view of the problem, leading to a more advanced way of thinking and acting in our world. "By their fruits you shall know them."

Seeing 4 X

As an energized individual progresses through life, the individual's seeing advances through four interrelational stages.

1. When children are born and begin to grow, "they are all eyes." They are searching to see more and more things in their immediate environment. The focus length of children's eyes is initially about 12 inches, the distance between the mother's breast and the mother's face. The faces and things in one's family environment make a deep impression upon a child and are remembered subliminally and pleasantly throughout life. In adolescence, kissing, hugging, embracing and calling one's beloved "Baby" pleasantly reflect this earliest stage. If any of these affectionate elements are missing in a child's earliest stage of development, it has been shown that the individual will having negative consequences on the way one later looks at life.

2. When the individual becomes an adolescent, the use of one's eyes becomes dualistic flight or fight. When a person feels shame before an authority figure or before one's group for a misdeed or for being of a lower status, the individual will cast one's eyes downward or askance, unable to look the other individual in the eye. This behavior is found in animals, like dogs and wolves, as well as humans. To intensify one's prayer or concentration on this level, people will concentrate by closing one's eyes so as not to be distracted by external things. Contrariwise, on this level, antithetically, an angry person may glare at an enemy, or give one's opponent the "evil eye."

3. In the adult phase, a person advances to looking another person inquiringly in the eye. Women who seek to attract attention from others, regularly highlight their eyes with mascara, eyeliner, eye shadow, and eyebrow darkener. A social consultant pointed out that too many people avoid close personal connection by defensively not looking other people in the eye. This habit is acquired in one's early self-defacing years. To develop the habit of "looking the other person in the eye," it is helpful for the individual to focus one's eyesight upon a spot between

the other person's eyes. Then, one's awareness will gradually expand to include the whole face and whole body of the other individual. This is done to socially bond with the other person.

4. Going beyond physically looking another person "in the eye," the viewer's transcending and spiritual vision can advance metaphysically through and beyond what is looked at. The *bindi* or dot worn on the forehead of Hindi and other religious groups marks the unseen "third eye." Here, the spiritual viewer "sees" the other individual's metaphysical God-given soul or logistic. That is why they greet each other with the salutation, *Namaste*, in which the God-aspect of me recognizes the God-aspect in you. To achieve this spiritual awareness, one should not close one's eyes as on the 2nd level. Here, in full physical sight, an individual gains insight into the spiritual, metaphysical dimension inside the other person or thing. This is the metaphysical insight that Plato pointed out in the famous parable of the cave. Here, the individual sees with physical eyes the physical dimension of things observed *and* sees with metaphysical insight their metaphysical dimensions. This is how a parent can look upon and deal correctively with an outwardly errant child, while inwardly filled with enduring, great love for the child.

Words and Seeds

For millennia people have recognized that words are analogous to seeds planted in infertile or fertile ground. Seeds hold the interrelational logistics of particular plants or animals in a materialized form of potential energy. This is like light photons which hold in the 2nd, intrarelational phase the logistic of a particular electromagnetic wave. Seeds can hold their respective logistic in a static state until it is placed in a compatible interrelational environment where it can grow. Similarly, the potential, metaphysical energies of written, signed, or spoken words contain internally interrelational information, which can be understood and incorporated by a reader or hearer years after the writing. Understanding an old text depends upon the logistical level of education in the reader.

Hieroglyphics were physical expressions of an unknown language for centuries. Learning how to read them in our modern age was in general a fourfold process. (1) Bits and pieces of hieroglyphs were viewed and record by curious scholars, who really did not understand them. (2) By comparison with other languages on the Rosetta Stone, an internally coherent understanding of hieroglyphic began to be decoded. (3) By the comparative analysis of hieroglyphic messages from different periods of history, changes and shades of meaning were recognized and appreciated in their respective historical contexts. (4) Finally, hieroglyphics are recognized as part of an evolved families of languages from earlier periods. In this process, the information objectively contained in those hieroglyphics was grasped by readers according to their progressive understanding of those texts. Again, "Quidquid recipitur in modo recipientis recipitur." Only when the logistic of the receiver is *of* the logistic of the energetic/material communicator is that informational seed able to be interpreted and incorporated in the knowledge and life of the receiver.

Types of Knowledge

Humans study very hard to acquire rational knowledge, which is highly prized and lauded in society. Sense knowledge and feelings are generally disparaged because they are thought to be more prone to error than rational knowledge. As a result, many people, including philosophers, think of rational knowledge as the only real form of human knowledge. Also, there is a human bias to think that only rational humans know anything, that all animals are dumb, and that material things know nothing. Prolonged studies of animals, however, reveal that there are many animals which know and communicate a great deal.

My physical heart increases and decreases its beating not by conscious knowledge but by subtle, unconscious chemical signals, as my body strives to keep itself in physical good health. Even on an atomic level, electrons must have knowledge of neighboring electrical charges to know how to travel around the nucleus of an atom in a unique, quantum, energetic way, as revealed in the Pauli principle. Distinct, atomic interactions are randomly free, oriented toward the lowest level of potential energy within a system to conserve potential energy and to become as interrelational as possible.

Rather than maintain the prejudice that rational knowledge is the only true knowledge, we must humbly recognize that many rational and scientific arguments have produced conclusions that where later proven false. IPhil recognizes that knowledge generated on each level—rational, sense, and material—is prone to be true and prone to be false. Knowledge on all levels must always be physically verified to determine whether it is really true. Sometimes one's logistically emergent intuitions are truer than the most carefully constructed rational arguments.

IPhil locates the origin of knowledge in the foundational character of potential energy itself, and this knowledge is time stamped.

1. Knowledge of potential energy is gained from experiences *now* from open existential awareness.

2. Receivers know they are receivers by the experience of change from their previous state. This demands a knowledge of the *past* in the form of carryover memory.

3. Since received energy has potential, there must be some logistical awareness of possible future states. That aspect of energy's knowledge is of the *proximate future* in the form of possible projections within an individual's logistic.

4. Logically consistent systemic logistics can evolve logically. Potential energy stimulates a subtle, metaphysical knowledge of an *advanced future* state.

Consider this classical Indian story. A person is walking down a path in a forest. The person recoils upon seeing a snake crossing the path in front of him. Upon closer examination, however, the person finds that it is not a snake but a wavy piece of rope. In this story, the individual's first impression was sense knowledge, but the individual's interpretation of it was not true. The individual's closer examination established the truth. This illustration about the rope (1st level) shows that there is difference between *knowledge* of a 2nd level initial conscious impression, and *truth* from a 3rd level verification. Continued verification of the truthfulness of a statement grows toward transcendental 4th level *certitude*. Until many observers verify an experience universally, certitude about the truth of a statement or concept will be limited by our current, limited, physical experiences.

IPhil recognizes that knowledge in and of itself is not necessarily true. Unfortunately, because of their limited experiences, adolescents strongly claim that what they *know* is the full truth. That is because the information they intrarelate into their ideas, incorporates all of what they have thus far consciously experienced and remember. Since metaphysical truth is the conformity of an idea with total physical experience, an adolescent's intrarelated ideas match their limited experiences, and they can strongly proclaim their ideas to be totally true. In strong defense of their self-esteem, an adolescent's absolute claims will result in a strong disassociation from anyone who disagrees with them. However, with the accumulation of more experiences, youths gradually come to recognize that many early ideas were not true to their now increased knowledge of the world. Early institutions, especially religious institutions, make the same mistake and must modify their claims, as increased evidence indicates to the contrary.

When we say something is "true," we commonly mean it is true to what we know experientially. However, knowing their experiences may be limited, loyal adults are often slow to admit that their partial knowledge is the whole truth. Adults humbly recognize that what they know from experience, study, and hearsay, is not the totality of reality. They are able to admit that what they currently know is only *probably true*, because our experiences are not universal but only partial.

Even if a revelation is said to come from the Almighty, what *we finite receivers* hear or see is only partial because hearers are significantly biased in a given historical situation. *Quidquid recipitur in modo recipientis recipitur* (What is received is received in the mode of the receiver). That is, the truth we learn from divine revelation, as from physical experiences, will always be the partial truth. To say or assume that any statement is *absolutely true* means that our reception of that information is not humanly partial but absolute, like the Almighty. In ancient times prophets claimed to speak "the word of the Lord." However, critical, historic examination of the prophetic statements shows them to be only partially right—not totally right—and that is how the people in their time understood them. Absolute interpretations of biblical statements by modern Fundamentalists are 2nd level and adolescent. In contrast, Jews turn to professional rabbis and scriptural professionals to interpret and apply ancient scriptural texts in mature, adultlike ways.

Scientists speak of a "best-fit model" of reality. It is "best-fit" because all gathered data and ideas currently fit best into that model. Intelligent individuals, aware of past erroneous scientific claims, recognize that a proposed model may not exactly match our physical universe in truth. Unfortunately, the Press and many lay people write and speak of the latest "best-fit model" as if it actually is the objective truth. They stretch the truth to simplify things that are interrelationally complex. They want to live in a 2nd level black-and-white world, which glorifies the new and denigrates the old.

Why is language dualistic? Knowledge and language on the 1st level are not dualistic. Rather, words are applied in a childlike manner to closely associated individuals, such as mother, father, dog, milk, etc. This is done in a simple accumulated way, analogous to the natural number system (N), Then, by logical evolution, in a 2nd level, adolescent way, the logical inverse is recognized (-N) as in the negative number system. Antithetical knowledge concerns what *is not* versus what *is*. Here, knowledge and language easily become idealistic, antithetical, and adversarial. As in all logical evolutionary advancements, this antithetical, dualistic characteristic is carried over into the advanced knowledge and language of all subsequent systems. An adult on the 3rd level recognizes and is willing to work with individuals existing and operating at various degrees between black and white extremes. This is analogous to the mathematical system of fractions (/N). Because there are multiple degrees of association between two idealized terms, these degrees are difficult to express in organizational language because of the forcefulness of carried over 2nd level thinking and dualistic language. Continuing the analogy of knowledge and language to evolving mathematic systems, 4th level transcendental experiences cannot ever be fully expressed rationally ($\sqrt{N}$). On the 5th level, some experiences go beyond what is physically rational to what transcends by logical evolution to what is a combination of what is natural and supernatural, material and spiritual analogous to the combination of real and imaginary numbers, as in (A + Bi).

Rather than fall into despair about recognizing that we know nothing with total objectivity, we need to be at peace about the limitations of our knowledge. We can act in confidence when we realize that the difference between what is known and what appears to be the objective truth is regularly "interrelationally insignificant." We must remember that in the end of any possible interrelationship, *knowing* the truth accurately is secondary to *making* things as interrelationally well-ordered and actual. Knowing the truth is not the end or purpose of human existence; loving interrelationally in a physical way is.

PART III: DEVELOPMENTS

Statements about the Future

Potential energy is oriented toward the future. In potential energy there are many different future possibilities. What can be said about the truth of statements concerning physical events and states in the future?

Statements about the future exist and are known in the mind; they exist in the metaphysical order, but not the physical order. Potential, future events are not yet energized or actualized in the physical order. This presents a problem. Knowledge is recognized as *true* when the evidence in the exterior world matches what is metaphysically known in the mind. Future metaphysical states of potential energy are not yet actual. So, because all future states are not yet energized or actual in the physical world, there cannot be a comparison between what is in the mind concerning the future and what is now in the physical world. If the measure of truth is the correspondence of what is now in the mind with what is now in the world, then all statements about the future in the physical order have *no physical truth-value*. That is, the criteria that establishes a statement as physically true in the physical order is not present in statements about the potential future.

Nonetheless, IPhil recognizes that statements about the future can have a *logistical truth-value*. Here, *truth* is found in the mind's conformity of metaphysical knowledge of specific possible states to a metaphysical logistic of its possibilities. When potential energetic events are of a logistic that endure through time, it can be said that projected future statements within that logistic are *probably logistically true*. The statement "I will have breakfast tomorrow" of itself has *no physical truth-value*, but because I regularly eat breakfast every day, the statement "I will have breakfast tomorrow" has a *probable logistical truth-value*. The lack of physical truth-value in the first statement comes from lack of physical evidence. The probable, logistical truth-value of the second statement comes from the enduring logistic of the individual as known from observed past behavior projected into the future. In a logical system, the founding axioms are simple in form, and they are very probably true from extensive observations. Statements which conform with the axioms of a logical system can then be said to be *logistically true*. Whether *logistically true* statements are *physically true* depends on the conformity of the axioms to physical reality. An individual metaphysical statement drawn from a specific physical event can be said to be *metaphysically true*. Still, a general metaphysical statement inferred from a limited number of physical experiences can be said to be only *probably true*.

A myth is a story that is the whimsical creation of an author or a poet that has little basis in fact. A mythology, on the other hand, is an "-ology" which is based on the logistical carryover found in logical evolution. Mythologies are well-ordered logistical extensions of known physical facts and are thereby analogously and partially true. The big bang theory, the first eleven chapters of Genesis, the premises of mathematics and physics, and the founding principles of IPhil are not statements of observed physical truth, but they are logistical extensions of physical truth and thereby analogously have much highly valued logistical truth-value.

Does God Know the Future?

The people in the ancient world felt very insecure for many reasons. They regularly consulted oracles and seers to learn about the future so that they could take evasive actions to minimize future misfortunes. It is much like people today investing in hedge funds following the advice of recognized financial pundits. In the Jewish world there were many prophets. The primary criterion for judging prophets as true was the accuracy of their future predictions.[4] The existence of that criterion of judgment implies that many prophetic statements were not truth. Like predictions of pundits today, many of the prophesies of the most famous prophets were not accurate in their details, though they were accurate in a limited way. That ambiguity was the common experience and expectation of all *prophetically true* statements in ancient days. It is exegetically wrong to assume that ancient *prophetically true* statements of ancient days are *physically true* today.

It was frequently said that God knows everything, even the future. However, many times it is recorded in the Bible that God made judgments about the future and then "repented" from that judgment. If God really knew how the people he tested would respond to those tests, then surely the test was superfluous, inefficient, and showed a lack of wisdom. In the test in the garden of Eden, God told Adam and Eve that if they ate of the Tree of the Knowledge of Good and Evil, they would die immediately (Gen 2:17). However, when they ate of that tree, they did not die immediately. Moreover, only by Adam's response to the LORD's question did the LORD know that he had eaten of the forbidden fruit. After they sinned and were expelled from the garden, Adam did not die immediately but lived outside paradise to the age of 940 years. Some might say that their *relationship* with God died immediately. However, outside the garden God still communicated with Abel, Cain, and others in personal ways. Likewise, Jesus of Nazareth predicted he would return in glory within one generation. Early Christians initially believed this prophesy literally. When that did not happen, they became confused and discouraged. Various authors then tried to rationalize that prediction by saying it would be fulfilled at some indefinite date in the future. In the Bible, most predictions about the future are not literally accurate in details, but only generally so. Everyone in the ancient world knew that. Only adolescent, 2nd level perfectionists take those hortatory biblical statements about the future literally.

Biblically, to "know" one's wife was to experience sexual intercourse with her. Real knowledge was understood as being more than mental; it was to understand from physical experience. So, in a biblical sense, it can be said that God did not "know" the future, because he had not experience it physically. Early Christian theologians

---

4. Deut 18:22.

broadly embraced Neoplatonic philosophy. Despite going against how the Bible described God's repeated interventions in the world, Catholic philosophic theologians strongly held to the claim that God is perfect, unchangeable Being. They maintained that if God's knowledge is perfect, God knows all things perfectly. They held that God as Being knows all things as being-as-*now*. However, how can God as Being, as perfect and unchanging now, *experientially* know the future as future? If God already knows all things past, present, and future simultaneously as now, there is no way that God, as perfect Being in whom there is no time, can know *experientially* the sequence of temporal events as they unfolded in time. A God who is simple Being cannot have full experiential knowledge of things as they are changing. There is no way that God as perfect Being can know whether an individual will repent from a sin that was committed the past. A perfect, unchanging God as Being cannot be a savior, yet the Bible regularly speaks of God giving salvation to those who follow his will, and punishment to those who don't. If God knows the subsequent disposition of a person's loyalty, why did the prophets so painfully call the hard-hearted to repentance?

To love others is to seek their *future* interrelational benefit. Therefore, the statements "God is Being" and "God is Love" are logically incompatible. Energy is oriented toward change, so the statements "God is Being" and "God is Almighty" are logically incompatible. Thus, God as Being cannot know the forward advancement of self-initiated acts of energy or love. IPhil recognizes that because of the forward orientation of the logistic of energy and love, the Almighty can only know the *logistical probability* of future acts of energy and love.

In heaven, will our awareness be conscious or unconscious? Experience shows that consciousness arises from unconsciousness when there is a dissociative stimulus toward some interrelational change. When I awake, I become aware of a change of mind. I subsequently am conscious of my awake state in relationship to making some change. If I am moving in a routine manner, I am not conscious of that routine until there is some reason for altering that routine. If heaven is a place of simple *being* in union with God as Being, then awareness of heaven's static peace would quickly drop to the level of unconsciousness. That means a person in a perfect, unchanging heaven would not be conscious of being in heaven. For me, that would be unsatisfying. I want to be conscious of being in heaven. In contrast, IPhil recognizes God not simply as being but as forward-leaning Primal Absolute Potential Energy. Consequently, union with God and with other individuals in heaven will likewise be forward leaning, and thereby conscious.

In summary, IPhil recognizes that by logical evolution, the Almighty is analogously like good parents, who by logical extension of past behavior, know how their children will *very probably* act in the future, but *not* know exactly how they freely, actually *will* act in the future until the deed is done.

Scientific Method

The beginnings of the scientific method can be traced to ancient Egypt, where there were documents describing specific empirical methods for approaching matters of mathematics, medicine, and astronomy. In the sixth century BCE, the founder of Greek philosophy, Thales of Miletus, refused to accept mythological explanations for the origin of natural phenomena. His observations and reflective inductions led him to the intuited assumption that every natural event has a natural cause. Aristotle made extensive, well-ordered physical studies of our world. However, in his ancient world, thinking and planning were the concerns of the aristocrats and their teachers, while direct dealings with the material things of our world were reserved to slaves and peons. So, in subsequent centuries, advances in academic knowledge were regularly in the mental, metaphysical order and not well-grounded in the material, physical side of life.

The first clear instance of an *experimental scientific method* seems to have been developed by Islamic scientists, like Alhazen in his *Book of Optics* in 1021 CE. By the fifteenth century, surgeons became critical of ancient texts on the human anatomy. By the end of the sixteenth century, modern science took a major step forward with the experimental methods and writings of Galileo Galilei. The modern scientific method was crystallized in the seventeenth and eighteenth centuries. Advancing beyond Aristotle's *Organon*, Francis Bacon wrote the *Novum Organum* (1620), in which he outlined a new systemic approach to science or knowledge, grounded in observations.[5] This improved upon the old philosophic process, which was built primarily upon intuited, idealized assumptions of peripheral observations. Then, René Descartes established the framework for the scientific method in his treatise *Discourse on Method*.[6]

The classical scientific method is basically a four-step process:

1. *Observation*. From a small group of experiences, an insightful observer intuitively recognizes consistent similarities. This observer then experiments more, to gauge whether that intuited pattern of behavior continues to be physically experienced.

2. *Hypothesis*. When the newly gathered evidence continues to confirm that intuition or modify it, the experimenter intrarelationally brings all that information together and puts forward a formal hypothesis, which he inductively proposes as a universal pattern of behavior in this type of phenomenon.

3. *Broad logical consistency*. The experimenter then broadens mental concerns about this specific behavior by determining whether this formalized hypothesis logically harmonizes with other known phenomena, information, and logical systems. That is, the proposed hypothesis must be logically consistent with other related, well-documented, and well-reasoned elements and systems. If the new

---

5. Bacon, *New Organon*.
6. Descartes, *Discourse on the Method of Rightly Conducting the Reason*.

hypothesis is not logically consistent with itself, with other hypothesis, and with logical derivations, further work must be done to bring all experience and generalized knowledge into logically consistency.

4. *Projective confirmation.* The new hypothesis expands the proposed system beyond past and present experiences. By logical extrapolations or interpolations of the proposed logistic, the new hypothesis is expected to make predictions, which further experimentation must also show to be true in an advanced way. Expanded experimental verifications increase confidence in the truth of the hypothesis.

These four steps in the scientific method match the fourfold progressive interrelational stages of the logistic of exchanged energy.

Is IPhil Scientific?

For interrelational philosophy to be accepted by modern scientists, it must be shown that IPhil is indeed scientific in a variety of traits. For a new theory to be respected in the scientific community, it is expected to meet certain criteria:

- Consistency. All its propositions and conclusions must be logically consistent, and none can be logically contradictory.
- Parsimonious. Using the metaphor of Occam's razor, the number of fundamental terms and foundational premises should be sparse or minimal.
- Useful. The theory must describe and explain observed phenomena in a way that advances understanding of the interrelationships of things in our world.
- Progressive. A new theory must refine or advance previous theories.
- Observable. Its premises and conclusions must be physical observable.
- Empirical. Its predictions cannot be assumed *a priori* to be true but must be physically testable and falsifiable.
- Provisional. It does not claim certainty, but it must be open to correction or advancement.

The expositions and arguments of this book, I believe, fulfill these conditions.

Chapter 16

Stages of Human Development

Hinduism: Stages of Life. Modern Developmental Psychology. Virtues. Personality Differences. Persons and Personalities. Four Levels of Love. Conscience. Unconditional and Conditioned Love. Buddhism and IPhil. Fourfold Interrelational Morality. Principle of Appropriateness. Justice and Mercy.

Phil holds that by logical evolution, all energetic things analogously follow the same foundational fourfold interrelational logistic. If that is the case, then this pattern of development should be found in all areas of human life. We see this in both the East and the West.

Hinduism: Stages of Life

Hinduism is the world's oldest religion, and within its literature are many practical and religious insights. Hindus generally recognize four main stages of life.

1. Student. This stage covers all of childhood until puberty during which youth not only receive material goods from their parents and elders, but also learn many life lessons, starting with living harmoniously within one's extended family.

2. Householder. After puberty, youth are expected to get a job, get married, raise a family, and learn to take orders from the leaders of their community. Stages of advancements into the role of leader in one's extended family, farm, or industry are blurred in this Hindu schema of personal development. Here adults are viewed as working tutors and servant leaders. In Hinduism, the role of honored teacher is not applied to ordinary adults; it is reserved to a choice few. This keeps adults humble and reticent to rise from their station as laborer.

3. Retiree. After many years of hard labor, the male adult is allowed to sit back with wisdom as head of their family, business, and community. The retiring person often withdraws to a hut in a nearby forest, away from the bustle of daily life.

4. Renunciate. Finally, the individual may follow spiritual pursuits, often becoming a celibate, beggar, and ascetic, striving for contemplative, spiritual union with Brahman.

In addition, Hinduism recognizes that individuals can have different ways in life.[1]

1. Kama. This way focuses upon the right use of pleasures of all kinds: food, sexuality, dance, aesthetics, etc.

2. Artha. This way focuses on the right way to acquire wealth and exercise power.

3. Dharma. This way focuses upon social duties and responsibilities for harmonious living with all. The term *dharma* is also applied in a logically evolutionary way to the other ways of right living.

4. Moksha. This is the way of spiritual liberation, which focuses on samsara, the cycle of birth, life, death, and rebirth.

In other words, ancient Hindus had a fourfold developmental view of the stages and ways of life, similar to the fourfold interrelational logistic of IPhil.

Modern Developmental Psychology

While most modern psychological studies are very narrow in their investigations, there are several longitudinal studies that cover the full human lifespan. Each longitudinal study concerns one particular aspect of the human personality. Consequently, these stages are comparable, but not in a strict parallel way. Sigmund Freud is noted for the triad: id, super ego, and ego. To these have been added: spirit. A modern presentation of these is found in transactional analysis: child morality, parent morality, adult morality, and integrated morality. The theologian Paul Tillich wrote of the states of heteronomy, autonomy, and theonomy. Lawrence Kohlberg studied moral development and noted these stages: pre-moral, pre-conventional, conventional, and post conventional.[2]

Jean Piaget researched in the field of education. His stages are more detailed and numerous, but they can be grouped into a fourfold way: amoral; fearfully dependent to opportunist; conform to persons and to rule; autonomous; principled to religious.[3] Jane Loevinger,[4] studying ego development, noted the following stages: presocial; impulsive fearful to expedient; conformist to conscientious; autonomous to integrated. Each of these studies have their own focus in the human personality,

---

1. Koller, "Dharma."
2. Kohlberg, *Essays on Moral Development*, and Kohlberg, *Cognitive Developmental Psychology*.
3. Piaget, *Moral Judgment of the Child*.
4. Loevinger, *Ego Development*.

but they show comparable stages of development, which IPhil holds originated in an evolutionary way from the four stages of the interrelational energy of which we are composed.

James W. Fowler in *Stages of Faith: The Psychology of Human Development* describes in detail the seven stages of faith most people go through in their lifetime.[5]

0. 0–2 years—Infant—Undifferentiated faith
1. 2–6 years—Intuitive-projective faith
2. 6–10 years—Mythic-literal faith
3. 11–20 years—Adolescent—Synthetic-conventional faith
4. 20–40 years—Young Adult—Individuative-reflective faith
5. 40–60 years—Mid-adulthood—Conjunctive faith
6. 60+ years—Maturity—Universalizing faith

These stages easily fit into the four stages of development recognized in IPhil:

1. Childhood—Stages 0–2. (0–10 yrs.) Faith is basically received.
2. Adolescence—Stages 3–4. (11–40 yrs.) Faith is increasingly individualized.
3. Adulthood—Stage 5. (40–60 yrs.) Faith is increasingly interrelational.
4. Senior—Stage 6. (60+ yrs.) Faith is spiritualized and transcendentalized.

Because of the fourfold logistic of the potential energy in my core and my bones: (1) My hope springs eternal. (2) I want more material things. (3) I want to get my life more organized. (4) I want to experientially discover something truly new.

When people observe something new (1st level), they normally immediately look for elements that are dissociative (2nd level) as well as the cause(s) of the imperfection(s). They regularly seek to blame someone or something for the imperfection and dissociative evil. They defensively want to keep their integrity and "save face" before their peers, blaming it on others. With maturity, 3rd level individuals go beyond complaining, and they look for ways to interrelationally correct and advance beyond the dissociative, imperfect situation. Experienced adult individuals build safeguards in their projects, because a constructed enterprise can be destroyed in a minute by some malevolent adolescent or random evil. Nonetheless, some problems defy 3rd level, rational improvement. Then, 4th level individuals turn for advanced assistance in projective faith, hope, and prayer to the Almighty, who is the holder of all energetic possibilities.

---

5. Fowler, *Stages of Faith*.

Virtues

Virtues are good habits. Good habits come from repeated good actions. Actions are good insofar as they are in conformity to one's logistic that comes from one's received interrelational potential energy, which is evolutionarily of and from the Almighty. With Socrates and Plato I ask: Can virtues be taught? In the interrelational logistic, there is a period of receiving, followed by a period of giving. "Being taught" is of the receiving period, and "exercising virtue" is of the giving period, so they are progressively distinct. An individual is "taught" progressively according to the logistic of one's potential energy by one's parents and teachers, then by professionals and political leaders, then by a progressive progenitor advancing toward an innovative, advanced order. Subsequently, an individual is internally pushed to self-initiate those directions physically to advance through repeated good actions and toward enduring virtues.

A vice is a bad habit that is dissociated from the above. There are also teachers of dissociative, evil deeds, including deceiving siblings, criminal organizers, despotic leaders, and a communal spirit of deception arising from like-minded, idealistically closed, dissociative groups of people. While associative or dissociative mores can be taught in various ways by teachers or role models, this information must be received by an open individual, then physically intrarelated in the individual's learning and worldview, and finally self-initiated by the observant student in the physical order. By repetition, associative or dissociative actions becomes routine and unconscious virtues or vices.

Personality Differences

Personality is not uniquely a human characteristic. Look at any litter of puppies, and it quickly becomes apparent that each pup has its own unique personality. On the receiving 1st level, each pup has slightly different physical characteristics, like its height, weight, color of eyes, ear shape, etc. There are 2nd level, intrarelational, interior psychological tendencies associated with its unique individual logistic, as evidenced by the individual's self-initiatives and self-control of bodily and mental abilities, including individualized ways of dealing with aggression and fearfulness. As the individual advances, 3rd level exterior sociological tendencies arise in interacting beyond one's peers with other community members and members of other species. The individual can be aggressive or timid in doing this. Then, 4th level self-initiating characteristics can express themselves in the self-initiative of the individual to break away from the pack and explore beyond one's current condition. Personality is really the totality of all levels of an individual's interrelationship with one's body, one's psyche, one's peers, and beyond.

Based on Carl Gustav Jung's *Psychological Types* (1921),[6] Isabel Briggs Myers and Katherine Cook Briggs developed what they called the Myers-Briggs Type Indicator, which recognizes four dichotomies within different aspects of the personality.[7] Each term has its own description and range of behavior.

Extroversion (E) — (I) Introversion
Sensing (S) — (N) Intuition
Thinking (T) — (F) Feeling
Judging (J) — (P) Perception

These four pairs of dichotomies result in sixteen possible personality types. Four letters indicate each personality type, one drawn from each of the lines above. Longitudinal scientific studies have shown that an individual's "type" remains basically the same over long periods of the individual's life. These four dichotomies express themselves independently, for they deal with the priorities an individual gives to different levels of the fourfold energetic logistic. For example, introversion emphasizes 1st level concerns of self-receiving and caution in a given situation; extroversion emphasizes 2nd level personal confidence to be outgoing in a given situation.

Persons and Personalities

What of changes of personality triggered by changes in the physical brain? IPhil recognizes that, just as there is a difference between one's potential energy and one's kinetic energy, so too there is a difference between one's potential, spiritual *person* and one's dynamic, physical *personality*.

Phineas Gage was an American railroad construction foreman, who survived having his brain pierced by an exploded rail spike.[8] That physical damage to his brain severely changed his 3rd level *personality*. Yet it did not change his holistic identity as a 2nd level *person*. Phineas metaphysically, spiritually remained the same *person*, even though physically and materially Phineas displayed a different *personality* because of his injury. Even if an individual is physically injured, the individual's received *metaphysical logistic* is closely associated but not identical with one's *physical DNA*, but they are different. Because of the physical damage to his brain, it no longer has the same material/energetic set of potentialities available to him. The change of his personality will remain altered until the physical parts of his brain are physically restored. Similarly, despite various physical injuries or debilitations due to starvation or various crippling diseases, an individual's personality can be severely handicapped, but the metaphysical logistic and identity of the individual as a person remains the same.

6. Jung, *Psychological Types*.
7. Myers and McCaulley, *Manual: A Guide to the Development and Use of the Myers-Briggs*.
8. Harlow, "Recovery," 327–47.

PART III: DEVELOPMENTS

Human personalities develop in stages according to the logistic of his person. There are certain personality traits that are recognizable from infancy, and more personality traits emerge with time. In a social situation, the personality of an individual can be displayed in a way that matches one's social situation, but interiorly the individual may experience and suppress very different personal feelings.

In fact, within an individual there can be different personalities. Fortunately, most intrarelational individuals know how to keep integrated the different sides to their personality in different situations to maximize effect. A youth will usually display one personality at home before parents, another personality before siblings and peers, another personality before teachers in school, another personality before a policeman, etc.

If the individual has difficulty intrarelating the different aspects of her personality, she may interiorly experience and even externally manifest a "split *personality*." That does not mean there are two *persons* in one body but that the individual was not able to achieve full 2nd stage intrarelational integration of her subpersonalities. The dissociation of different interior personality traits can be painful, and the individual will seek to integrate them to find peace within herself. In contrast, an individual may be so much in control of her different interior personalities that she can very effectively show one side of her personality in one social situation and another side of her personality in another social situation to achieve greater social advancement in each different setting.

Idealistic essentialists believe that there is only one, ideal, right way to interrelate with everyone. It is common for 2nd level adolescents to condemn themselves and others for acting differently in different situations. For them, there is only one way to act in every situation: the singular, universally right way. They condemn themselves or another person for being a hypocrite when one's performance is not always objectively the same way in different situations. When a comedian cracks family-friendly jokes at a parish picnic and then cracks raunchy jokes in a night club, a black-and-white essentialist will condemn this comedian severely for allowing the performances to be different in two different places, especially if the comedian's humor does not live up to the Puritan expectations of the essentialist.

It is said that beauty is in the eye of the beholder. Yet, some art objects and musical pieces are considered beautiful by many. IPhil recognizes that individuals have a logistic that is uniquely individual within a community, *and* a society has a common, communal logistic with a common sense of order, taste, and beauty. IPhil goes beyond a 2nd level, dissociative, individualistic either/or view of taste and beauty, for humans also have a 3rd level logistic and associative both/and view of order, taste, and beauty. In this way, I share a common sense of order, taste, and beauty with other members of my society, *and* I have a sense of order, taste and beauty that is uniquely my own.

When a pet does something unacceptable to a 2nd level idealist, the highly critical individual will angrily chastise the pet by yelling, "Bad dog!" In doing this the idealist equates 3rd level interrelational behavior with 2nd level individual identity, leaving the pet with a bad feeling about its self-identity and self-esteem, hobbling future interaction of the pet with the criticizer. When a child or youth does something the parent does not like, too frequently an immature parent will yell, "Bad boy!" or "Bad girl!" This is another example of confusion between 3rd level interrelational physical behavior and 2nd level metaphysical individual identity. Unlike self-centered, 2nd level individuals whose primary concern is the development of everyone their way, which is the ideal, "right" way, mature 3rd level individuals recognize that they live in a pluralistic community in which individuals have different personalities, abilities, and needs. The more efficient way of dealing with people with differences is not antithetically but synthetically for the sake of building individuals into interactive members of a pluralistic group, organization, or community.

Four Levels of Love

From the Age of Chivalry, bards have sung the praises and woes of romantic, passionate love. Many vocalists have sung variations on the theme of: "Love makes the world go 'round." Popular movies, books, magazines, songs, and television dramas usually have some form of romantic love theme. "Love is in the air." Still, people say they love ice cream; they love their pets; they love their favorite subject in school, and they love just lying around. In today's parlance, anything that is sensually pleasant becomes an object of one's love. However, the way we love ice cream, our friends, our country, and our God are interrelationally very different.

In contrast to our egalitarian American conventions that demand that everyone and everything be treated and described the same way, the ancient Greeks had four different words for love, each of which focused on a different object of one's love, for interrelationally we do not love everyone and everything in the same way. C. S. Lewis in *The Four Loves* discussed the four types of love identified by the Greeks.[9] These match the four aspects and stages of the interrelational logistic described by IPhil.

1. *Stergo* (familial love). This is the mutual love of parents and children—the natural affection and material commitment toward family members. This term also describes the practical, daily, concerned love of a caring leader toward the people in his realm, and the love of loyal subjects toward their leader.

2. *Eros* (passionate love). This involves great feelings of attachment to individuals and things. These passionate feelings are not just sexual but can be highly emotional toward any beloved. Socrates talked about the passionate *eros* he had toward philosophy and truth. Professional musicians have *eros* toward their artistry.

---

9. Lewis, *Four Loves*.

3. *Philia* (friendship love). This is often referred to as "brotherly love." This love is expressed through mutual care, assistance, and bonding with a companion. Comrades, buddies, and compatriots in civil concerns exemplify this love. Beyond self-centered *eros*, marital love is of this order—and more.

4. *Agape* (altruistic love). This love emphasizes giving to another in need. Here a person sacrifices oneself for the benefit of another, with little or no concern for oneself. In the New Testament, the statement "God is love" is not a cuddly, familial love but a healing, developmental, altruistic, self-sacrificing love. The logistic of agape-level love is an evolved form of the reciprocal, fourfold, mutually benevolent logistic of the primal potential energy of the Almighty.

Even though in the English language people continue to lump all love together into one big melting pot, IPhil recognizes, as with the Greeks, a person expresses different types of love in different situations, depending upon one's interrelational connection with the object of one's concern. That is why there really are four formally different types of love. Nonetheless, C. S. Lewis warned that each of these loves can be idolized, and each can become exclusive and evil. IPhil recognizes that all pleasing associations may be physically good, but in a fourfold way they also need to be applied differently in different situations toward different interrelational needs and greater interrelational development.

Aristotle described *love* as "two bodies, one soul." Putting this into IPhil terminology: Love is the communal logical evolution of two 2nd level, logistical individual persons into one 3rd level corporate person. In that logical evolutionary process, the founding individuals remain independently free, *and* they become one advanced bonded interdependent corporate reality, advancing our unifying society and world. Individuals have their own individual logistic-souls, *and* a love bond between two or more individuals establishes a communal logistic, which has its own communal logistic-soul. While infatuation may be instantaneous, mature love is not. Deeply caring for and dedicating oneself usually begin as a subliminal attachment, grow into conscious feelings, advance to physical involvement, and climax in physical sacrifices to assist the other person or thing interrelationally in overcoming their deficiencies and in achieving their maximum potentials.

Conscience

Conscience in IPhil is initially usually a 1st existential level intuition of a possible action that is more dissociative than associative within one's receiving logistic. On the 2nd intrarelational level, a self-centered individual can experience a twinge of conscience when the individual comes close to an action or interaction that is dissociative from one's individual logistic. In contrast, on the 3rd interrelational level, judgmental intuitions of logistically dissociating behavior are experienced as shame within one's local group or as guilt within a law-abiding community. A twinge of conscience can lead to a self-initiated or community-initiative correction. On the 4th level, one's intuition is the foundation of one's transcendentalizing faith. Here, one's conscience rises to the level of a possible positive, associative or a negative dissociative relationship with the Almighty.

When a wrongful act is done, normally one's conscience immediately makes the individual feel guilty and pushes the individual toward initiating some logically inverse, corrective action. In this way, the evolved spirit of one's potential energy immediately pushes the errant individual toward correcting that dissociative evil. Because of freedom, however, a person can respond or ignore that the corrective impulses of conscience. Sometimes after many years, an individual's guilty conscience can push the individual to confession and atonement of some previous wrong.

Unconditional and Conditioned Love

Some people identify unconditional love with altruistic love, but IPhil does not. Unconditional love originates in the metaphysical order, and it is identified with the antecedent, inherent, metaphysical tendency of one's potential energy to be as interrelational as possible. Logistically defined potential energy is enduring and forward-urging, and it overflows into the physical order, from one's spirit through one's body toward another. Thus, unconditional love springs from the depths of one's soul, core, or "heart." Metaphysical, spiritual unconditional love *can* be expressed in all four interrelational ways: (a) in the unconditional love of a parent for a child; (b) in the unconditional love toward a comrade-in-arms; (c) in the unconditional loyalty toward one's nation or country; and (d) in the unconditional charity toward another as a related child or creature of God.

Then, one's interior, unconditioned metaphysical love becomes physically conditioned by limiting environmental factors. That is, one's metaphysical, maximizing, *unconditioned* love quickly becomes *conditioned* by one's current condition in the physical order. Thus, love is neither unconditional or conditioned, but fullness of love is both unconditional and conditioned. Our inner energy pushes an individual's metaphysical, spiritual, unconditioned love toward physical, pragmatic, conditioned love toward another. Love-hate relations between teenagers and their parents can often be understood by the enduring positive logistical metaphysical relationship between parent and teen, *and* by an immediate resentment over physical restrictions by the parent upon the teen. In the other direction, when a child has a history of misconduct, the parents can have *unconditional love* in their hearts for that child, and at the same time they can have *conditioned love* in their bodies, sometimes expressing "tough love" toward a delinquent child in the hopes that strong, painful, natural consequences will cause the child to reflect upon and physically change misbehavior into good conduct.

The urge and the call to unconditional love have been with us for millennia. In the Bible, the call for *unconditional love* is directed toward the Lord God, whom one is to love with all one's heart, with all one's soul, with all one's mind, and with all one's strength. Yet, how a person does that is *conditioned* by one's interior ability and exterior circumstance. In Hinduism and Buddhism, the Sanskrit word *bhakti* refers to the concept of unconditional love, especially unconditional devotion of a devotee in worship of a divine. However, each person will show that devotion according to their material ability and circumstance.

Buddhism and IPhil

Today, branches of Buddhism have supernatural elements and can be properly called a religion.[10] Originally, however, Buddhism was primarily a psychology, designed to bring its followers to a state of inner peace and happiness in one's life. In the sixth century BCE, Siddhartha Gautama, the Enlightened One, or Buddha, formalized the Four Noble Truths, which have inspired many spiritually minded people for centuries. These Four Noble Truths reflect the successive stages in Gautama's life.

1. Gautama was born a prince in Lumbini in modern-day Nepal ca. 563 BCE. Because of a prophecy, he was highly protected and well-provided for as a child within his parent's ancestral palace.

2. When he was a teenager, he left the palace, and he was shocked to see various forms of evil in the world outside of the palace. One person was sick, another person had died, and another person was getting old and feeble. Struggling with the problem of evil, Gautama learned the First Noble Truth: "Life is *dukkha*" or suffering.

3. He decided to leave the palace and meditate upon the problem of evil. He became a monk practicing extreme asceticism. However, he soon found that long hours of thinking, prayer, and severe fasting did not solve this problem. In this search, he did discover the Second Noble Truth, namely, that the cause of one's *dukkha* or suffering is a craving to resolve an experienced problem. Intensely attempting to mentally solve the problem of *dukkha* by craving for a solution results in further suffering by various forms of thinking, and it yields no relief. It displays one's inability to solve rationally the problem of the *dukkha* associated with the painful dissociations we experience in life.

4. While meditating under a bodi tree, Gautama became enlightened with the Third Noble Truth, which provides a remedy for the dissatisfaction and the pain arising from *dukkha* through the doctrine of *no-atman* or "*no-soul*." This is similar to the 4th stage of IPhil's logistic in which the individual "lets go" of intellectually trying

---

10. Tucci et al., "Buddhism."

to solve the problems of life. This is also like the wu-wei principle of non-action action in Daoism, which finds true freedom by breaking from the many social rules of Confucianism.

5. Once inspired by this great, logically evolutionary insight, the Buddha became a teacher and the leader of many monks and followers. His enlightened insights led him to formulate a set of principles which lead dedicated Buddhists to a new, higher level of right, peaceful living. The Fourth Noble Truths leads to living an advanced interrelational life by replacing one's previous, negative, painful focus upon trying to understand and resolve life's *dukkha* and the problem of evil, by positively focusing on "right living." Buddha described this new level of "right living" in the Noble Eightfold Path. This is a set of eight interconnected behaviors which when developed together lead to the cessation of *dukka*. This Noble Eightfold Path consists of: right view (or right understanding), right intention (or right thought), right speech, right action, right livelihood, right effort, right mindfulness, and right concentration.

The psychological insights of Buddha's classical Four Noble Truths and Eightfold Way have been increasingly appreciated in the West. In terms of Western medical terminology, the Four Noble Truths present the four stages of: (1) symptoms of the problem; (2) etiology of the causes of the problem; (3) diagnosis of a solution of the problem; (4) treatment of the problem; followed by (5) living beyond the problem. This series of steps brings a hurting patient to psychological contentment with a new way of life, even when the physical problem may not be resolved.

Like the white paper on which books are printed, in traditional presentations, the 1st stage of life is regularly taken for granted and not given intellectual attention. When it is included in the above order of the Four Noble Truths, many parallels can be seen between traditional Buddhist wisdom and IPhil's fourfold logistic of energized living, leading to a more advanced 5th state of life.

One of the great enigmas of Buddhism is its doctrine of "no-soul." The very word "no-soul" indicates that this concept is a logical inverse. Here, the notion of "soul" is not the metaphysical principle of life but the psychological faculty of conscious thinking. In this way, the principle of "no-soul" and the activity of "right thinking" are not metaphysical, diametric opposites but psychological, dialectic contraries. Conscious concerns about self or others are oriented toward change, and this is always psychologically distressing. IPhil recognizes that dissociations moves 1st level unconscious awareness to 2nd level conscious concern and self-knowledge. Conscious distress over dissociations is the root of *dukkha*, which Buddhists seek to remove from their lives. This ideal seeks to advance spiritually from a state of active, complex, conscious concern to a state of passive, simple, peaceful awareness of the events in one's challenging, distressing world. Through meditational practices, the physical world becomes for them only as a sensed image which has no real conscious, distressing impact upon the individual. This happens when inner metaphysical awareness becomes detached from exterior concerns about our physical world. A mature Buddhist is like a person at peace inside one's house looking out through a window at the distressing, chaotic affairs of the outside world. It is like the person who sits back and just observes a movie passively. It may be said that the main understanding of "conscious soul" in Buddhism is that of a conscious, concerned ego. Consequently, the state of "no-soul" is inversely a semiconscious, unconcerned *awareness*. This understanding of soul is different from IPhil's understanding of "logistic-soul," which will be discussed later.

Fourfold Interrelational Morality

Most adolescents and essentialists have an idealized, static, 2nd level view of morality and ethics. In ancient days, the moral standard of individuals within a community was specified by one's ancestral birth (*natus*). Because of their limited, antithetical communications with other cultures, people insisted, out of ancestral loyalty, that their traditional local moral norms were the one and only standards for right conduct. The word "tradition" literally means "handed over," and because it was handed over by the ancestral family and religious leaders, traditional moral standards were honored and considered sacred. Because of their limited experiences, local cultures and closed groups strongly and unwaveringly insisted that their standard of morality was the best, ideal, and even God-sent.

Dogmatic and philosophic religious beliefs (orthodoxy) pertain to established events of the past and are of *being*, while morality (orthopraxy) pertains to *becoming* and advancing, active interrelationships now. That is why historic religious credal tenets do not change through time, while religious moral prescriptions advance through time.

Idealistic, religious, and juridical leaders preferred a clear 2nd level or white-and-black moral standard. They resisted dealing with confusing "grey areas" of morality. For example, the commands not to lie, steal, kill, bully, commit adultery, etc., were usually spoken about and applied in the same, literal way on all occasions. Reasonable variations were publicly and instructively set aside. Consequently, the moral code had clear, well-defined boundaries, so it was easy to judge whether a community member was an obedient and faithful member of the community or a disobedient wrongdoer.

In the language of the Lakota people, "to be different" is *tokeca*, and "to be an enemy" is *toka heca*. Both ideas are very close in thought and speech. To act contrary to community traditions was closely associated with acting like an enemy to the community. If there were questions of right conduct in new situations, only authorized legal experts or elders could officially interpret the traditional laws of right moral conduct. However, our modern world is quickly becoming multi-cultural, and people need to move from a 2nd level official application of traditions to advance to a 3rd level, progressive, communal logistic of one's advancing, multileveled society that respects and honors the old, *and* interrelationally goes beyond it in a logical evolutionary way.

As people begin to do this, there regularly is considerable diversity of opinion as to what the moral standard or "natural law" is for all. IPhil predicts that, as the standard of morality switches from 2nd level religious and cultural traditions to a 3rd level public recognition of the universal logistic of potential energy, progressive moral standards, it will become increasingly well respected, objective and universally accepted. We already do that by not imposing the same moral standard on children as on adults. We qualify the moral standard demanded on adolescents in society, from what is demanded to mature adults. IPhil recognizes that there is a carryover of 2nd level standards into a 3rd level morality, and this 3rd level of morality is founded on many interrelational experiences and the universal push of potential energy to be more interrelational. Phil recognizes that the logical evolution of moral standards is pushed and guided by the potential energy and communication Spirit of the Almighty. The worldview of 3rd level of morality goes beyond moral judgment of individual deeds to promote the progressive cooperative, interrelational advancements of diverse groups in society.

Furthermore, because energy is a fourfold interrelational reality within all things, from IPhil's perspective, there should be a foundational comprehensive code of moral conduct in the physical order, just as there is a single communal logistic in the metaphysical order. This morality should be fourfold, dealing with people's physical health, their psychological health, their corporate community cooperation, and their individual and community spiritual/metaphysical relationship to the Almighty.

IPhil recognizes that from the primal communal logistic of PAIS, and by logical evolution of the communal logistic in the evolutionary Tree of Life and Knowing, there is a universal, foundational comprehensive code of moral conduct for all human societies as well as local and individual standards. Like the Roman juridical recognition of a universal *Lex Gentium* (law of the nations), IPhil recognizes that there is a common, foundational, metaphysically universal moral code within the metaphysical, communal logistic of the humanity. Also, there will be transient, local moral norms of conduct in different cultures and for different individuals—valid but subordinate to the universal code associated with the communal logistic shared with all humans in all cultures. Being logistically universal for all humans, this universal code of conduct is metaphysically the human communal logistic, which is known only inductively and verified by metaphysical reflection on the physical experiences of all humanity and all its history. Because of the physical limitations of human investigators, this universal moral norm is metaphysically partially known only through Spirit-guided wisdom.

Augustine of Hippo wrote: "Wrong is wrong—even when everyone does it. Right is right—even when no one does it." The first part of each of those two true statements is in the enduring, metaphysical order, while the second part of each of those two statements is in the free, physical order. The truth of those statements comes not from socially approved mores of different social groups but from the interrelational logistic an individual received from the Primal Absolute Potential Energy (PAPE), by which everything in our world is sustained and from which it has evolved.

How can a person advance from a 2nd level morality to a 3rd level morality? By recognizing that the 2nd level idealization of who or what is right or wrong is regularly exaggerated and does not totally matching our diverse, progressive world. This is done by recognizing that some traditional judgments are intellectually prejudiced, by recognizing that some "right actions" are sometimes progressively wrong, and some wrong actions are sometime interrelationally, progressively right. We recognize what is right or wrong in a given culture or a given situation by recognize the interrelational physical fruits of one's actions and not one's intellectual standards. (See the parable of the Good Samaritan in Luke 10:25–32.) A 3rd level person will strive to judge what is right or wrong not according to some fixed ideal norm but will strive to live according our God-given, evolutionary individual and communal logistic, which is always interrelationally progressive for both individuals and for the common good.

Voltaire quoted[11] an Italian aphorism when he wrote: "The perfect is the enemy of the good." IPhil recognizes that this statement is true in the metaphysical, potential order. However, in the physical order, every partial good is an interrelationally advanced, evolved actualization from many non-actualized potentialities. As it is said, "Actions speak louder than word." A person is not *interrelationally* judged not by their thoughts, their plans, or their words, but by their deeds. Different authors have written that anything that is interrelationally worth doing is worth doing poorly—rather than not at all. In IPhil, an action does not in itself have an essentialistic, 2nd level value. Every physical action is interrelationally 3rd level, and it is from that perspective that all physical actions draw their moral value, regardless how small. Rather than striving for the objective best, it is better to strive for one's personal best and be at peace with one's partial performance, rather than not striving at all and giving up in face of the challenge.

Principle of Appropriateness

Are certain statements appropriate or inappropriate to a given situation? Whence come their truth-value? Or are they all a matter of opinion? Is appropriateness a moral value?

In a given communal situation, certain behaviors are intuitively considered inappropriate or socially wrong. I frequently hear an older sibling correct a younger sibling by saying, "Act your age." Adults frequently complain, "That remark is not appropriate here." In one context, certain behaviors may be appropriate, but in *this situation*, this particular behavior is "out of place." This moral perspective goes against the essentialistic view that an action that is appropriate in one context will always be appropriate

11. Voltaire, *Dictionnaire Philosophique*.

in all contexts. The principle of appropriateness also goes against the self-centered, independent adolescent belief that all statements and actions should always be allowed everywhere. For example, it is interrelationally immature to demand that every person should *always* tell the full truth, even when the persons hearing that truth will probably act in a destructive, antithetical way with knowledge of the truth. When a person asks you an intimate question about your sexual life, is it appropriate to reply in a form of evasion such that the hearer will probably infer a wrong answer?

Avoiding extremes, IPhil recognizes that it is wrong to say that particular physical acts are either morally determined or morally relative. Some dissociative variations from the moral norm are interrelationally insignificant and are tolerated if they are quickly covered in an interrelational way. Yet, other dissociative behaviors from the social norm are interrelationally significant when it would take a significant amount of time to correct that dissociation. All energized behavior takes place within an individual and a society, and their individual and communal logistics set the parameters within which behavior is interrelationally beneficial and therefore moral—or not. To call someone a derogatory name would be an enduring dissociative evil in a political setting, yet it may be allowed as satire and as a joke among friends. It takes experience and wisdom to know what is truly appropriate in a given situation and what is not.

Dissociative, painful behavior, such as "tough love," *can* be logistically right and appropriate when a person is misbehaving, but "tough love" is not generally the appropriate norm for raising a youth. Contrariwise, it is inappropriate to dote upon and spoil a child, or to make an environment so sterile that the child's immune system does not develop adequately. Depending upon an individual's progressive development, there are times when tough love or telling a raunchy story is appropriate, and there are times when it is not. So, which actions are interrelationally appropriate or inappropriate in a given context? I refer again to the classic norm: "By their fruits you shall know them." If a behavior distresses the formation and development of an advancing interrelational individual or community, it is inappropriate. If it fosters harmony and unity within a progressive individual or community, it is appropriate. In a mixed community or social event, different behaviors will be interpreted and taken differently by different people. Whether a socially dissociative act is appropriate depends upon whether that action has inconsequentially or consequently evil effects to most of the people there, who can moderate the displeasure of the few who are offended. The faster a disintegrating, inappropriate act can be corrected, the more it can be tolerated and easily dismissed as communally insignificant. If used sparingly, deliberate, inappropriate words or actions can occasionally have a positive, long-term significant effect—like Jesus' clearing of the sellers from the temple, or the Indian elder's telling a raunchy story at a religious meeting. The interrelational context and the probability of social advancement are important.

The principle of justice is the community cousin to the individualized principle of appropriateness. The principle of appropriateness subjectively seeks to match the performance of an interrelational deed to a locally established social or moral imperative. In contrast, the principle of justice objectively seeks to match a societal punishment to a misdeed, or a societal reward to a right deed. Consider these cases: (1) A six-year-old lies to his mother, telling her that he did not take a cookie from the cookie jar. (2) A team member lies to her coach, telling him that she did not share any of their game strategies with their opponent. (3) A soldier lies to his sergeant, telling him that he did not tell battle plans to the enemy. (4) A person lies to God by telling a falsehood when under oath in court. Though these are all lies, the interrelational moral weight of each one is different. A just punishment is different for each lie, not because of the lie itself but because each falsehood impacts different levels of interrelationships. This makes common sense to an interrelational thinker. However, to an idealistic, essentialist thinker, every lie is a lie and is intellectually deemed worthy of the same punishment.

Adolescent essentialists want judgments to be "fair." They want the same penalty to be applied to the same physical misdeed, regardless of its interrelational consequences. They focus only on the deed itself. Therefore, they cannot understand how a parent can make different judgments in different social situations. In contrast, IPhil recognizes that the moral value of deeds lies primarily in their interrelational effect. Each lie is inappropriate according to its own social setting, and corrections should be appropriate to the maturity and interrelational responsibilities of the offender. That is, IPhil recognizes that for justice to be done, a punishment should not just match the wrong deed, it should also take into account the developmental situation of the wrongdoer and the interrelational impact of the wrong deed. This why sanctions in a mature society are meted out not by legalistic judges who only see the letter of the law, but by wise judges who see the bigger picture. It is juridically appropriate that courts have wise judges to mete out penalties to produce the greatest "common good" in the long run. This is commonly practiced in society today, and IPhil provides a philosophical foundation and justification for that practice.

People forget that most of the moral laws in Hebrew Scriptures were written to be learned and practiced by adult men who were thirteen years old or older. Only a few of those laws applied to children, women, or foreigners. Today, biblical Fundamentalists want to apply the same code equally to everyone, and they do not take into consideration the many changes that have happened since the original writing of those laws. That is why Jewish commentaries on the Torah by educated rabbis have been so important in Jewish history. A modern moral code must consider the non-homogeneity of today's society. Moral imperatives and commandments in Sacred Scripture may have been very appropriate in the historic setting in which they were made. In ancient times, stoning of a wife or woman caught in adultery may have been judged appropriate within a tribe in the desert where there were no jails. But today, several thousand years later and in another geographic context, literal applications of these original commandments are considered inappropriate.

Justice and Mercy

It has been said that Christians must always forgive sinners. That is not true. Jesus said, "If your brother does wrong, correct him. If he repents, forgive him. If he sins against you seven times a day, and seven times a day turns back to you saying, 'I am sorry,' forgive him" (Luke 17:3–4).[12] Notice the qualifying clause in the sentence "If he repents, forgive him." Whether a Christian forgives a sinful person depends on the offender's repentance, metanoia, or change of mind and heart. If a person is not repentant, the essential condition for mercy is missing. When there is no 4th level metanoia or repentance, it is proper for a Christian to require a 3rd level just penalty for the misdeed (see Matt 18:32–34). In Hebrew Scriptures, God regularly punished his people Israel for their disobedience. Still, God always repented of his painful judgment when Israel repented of their misdeeds and turned back to the Lord. If God or Christians were always forgiving without the requirement of repentance, there would be no need for any right conduct in our world.

The command to "turn the other cheek" to one's opponent does not mean to forgive one's opponent for the wrong, but rather to continue treating one's opponent civilly and equitably as before. Moreover, if a person has offended another unwittingly or unconsciously, there is no need for a conversion of heart, so the restitution for the objectively wrong deed would be less (Luke 12:47). Jesus did not turn the other cheek in his cleansing of the temple, in his prophetic "Woes," or in his repeated condemnation of the legalism of the Pharisees.

It is a common misconception in modern moral theology that only conscious, deliberate misdeeds are sinfully against God's will. Yet the annual Jewish observance of Yom Kippur is for the forgiveness only of unintentional, indeliberate sins. Hardhearted, deliberate sins must be forgiven in another way. Inadvertent mistakes are physical, 1st level dissociations, and materially they can be covered fully by the associative deed of another individual. When a child breaks a neighbor's window by accident, that material, dissociative damage needs to be covered. However, since no change of heart is needed by the child, the indeliberate offense and material sin of the child can be covered by a parent or another.

Deliberate wrongs, however, are self-initiated and from a person's metaphysical, spiritual heart. Deliberate 4th level wrongs are the responsibilities of the "hardhearted individual," and even when the material damage is covered physically by another, merciful forgiveness of a deliberate offense on the metaphysical level can rightly occur only when the offender makes a 4th level change of heart. Until that time, the one offended is normally required to deal with the wrongdoer at a civil 3rd level in a court of law.

The law of justice and equity in the Hebrew Scriptures is 3rd level in its demand for "an eye for an eye, and a tooth for a tooth." This moral injunction was an advancement from the 2nd level tribal "law of revenge," where an offense was paid back many times

---

12. Cf. Matt 18:15–18; 21–35.

over. Biblical justice sought to match the punishment with the crime, at least among fellow Jews. A third-party judge was often asked to referee a fair retribution. In a 4th level morality, if an offender goes through a metanoia change of heart, the individual may be shown mercy—as long as that individual maintains that change of heart in one's actions. This is what the practice of probation is in our current legal system. However, if the commuted individual breaks that probation, displaying a regression of heart, that probation and that mercy is terminated, and the just sentence again takes effect.

Modern society is strongly individualistic, and the above discussion of mercy applies to individuals. Nonetheless, like energy, mercy also has a communal dimension and a communal logistic. Groups and nations can collectively commit social sins. So, a discussion on mercy (*hesed* in Hebrew) on a community level is needed. An example of this found in the next chapter.

Chapter 17

Societal Matters

Simplification and Diversification. Democratic Development. Taxes. Preferential Option for the Poor. Equality and Inequality. Rich and Poor. Stockholders and Executives of Corporations. Randomness and Luck. Interrelational War Theory. Crimes against Humanity. Appreciative Inquiry. Principle of Subsidiarity. Resolving Conservative vs. Liberal Stand-Off. Concluding Thoughts.

Energy is interrelational. That which is interrelational has terms and a nexus connecting the terms. Terms have their individual fourfold logistics, and the connecting nexus between terms has its own fourfold communal logistic. The nexus between terms keeps the terms both distinct and close. Consequently, there is a tension between individual terms via the communicating communal nexus. The bonding of individuals produces enduring, logically consistent corporate and national entities, which can interrelate with other enduring, corporate entities and nations in a fourfold interrelational way. This chapter considers several dissociative-associative relationships in energized communities and societies from the perspective of IPhil.

Simplification and Diversification

IPhil finds the origin of simplification and diversification in the existential *being* and in the progressive *becoming* of potential energy. These are found in all aspects and phases of the individual and communal logistics of energy. Diversification is found on the thetic 1st level in the randomness found in the chance material formational of an individual. Yet, a child shares common DNA traits with one's ancestors, among whom the child usually finds security in a caring, bonding family. On the 2nd intrarelational, antithetical level, adolescents push for group conformity and also pushes in an absolute black-and-white way away from another group. Groups simplify their intellectual categories to deal more easily with many different things. Still, a group glorifies its leader, who is ranked higher than and is an exemplar of the other members of group, leading them to greater achievements before outsiders. On the 3rd level, diversity of groups within a community is recognized, respected, and is seen as an opportunity for corporately drawing many together unto a unifying common good.

On the 4th level, innovative, evolutionary progenitors are individually outstanding, but they need a supportive group of compatriots to produce an enduring, advancing, logically evolved interrelational system. In this way, IPhil recognizes that individualism and egalitarianism have their appropriate time and place within the logistics of every energized, individualized unifying and evolving community.

The interrelational logistic recognizes that leadership progressively advances from family love and protection, to group instruction and direction, to community law and order, to virtue and sacrifice for the benefit of all. For Confucius, advanced leaders gain followers not by violence or by rule of laws but by virtue. Confucius strove to resurrect the traditions of benevolence, propriety and ritual, thereby promoting peace and harmony through the country. Five centuries later, Jesus taught the Twelve that leaders are not to make their authority felt but they are to be "servant leaders." This is done through charity and sacrifice for others and for the advancement toward a universal, interrelational, Godlike world of peace and love. IPhil goes beyond 2nd level antithetical authoritarian leadership, to 3rd level synthetic, interrelational leadership, to 4th level progressive and evolutionary advanced respect between all in All.

Democratic Development

The origin of democracy was in Athens ca. 507 BCE by Cleisthenes. In chaotic times, he relied on a broad band of groups to rise in power. Rather than proclaiming himself as king, he refused, like George Washington, to be made king, and he drew together the landowning citizens of Athens to form a democracy. Unfortunately, their democratic rulings often were for the benefit of self-interested factions, rather than for the broader needs of the polis. They made numerous wrong military decision, and Athens and its democracy fell to the Spartans in 404 BCE.

In the Middle Ages, countries and the Catholic Church were ruled by military monarchs. After a civil conflict in England, King John at Runnymede on June 15, 1215, signed the *Magna Carta Libertatum*, a charter of rights, which relinquished to the demands of nobles. The House of Lords was established. Later, after severe conflicts, the rights of commoners were recognized, establishing the complementary House of Commons. Gradually, the range of democracy in England expanded with the extension of the franchise to vote in their representative democracy. The history of the development of representative democracy in the United States has expanded in a similar way from giving voting rights only to property owners to gradually giving the franchise to vote to non-owner adult men, to freed slaves, to adult women, to American Indians, to citizens of younger age, etc. There is a significant group of leaders who oppose such expansions of democracy. Increased divisiveness in the modern United States representative democracy is leading to its 4th stage demise. In this way, the history of democracy is following the four-phase development of the fourfold logistic of potential energy.

In advancing from a 2nd level authoritarian society to a 3rd level communal society, it is tempting to pursue a pure, democratic model where every individual has an equal vote. Unfortunately, the majority soon becomes dictatorial, demanding conformity to their ways and beliefs. The tyranny of the will of the majority and the rich regularly soon suppresses the poor and various minorities.

The 2nd level of the interrelational logistic recognizes the rightness of having charismatic leaders of different groups within the community. The progressive advancement of the interrelational logistic into 3rd level justifies the formation of representational democracies. The maturation of communal societies leads to incorporation of small, minority groups. Such 3rd level representational democratic societies usually have a strong middle class, because they experientially know how to better interrelate and innovate. Unfortunately, a society's lower class and the upper class are unusually more concerned with their own personal survival. To advance further it is necessary to foster 4th level values that are built upon 3rd level mutual respect and appreciation and are directed toward higher, spiritual values for all.

Taxes

All individual and communal systems must receive and possess energy to exist and endure. In societies with communal logistics, its physical energy is usually received in the form of money through some form of taxation. Taxes are collected by governments to pay for the expenses of running a society in various interrelational ways. There are several different ways of viewing taxes.

1. Childlike individuals passively accept taxes imposed by leaders, just as they dutifully respond to responsibilities ordered by parents. They accept these as a part of life and simply go with them, assuming they are for their own benefit and the common good.

2. Adolescent individuals view all taxes and government regulations negatively—as dissociative evils that take away their income and freedom, which they view as rightly their own alone. Because they want full control of their income and resources in an exclusive, self-centered way, they oppose all government regulations, find ways around taxes, and circumvent regulations in the same way adolescents oppose and circumvent rules imposed by parents.

3. Adult individuals positively accept and recognize the communal values of taxes as directed toward the common good. They realize that the organization of many people with different talents can do more than can one person alone. For such a corporate person/society to survive and operate effectively, it is necessary to "feed" the corporate person of one's government through various forms of taxation. Responsible, mutually caring citizens recognize and publicly promote those taxes by which improvements can be made in infrastructure, advancing communications,

and transportation of various things needed for the efficient operations of the community. They financially support health, education, and welfare on all levels of society, recognizing how they improve the prosperity of all.

4. Spiritual individuals contribute more than their share of taxes by their support of diverse charities directed toward the needy in society. They do this in the hope that their generous contributions and activities will lift the lowest members of society, and thereby interrelationally lift the entire community and the world.

Preferential Option for the Poor

Individuals who are 2nd level are highly intrarelational, seeking their own development rather than others. They think that if there is to be any preferential treatment, it is to be shown to *them*. If preferential treatment is shown to others, they shout, "Unfair!" "Unjust!" To them, showing a preferential option to the poor is contrary to their self-centered, egalitarian worldview. Privilege, they think, should be shown to those of privilege, who worked hard to get where they are. They strive to make the poor and lowly serve them, rather than the reverse. Their upper-class exclusivism promotes the judgment that those who have less rightly deserve to receive less, according to their lowly social station. Caste and class privilege are to be defended, and all attempts by lower classes to topple that materially favored hierarchy are to be confronted legally and suppressed by law enforcement and the military.

In contrast, showing an adult, 3rd level preferential option to the poor is logically inverse to the above adolescent 2nd level mode of thinking, moving from intrarelational, self-centered values to interrelational, other-centered values. Extending a preferential option to the poor comes from the progressive realization that the poor have few resources with which to compete in most societal situations. The poor are like infants and children who are not yet mature enough to compete with those individuals who are financially secure and politically powerful.

In sports, there has long been a preferential option for those who are younger, smaller, and less experienced. Athletes of different ages, weight, experience, and physical handicaps are not indiscriminately made to compete with each other. They are put into leagues where people of the same ability compete. Making athletes of the same caliber compete against each other is said "to make a level playing field." When poorer bowlers are handicapped by giving them extra pins based on their past performances, poor and good bowlers are able to compete against each other in a way that is really a competition against their current normal. This evens out the playing field so that the less talented bowler has as much chance of winning as the better bowler.

Similarly, showing a preferential option for the poor gives the poor a handicap so that they are able to compete with persons more advanced in their resources, education, and talents. This handicapping procedure fosters not only an even playing field for the poor and the average person, it also supports their greater participation in society. Just as it is appropriate for parents to show physical and material preference to their children according to their current personal abilities and needs, so too leaders in a community need to bestow educational and material preferences to poor and struggling citizens in their societal needs, until reciprocity of giving and receiving are possible in society. Assistance is not given to the poor only because they are poor. Generosity and preference to the poor are given with a social objective—with an expectation that they become more interactive and benefiting (3rd level) members of society. If individuals are physically (1st level) or psychologically (2nd level) not motivated to advance as contributing members of our community, they should not be shown preference but rather given only sufficient support to live a physical and psychological healthy life.

Self-centered, 2nd level leaders and individuals oppose such supports, for they strive to take more from the community than they give back. In this way, their lives are not consonant with the interrelational logistic of the potential energy that is the logistical foundation of our world as consonant with the logistic of the Almighty. They greedily want as much income as possible for themselves and their own, with as little as possible to go to others, so that their social superiority to others may selfishly grow. Consequently, it is sometimes necessary to oppose legally, militarily, or violently such oligarchs. In these ways, the poor will have a better chance for progressive advancement of the individual and communal aspects in their personal lives. These chances can also give their society a better chance to expand the talent-base of a growing middle class into a healthier economy and political society. Large property owners in third world nations claim the nationalization of some of their properties is not just. That is true if the laws on the books are biased to the privileged classes in that civil society. However, such biased civil laws are not consonant with the equitable, reciprocal logistic of the Almighty. Therefore, these prejudicial laws are immoral and sinful before the Almighty, and correcting those prejudicial laws and their societal structures may rightly be opposed and replaced—in a progressive fourfold way—beginning with inner conversion of the prejudicial elites, civil efforts to change these laws, and ultimately physical opposition to the individual values and social structures that have led to these inequalities.

Equality and Inequality

The notion of "equality" was emphasized in the social theories that arose during the Age of Enlightenment. In the American colonies, the Declaration of Independence, written in 1776, begins by emphasizing equality. "We hold these truths to be self-evident: that all men are created equal." In the French Revolution (1789–99) the battle cry was: *Liberté, Egalité, Fraternité*. In their historic contexts, these calls for equality arose antithetically and inversely because of the magnitude of the inequality the people were experiencing at the time from the rich and elite class of nobles and kings, leaving the working poor suffering greatly. However, only a few years later after the American Revolutionary War, the approved US Constitution did not recognize all men as equal, for slaves were counted as partial persons with no voting rights. In France and in the Americas, the disparity of rich and poor remains a social reality even today, and it is getting worse. Second-level inequality is a lasting reality in our world despite our constant call for equality. Why this discrepancy?

Where is the first appearance of inequality? Starting with the Primal Absolute Interrelational System (PAIS), dissociative inequality precedes equality—but only for a very short time. In the evolutionary advancement of PAER into PAIS, the first active participant is the all-giver who has all potential energy. Then, the all-giver gives all potential energy to the all-receiver, transforming it into the all-receiver-giver. In fact, at any given moment in the absolute exchange, only one interactive term has All interrelational energy, while the other interactive terms have zero interrelational energy. Nonetheless, if one looks at the interrelational process of PAIS as a whole, there is symmetry and equality of exchange within the interrelational process. By logical evolution, it is expected in the operations of finite individuals and communities that inequality will be recognized in each 2nd level phase, while equality is the expected goal in each 3rd level phase. Unfortunately, with the exception of religious communities, most human communities display major dissociative material and monetary evils through the greed of 2nd level individuals and groups.

Looking at the four stages of development of the game of catch, IPhil recognizes that *equality* is first found only in the 3rd stage of the advancement of the interrelational logistic. Only then is the symmetry of the logistic of potential energy realized in the transformation of the receiver into a receiver-giver, oriented toward establishment of a system or society in which all members play an equitable interactive part according to their abilities. Consequently, the origin and legitimacy of material *inequality* is found in the 2nd stage of all energetic developments, and *equality* in a system inversely arises in the 3rd stage of development of a system. During the interrelational process, the painful dissociative character of inequality is the *occasion* and *opportunity* for advancing to the more associative and pleasant state of equality.

When a mother bird feeds her chicks in the nest, she can only feed one at a time. As a result, the aggressive chick will receive food first and will grow in strength slightly faster than the others in the nest. In every family, there will be certain variations in food, nurture, and information in the different children from their conceptions onward. Because children are not born at the same time, and because they inherit different strengths, this leads to material and intellectual inequalities among siblings. The eldest child who receives first tends to be more mature and takes on greater responsibilities at an earlier age, but this is not always the case.

I regularly hear siblings complaining that their parents love their "favorite" more than they. While that may happen, loving parents can honestly respond that they love all of them equally—not in a physical, material way, but in a metaphysical, spiritual way. Good parents are oriented to respond to each child according to the child's current material needs and stage of development. Older children grow faster, and they need more food and new clothes more often. On the other hand, a sick child has greater medical needs than those who are well, so parents will necessarily spend more time with them in doctors' offices and in the hospital. Youth are usually 2nd level idealistic, demanding equality in all things, and they tend to complain when they receive less than others. They see the others only from their own perspective and not in relationship to the developmental stages of the other siblings, and to the needs of the family as a whole. The unconditional, spiritual love of the best parents is equal insofar as they strive to care for each child according to its current need. Optimally, that is true even after the youth leave the home. They may not receive equal material support from their parents, but the parents are inwardly pushed to continue to give love, advice, and support each according to their need.

From IPhil's perspective, it can be said that from an individual, physical point of view, God does *not* materially love every individual equally. That is easily seen in the difference of ability found in different individuals and in different species. Yet, IPhil recognizes that the Spirit of interrelational energy pushes all metaphysically to a better life according to their received individual logistic. In this way, God's love is logistically and spiritually equal according to their respective individual logistics, while they are materially unequal according to their different bodily abilities and their different physical environments. Distress is usually experienced by individuals in their 2nd level recognition of the dissociation of one's own material condition with the material situation of more prosperous individuals. True personal happiness is found in the 3rd level matching of one's physical reality with one's metaphysical logistic and in the 4th level "letting go" of one's desire to be like someone else other than the best individual one's logistic allows one to be.

Rich and Poor

Money is a medium of exchange; it is analogously equivalent to energetic and material power. Like energy, money has no real value if it is not exchanged. Rather than hoarding money and assets, the exchange of energetic resources is what gives an economy and a country prosperity and endurance. Consequently, in the 2nd stage of development, it is interrelationally right and for the common good if rich individuals gain and enjoy accumulated wealth—but only for a short time. After a logistically appropriate time, it is important for rich individuals to "let go" of and disperse their acquired wealth to enrich others so that others too can become active participants "in the loop" of a 3rd level society. Prolonged hoarding and accumulation of assets by intrarelational, 2nd level, self-centered individuals cause the rich to get richer and the poor to get poorer. This growing inequality results when the rich seek to hold on to their wealth too tightly, for the sake of extending security for themselves and their progeny. Like the stockpiling of energy, hoarding and building up wealth are usually directed toward the development of one's personal empire. However, a *polis* involves all the people in a society, and politics should be directed toward greater financial and material exchanges for the interrelational common good of the whole society. The more energy, money, and assets are actively exchanged between an increasing number of members of a society, the more that society tends toward its maximal interrelational character as a system and its logically consistent endurance. The larger and stronger the middle class become, the more reciprocal, efficient, and prosperous the society will be as a whole.

Thus, IPhil recognizes that in a developing society, (1) while all begin basically poor, (2) there will be a few who break away from the pack and become rich and powerful. But (3) in the growing exchanges of a healthy society, all interactive members should also gradually grow in wealth and enjoyment of the fruits of their labors. If they don't, corrective inverse actions need to be taken toward establishing equity within that society. This makes the entire society richer. Furthermore, (4) it provides all members of that society to use their surpluses for advancing, transcendental activities. When the poverty of individuals and classes become so great that they are severely hindered in their ability to develop personally or become meaningfully involved in society, IPhil recognizes that a high level of poverty to be interrelationally wrong and logistically sinful. Note that IPhil recognizes that there is nothing wrong with *becoming* rich. What is wrong is *staying* rich at the expense of the interrelational life of others. As is said, "Money is not the root of all evil, the *love* of money is." The *love* of money pushes a person to hold on to money longer than is appropriate, according to the interrelational logistic.

Andrew Carnegie, John D. Rockefeller, Melinda and Bill Gates, cofounder of Microsoft, Bernie Marcus, cofounder of Home Depot, Mark Zuckerberg, cofounder of Facebook, Oprah Winfrey, and many more wealthy people are examples of entrepreneurs who can be called "moral capitalists." Their early, 2nd level, personal ingenuity and efforts brought great achievement and wealth into their lives. Later in life, their subsequent 3rd level philanthropy helped or continues to help many needy individuals and institutions to advance. Several people have suggested a general rule: The first half of one's life is for making money, and the second half of one's life is for giving it away. It is a common practice that people "let go" of their strictly personal accumulated properties and treasures as they approach the end of their lives. Rather, this individualistic "letting go" is advanced to communal dedication to the holistic benefit of society—of which they remain an active, interrelational part. Here, the rich maintains an active part in the both/an advancement of society, with them as leaders in improving the structures of society. This practice matches the receiving-giving cycle of the energy exchanges recognized in IPhil, starting with PAIS. This rule of life then matches the logistic of PAIS and therefore is holistic and holy, like the Almighty. This is the way God meant our energetic world to be. To go against it in an insecure, greedy, and self-centered way is enduringly dissociative and evil. Within IPhil's communal interrelational logistic, 1st level poor and suffering individuals and groups would be justified to rise up and antithetically oppose those individuals and groups who are 2nd level self-centered and greedy.

Stockholders and Executives of Corporations

Unfortunately, in our modern world, as corporations grow, they tend to fragment into their constitutive parts: stockholders, management, workers, and consumers. In this fragmentation process, these parts regularly lose a sense of their interdependence and their responsibilities to one another. They lose sight of their corporate consciousness and long-term goals. They turn their focus to 2nd level self-interest. They are more interested in one's bottom line and one's immediate financial self-interest. The spirit of materialistic greed and self-interest pervades an organization on all levels.

Essentialists on the 2nd level of social development advocate and tenaciously hold on to a self-centered, idealized definition of capitalism, where the desire for capital or money is the sole motivation in an economy. History shows that Adam Smith's understanding of economic order was inherently open and flexible. Karl Marx, on the other hand, saw capitalism as divisively antithetical, where the bourgeoisie oppressed the proletariat. In 1970 when corporate social responsibility was much in vogue, Milton Friedman wrote a reactionary, often-quoted article in the *New York Times* in which he said that the chief executive officers (CEOs) of corporations are employees of the stockholders. They are hired to solely give the greatest financial return to the stockholders. For an executive to do charity work in the name of the corporation is wrong, for stockholders are interested solely in a greater return for their investments. Taught in most MBA programs, this became the conventional view of capitalism.

Recognizing that a corporation has contributing stakeholders as well as stockholders, there has been a gradual, inverse movement toward including more parties in corporate decision-making. In the 1990s, a Minnesota initiative established an international network, the Caux Round Table (CRT) for Moral Capitalism, focused on taking care of all stakeholder. In 2019, two hundred CEOs of major American corporations of the Business Roundtable conceptualized a new theory of capitalism. They postulated that each of our stakeholders is essential. They are committed to deliver value to all of them—for the future success of our stockholders, our companies, our communities and our country.

Pure laissez-faire, free-market capitalism is adolescent. From IPhil's perspective, Friedman's concept of capitalism, which puts the financial interests of the stockholders over anyone else, is self-centered and immaturely 2nd level. Free-market capitalism seeks immediate self-gratification for the short-term growth of wealth and power of the stockholders. A good quarterly economic report and dominance over competitors bring their stockholder's great self-satisfaction. Its investors idealize free-market capitalism, thinking that it is godlike and will last forever. However, history shows that the life spans of most capitalistic empires were short, unless they were defended by effectively enslaving their workers by keeping their income down, by keeping the levels of their education and political influence of their workers down, and suppressing labor leaders both legally, physically, and militarily. Free-market capitalism is like a virus that eats at the hearts of its members and attacks the all-inclusive spirit of a democratic society.

In contrast, through a reactionary inverse operator and the evolution of society, 3rd level humanistic, moral capitalism recognizes and supports progressive, subsidiarity interrelationships between the environment, workers, corporate businesses, and the state, according to a communal energetic logistic that is developmentally more interrelational. Moral capitalism is capable of greater, long-term, systemic progress for all: the environment, workers, corporations, nations, the building of a mutually respectful, universe, and interrelational glory in the Almighty.

Today, moral capitalism calls maturing stockholders to "let go" of the desire to control CEOs in all things, solely for their own profit, giving CEOs the respect and the responsibility of a central neurological hub for maximizing the interrelational involvement and benefit of not only stockholders but also all corporate stakeholders. Customers, workers, management, and financiers share in the rights and responsibilities of maximizing the interrelational exchange of resources and money—according to the changing historical situation and the abilities of each. Not all participants are at the same stage of development. To facilitate this diverse, progressive situation, IPhil presents the orderly fourfold understanding and process of the interrelational logistic of energy to guide the maturing and maximization of benefit of all in a corporate society.

Randomness and Luck

Through the centuries, randomness has been viewed different ways. In ancient times humans lacked control of the dynamic of randomness in the world around them. Striving to understand and control the affairs in one's life, people blamed the uncontrolled, spontaneous events in their lives on invisible evil spirits or the Fates. Striving for a sense of security, some people held that all chance events expressed God's control and will. How does IPhil view this matter?

When a gambler plays craps, he "lets go" of control of the dice, and this *releasing* of the dice to act randomly in this setting is according to the "free will of the gambler." Similarly, in the Almighty's "letting go" of individuals so that they may act freely, that freeing act and its random result is according to the diverse potentials of the interrelational logistic of Primal Absolute Potential Energy (PAPE), so freedom and random expressions of that freedom are according the "will of the Almighty," which is not eternally deterministic but free within the logistic of a given system. For example, in the Bible the casting of lots, and the high priest's use of Urim and Thummim, evoked

God's blessing—rather than control—of the random physical actuation of potent possibilities considered of basically equal value. Elsewhere, some people attribute random events in life to foreordained Kismet or the goddess *Fortuna*. Gamblers and risk takers call for the assistance of Lady Luck, as they hope for financial gain in their gambling. When randomness produces positive results in one's life, some people call it "Serendipity," for it leads to serene and happy feeling.

From this list, it is clear that throughout the world, because of the frequent dissociative or evil effects, humans on the 2nd level try to blame someone else or something else for chance and random events. IPhil finds dissociative and associative random events to be an expression of the diffused 1st level state of potential energy.

There are many interrelational things that happen best in a random way—such as the flipping of a coin at the beginning of a football game. This ritual energy is actually sacred. The deliberately freeing flipping of a coin at the beginning of a football game is analogous to the deliberate, freeing action of the all-giver to the all-receiver in the establishment of the Almighty in the Primal Absolute Interrelational System (PAIS). Similarly, when conscious, rational thought cannot reach a decision in a particular matter—as when there are two equally qualified candidate for an office—it is within the interrelational logistic and will of the Almighty that a decision be made by chance, by lots, or by drawing straws.

IPhil recognizes that the potentiality of potential energy pushes the holder of existential energy out of its current existential state to a subsequent existential state that is progressively new. Likewise, the potential of an individual pushes the energized individual toward freely initiating a state that is progressively different from the current logistic of the individual. This results in a self-initiated actualization of a logistical possibility beyond the current logistic of one's antecedent logistic—such as deciding who will kick off and begin a football game. That random event is not unto a chaotic universe but is oriented toward a more advanced logistical order. Whether that chance event will actually produce an enduring, logistically consistent system, needs to be seen. "By their fruits you shall know them." In a game of chance or in any energetic system, the potential energy of that system pushes that system toward self-initiating a potential state that is truly free from the determination of all the other elements in that system and therefore truly random.

Who causes random events? That is a 2nd level question, and it expects a single answer that is a 2nd level integrated individual. Rather, the material cause of random events is 1st level, diffused potential energy that enables its receiver to be free in its self-initiatives. When a thrower lets go of the roll of dice, it is the current potential energy of the dice that is free to actualize any of its potential states. Remember that potential energy is a no-thing. So, it is the no-thing of the potential energy of the dice—and not the materials of the dice—which efficiently causes the subsequent random physical event. IPhil recognizes that all events in our world are energetic, and energy has a progressive four-stage logistic in the physical order. In each physical stage are found all four metaphysical aspects, of which the fourth is self-initiative.

There are several ways to deal constructively with randomness in life.

1. Concerned simply with one's immediate development and environment, non-reflective 1st level individuals, in infant-like ways, experience many random events unconsciously. Regardless, their received logistic is already oriented toward integral development. Here semiconscious youth and naïve individuals simply deal with chance events as they come—not getting angry at them but accepting them as a natural part of daily life.

2. Adolescent individuals seek to put order in their lives by trying to control as much of life as possible. They see uncontrolled, random events as their enemy. When they recognize that not everything is under their control, they highlight the negative aspects of randomness and what could go wrong. Then, they deliberately activate self-initiatives to counteract destructive randomness as much as possible. Seeking to cope with their random negative experiences, but having difficulty trying to control of their lives in this matter, they imaginatively identify a "scapegoat" when things don't go their way,

3. A more secure adult tolerates the possibility of random events. Recognizing that random events can be either good or bad, an adult individual strives to incorporate random events into her current situation in beneficial ways. In a medical research project, a researcher accepts that there will be many negative results before finding one result that marks a breakthrough. Dismissing many dissociative random events, the adult individual is very patient in looking among many random events and samplings until a positive result is found.[1]

    The psychologist Richard Wiseman recruited subjects who thought of themselves as either unusually lucky or unlucky. He discovered that the self-described lucky ones share a set of behavioral traits that maximized their good fortune. They were receptive to new experiences and invested time in expanding their social and professional networks. When things went wrong, they didn't complain but reminded themselves that things could have gone worse.

4. The writer Kala Starr describes what she calls "structured serendipity." Here the individual structures her day so that there are times in which more random ideas and events will naturally be generated—such as, browsing through a bookstore for an hour. Here the open, alert mind can associate with other ideas and occurrences which, in that fertile environment, can produce new interrelational connections, and which *can* become a "seed" for new, beneficial ideas and plans. Constructive brain-storming sessions and trial-and-error experiments in a laboratory promote and structure serendipity. That happens because

---

1. Burkeman, "How to (Truly) Improve Your Luck."

many random events will more probably generate one or two positive results. Increasing the number of random ideas in a structured environment increases the likelihood of a fortuitous discovery. Persistent young actors looking for a big break go to as many auditions as possible. In this way, they "make their own luck." A successful thinker and writer will be alert to the many random events that pass by each day and will grab and store in memory those ideas and experiences which have greater potential.

When a consciously thoughtful individual is unable to solve a problem rationally, the person can structure serendipity in the random operations of one's unconscious by pre-setting one's imagination with a question before going to sleep or taking a walk. By consciously priming the mind with a dissociative question or problem before going to sleep or taking a walk, the individual allows the random, very busy unconscious mind to play imaginatively in unorthodox, random, visual ways with the problem. In a state of rational rest, the individual is more likely to have a sudden conscious "aha" or eureka moment from one's searching unconscious mind, as Archimedes did when he soaked in a public bath.

For me, the few moments after waking are very important. During that transition time of being only half awake, my unconscious mind can give my conscious mind new insights or solutions to a question or concern my unconscious mind has been thinking about during the night. My subconscious mind is 1st level, and dreaming is highly symbolic and sentient. It can randomly discover a new logical seed that can become the progenitor of advanced ideas. There is good reason why the famous, prolific inventor Thomas Edison had a cot in his laboratory where he could let his highly rational mind sleep in catnaps so that he might unconsciously discover new solutions to problems while he slept. There is good reason not to act on a decision rationally made until one "sleeps on it," for the unconscious mind can expose additional, extraneous aspects of that conscious, rational question.

It is a common mistake of 2nd level essentialistic thinkers to hold that their conscious, deliberate mind is the only place where rational, advancing thought can happen. Likewise, it is a mistake to claim that an idea gained during sleep is a God-given or a spirit-given answer. Every metaphysical idea from one's unconscious mind, as from one's conscious rational mind, needs to be checked out against physical reality to establish whether and how it is really true.

Interrelational War Theory

Idealistic, essentialistic, 2nd level philosophies and moralities emphasize several major moral maxims: "Thou shalt not kill," "Always turn the other cheek," "Injustices are best confronted in nonviolent ways," etc.

The commandment "Thou shalt not kill" (Exod 20:13) is a good commandment, but it is not a universal commandment. In the very next chapter (Exod 21:14), God orders the Jews to take a malicious murderer who may be clinging to the altar of the Lord and kill him. In the Sermon on the Mount, Jesus directs his followers when hit on one cheek, to turn the other (Matt 5:39). There is wisdom in this saying, but it is not universal in its application. Jesus did not practice this maxim when he faced Annas. Because of his answer to Annas, he was slapped by a guard. Then, rather than turn his other cheek, Jesus protested against that strike (John 18:22). Reacting to injustice in nonviolent ways is a good policy, but it is not a universal one, for Jesus took a whip to the sellers in the temple for their violation of the command that God's house is to be a house of prayer, not a market.[2]

One of the main reasons the famous nonviolent protests of Mahatma Gandhi in India and by Martin Luther King Jr. in the United States came to successful conclusions was because they took place in countries where some political leaders had been imbued with Christian values. After watching how the people in those demonstrations were willing to suffer and die for the cause of equity, British leaders in India and the President of the United States had a change of heart—a metanoia—toward long-standing social injustices. These leaders acted and pushed through legislation that recognized the justice and righteousness of the protests of these abused classes. Unfortunately, that corrective response is not always the case, for sometimes the leadership of a country remains hard-hearted and unsympathetic to the needs of the people. The king and nobles in France did not respond to the cries of the poor, and their lack of sympathy to the mass protests of the people led to the French Revolution. The Christian Russian czar and nobles did not respond well to the cries of the poor, and their lack of an effective response to the protests of the people led to the Communist Revolution. Revolutions are usually inverse, antithetical actions, which are rightly retaliatory and destructive of their non-responsive rich and powerful exploiters.

World War I started when Austria-Hungary declared war on Serbia. Complex alliances in Europe drew many countries into that war, and each country strove to protect its holdings and ambitions. After the war, in the Treaty of Versailles the Allied winners demanded full, just restitution for the war damages they had incurred. These material and economic sanctions were excessive, putting the German people in difficult straits. In response to these oppressive, debilitating demands, the German people became deeply angry, and they readily accepted Adolph Hitler as leader because he filled them with a sense of moral strength and retaliatory hope. He promised a better life for the German people, whom he called the best people in the world. They backed him in another war against the Allies. With hindsight, it is clear that the demands of the Treaty of Versailles for 3rd level "just retribution" did not promote enduring peace but rather set the stage for a reciprocal, 2nd level war of revenge.

2. John 2:15.

In contrast, after the termination of World War II, learning from the lessons of history, George C. Marshall proposed, the US Congress approved, and President Harry S. Truman signed the Marshall Plan. Rather than further pillaging Germany by demanding in justice full restitution for war damages, the Marshall Plan of the United States gave $13 billion in economic assistance to help rebuild all the economies of Western Europe. This plan included the restoration of the economy of German, our former enemies. General Douglas MacArthur acted similarly with Japan. As these economies grew, they became leaders in the development of enduring peace and prosperity, not only for their own countries but in a cooperative way with all war-torn countries in the West and East. Subsequent mutual respect and economic prosperity provide an example of the positive consequences of loving one's enemy.

Thus, history shows that the 3rd level international strategy of restitutional justice after World War I did not establish enduring peace and prosperity, but rather led to another world war. In contrast, the 4th level, merciful interrelational strategy of the Marshall Plan and MacArthur leadership, mercifully assisted repentant enemies, led all to an extended period of international peace and prosperity. These are classic examples on a societal level of the maxim: "By their fruits, you shall know them." History shows that a 4th level *merciful* approach to dealing with one's repentant enemies after World War II bore more enduring economic prosperity and international peace than the 3rd level *just* approach taken after World War I.

Many people think a war is over when the military conflict is over. For them, peace is only the termination of physical conflict. Fullness of peace, however, is found in the achievement of enduring interactive harmony after a conflict. The traditional Just War Theory focuses primarily on events that justify going to war. In contrast, IPhil holistically recognizes that the termination of military conflict is not the real end of a war but only the conclusion of the 2nd stage of a war. In the following interrelational war theory, IPhil recognizes that there is much more to war than starting or ending the actual physical conflict. The resolution of war is achieved only with the reestablishment of enduring, harmonious economic and social interrelationships.

Let me begin IPhil's exposition of its interrelational war theory by making an analogy. In an operation, a surgeon must cut the patient open to remove an infection that is affecting and threatening the whole body. The surgeon cannot be timid, because cutting out only the obvious, minimum manifestation of the ailment may leave unseen roots, such that the problem probably will return in the patient's weakened body. Still, the surgeon cannot remove everything, for that would kill the patient. In a living 1st level patient, once a physical evil is significantly dissociative and its aggressive 2nd level antithetic actions affect the whole body, it is necessary to remove the infected trouble fully, for otherwise, the infection will continue to spread throughout the body. Similarly, when military aggressions are dissociative, there is need for decisive, military extraction of the dissociative agent(s). Also, the surgeon will perform the physical extraction as quickly as possible. Then, in a well-ordered,

3rd level way, the surgeon sews the person up to make the patient intact again in a healthy way. But the success of the surgery needs to be followed by 4th level rehabilitation in the hospital. Finally, the patient returns home to begin living in a 5th level new way that accommodates the healing and directs the recovered patient to a new interrelational activation of his current potential.

IPhil's interrelational war theory is different from the long-standing just war theory, which was initially crafted by Augustine as the Roman Empire faced enemies that would kill Romans and pillage Rome. This just war theory was subsequently augmented by Thomas Aquinas and others. This theory focuses on what can justify a righteous nation going into war, even as Christ's teachings and Christian tradition opposed killing of one's enemies. Essentialistically, the killing in war is recognized as an inherently evil activity. So, how can Christians ever declare or participate in war?

In the early decades of the Christian community, Roman solders withdrew from the Roman army because they considered the profession of a soldier to be logically inconsistent with being a Christian. Crafters of the just war theory presented their conditions carefully. Nonetheless, these writers were foundationally essentialists, and their arguments appear to be practical, rationalized exceptions, with little solid philosophic foundation. Consequently, along with other critics, I find their justifications for war to be logically inconsistent with the essentialistic foundations of their philosophy and beliefs. "Thou shalt not kill" was recognized by them to be a universal moral imperative. Still, they allowed it in their just war theory. Various authors have added more corrective mental reservations to this theory, but I find the traditional just war theory does not stand well in light of their foundational essentialistic 2nd level principles.

In contrast, the interrelational war theory that I propose here is 3rd level, and philosophically it flows naturally from the evolutionary principles of interrelational philosophy (IPhil). It is based upon the fourfold logistic of the energy which drives our world forward. Therefore, it can be said that it is an expression of the "natural law" as logically evolved from the fourfold logistic of Primal Absolute Potential Energy and the Almighty.

1. The interrelational war theory begins on the physical level, recognizing that diseases, droughts, impacts by meteorites, etc., can threaten the very existence of a nation and civilization. For example, scientist have found that the probable cause of the destruction of the great Mayan empire was prolonged drought. Here we are "at war" with random, destructive, 1st level acts of physical nature. Humans are often powerless before the magnitude of such physical evils. Here, immediate reactions are mostly defensive to reduce the impact of disasters, striving to preserve and strengthen people's lives and community. Defensive strategies are put into place so that a subsequent disaster will have less effect.

2. With the rise of strong, charismatic leaders, powerful groups can aggressively attack and pillage other nations to expand their 2nd level dominion. Initially, small economic and military infringements may be judged inconsequential, and they can be easily dismissed or averted. But there comes a point at which the aggressions of one nation or league of nations against another are societally consequential. At this point, such aggressions need to be recognized as a state of war by an aggressor against the attacked group or nation(s). Here is where the difference between idealistic essentialistic thinking and realistic interrelational thinking go in different directions. Seeking to reduce the dissociative harm, the defending nation, like the aggressive surgeon, IPhil recognizes that the defensive military should be decisive and strong in its attack against the enemy, striving in a 2nd level dissociative way to terminate the aggression as fast as possible to reduce the amount of dissociative evil within this war.

In a reciprocal response against a 2nd level aggressor, a physical 2nd level war initiative must be made even by the 3rd level society. That is, 3rd level defenders rightly can defend themselves in an antecedent, 2nd level, equitable, military way. This is justified because in a secondary, carryover way, the antecedent logistics of a 2nd level nation remains within the advanced logistical foundation of the logically evolved 3rd level defender. That is, by the carryover characteristic of interrelational logistic and logical evolution, a 3rd level society is justified to reactivate 2nd *physical external* responses when appropriate—while still retaining a 3rd level *metaphysical internal* mindset. That is, during a 2nd level physical conflict, the 3rd level nation will continue to hope that the enemy will eventually become a 3rd level society that can interrelate with them respectfully. This is similar to the way a college-educated mathematics teacher will teach youth in high school at their high school level of competency, even while the teacher retains but hides her college-level competency until the student is ready for that.

Essentialistic thinkers maintain that moral (Christian) individuals and nations must only act in a non-aggressive (Christian) way. In their rigid idealism, any antithetical response to an aggressor is deemed immoral in their idealistic, only-good, Christian culture. In this setting, as Voltaire noted, the best becomes the enemy of the good. In contrast, IPhil maintains that 3rd level individuals and nations can temporarily "cover" their synthetic, 3rd level moral values when dealing with an enemy who is acting only in a 2nd level antithetical way. IPhil philosophically recognizes through the progressive logistic of potential energy that dissociative war is necessary and justified in dealing with an antithetical, aggressive opponent in a reciprocal antithetical, aggressive way—for a time.

3. In dealing with an attacking lion, a human person cannot stop that lion by talking to it rationally. Rather, an attacking lion needs to be dealt with on the same level of aggression as the 2nd level aggressive lion. Nonetheless, after restraining that lion, by treating it well in a 3rd level way, that lion can be turned into a pet by the

caring victor. Then, the lion will probably respond cooperatively to the words of the victor—in ways that are for both more interrelational. Similarly, a 2nd level, self-centered, attacking human enemy, who initially does not respond to 3rd level diplomacy, needs to be dealt with in a reciprocal, 2nd level, aggressive manner. Then, once the war is successfully over, the latent 3rd level motivations of the victor will be "uncovered," and the victor will start treating the conquered enemy in a 3rd level way in the hopes that the enemy will respond in a respectful, reciprocal way. During the 3rd level defender's 2nd level response to a 2nd level aggressor, the 3rd level defender will hold metaphysically in his heart the hope for future, 3rd level, cooperative, international association with the former enemy. This 2nd level response by a 3rd level defender is not a betrayal of the higher values of the defender, during 2nd level physical reactions to the 2nd level physical aggressor. The progressively developed defender metaphysically has 1st, 2nd, and 3rd level logistical interrelational values. The 3rd level defender will use those logistic behaviors which are more appropriate to the given physical situation. Hopefully, the physically terminated dissociative aggressions of a 2nd level enemy can be subsequently elevated to the 3rd level through a program, like the Marshall Plan, thereby uniting the former enemy and the victorious defender with 3rd level associative values.

4. Once the military phase of the war is over, the interrelational logistic and the logical evolutionary theory point the way toward advanced international developments beyond the past international war situation. To start with, leaders steeped in the values of interrelational philosophy can work with the agreeable, reformed bureaucrats of the old regime to restore and build in an interrelational way an advanced civil order that works in harmony and cooperation with the winning side. This is one of the reasons why the expansion of the Roman Empire led to the enduring *Pax Romana*. Agreeable bureaucrats and citizens embodied the logically consistent logistic of the aggressive society. After conquest, they can then go beyond that former logistic to establish and organize themselves and their state easily into the new, advanced order of the cooperative conqueror. Here, the population of the former enemy is treated as respected, potential partners in a well-ordered, expanded, mutually supportive alliance. Here, the principle of subsidiarity promotes appropriate empowerment and mutual respect between winners and losers.

In summary, what is traditionally called the "justified war theory" is not really *logically* justified, while the "interrelational war theory" is *logistically* justified.

Recognizing that some participants in the war have been seriously debilitated by this war and are not able to be strong participants in the expanded league, IPhil acknowledges that it is important that leaders of the winning nations of the league not further debilitate the conquered enemy. Rather, experience shows that greater loyalty and cooperation are experienced when the winners in a conflict materially and morally assist all injured members on both the winning and losing sides. If members of the conquered nation show a desire to do better, the wealthier nation(s) of the newly established league *can* in mercy, faith, and hope assist the weaker nations so that they can regain their interrelational strength—for the sake of long-term advancement of both themselves and the expanded alliance.

Here, economically helping one's defeated enemy will transform a once aggressive enemy into future energetic contributors to the prosperity of all members of the post-war league. In this way, "loving one's enemy" invites and transforms the defeated enemy into committed partners. The long-term success of the Marshall Plan has shown that the evolutionary war theory can work today.

Crimes against Humanity

After World War II an international court was established in Nuremburg to try German leaders who were said to be responsible for tremendous atrocities against humans during that war. Many of those indicted proudly and defiantly claimed they were innocent because they were only following orders. With respect to the moral standards of their German culture and government, whose leader they obediently followed, they judged themselves as acting in a morally righteous manner. The international court, however, claimed that there was a higher moral norm. The human race communally has a higher moral norm, and according to that standard—or logistic—they were guilty of "crimes against humanity."

IPhil recognizes that the groups and societies which humans form are parts and subparts of the greater society of all humanity. It is true that individuals are the grounding terms of friendships, marriages, teams, groups, states, and nations, which are subgroups of worldwide humanity. Each individual does have one's own individual logistic, and the consciences of individuals give guidance and judgment according to their individual logistics. Here is grounded the defense says, "I did no wrong because I was following my conscience."

Nonetheless, war is not just a series of individual or group interactions. War between nations is a dissociation on an international level, involving the communal logistic of humanity. Historically, one's human communal logistic is received in a carryover process from the first human parents. By the carryover process found in generation after generation, all their human progeny down to today have not only their own individual logistic but also the same communal human logistic or human nature. By one's individual and communal logistic, guidance for and judgment of one's actions are made not only according to one's individual logistic but also according to one's communal, human logistic. Although those indicted were following their individual consciences with respect to the orders of their German superiors, in an international setting as a world war, every human individual has a higher interrelational responsibility for living and acting according to interrelational standards of conduct that are present in one's human nature and communal logistic.

Party loyalties, group loyalties, racial loyalties, and national loyalties are prejudicial unto themselves, and they tend to be against those who are not members of one's group. Loyal associations to one's group are inwardly directed for the sake of defense and greater intrarelational security and prosperity within the group. These associations make a person in a group feel good. These group loyalties, however, are still 2nd level and have exclusionary walls. Those outside of one's group are regularly looked upon with distrust and fear, for by breaking the walls of one's exclusive groups, one's current social order, peace, and happiness are threatened. Loyalties to one's group then encourage negative, dissociative, hateful reactions against those outside of one's group. Consequently, 2nd level group loyalties generally are prone toward crimes against humanity.

Appreciative Inquiry

To establish a company or a nation, its founders must gather resources to establish its viability. As the organization or nation is forming, it must protect itself from outside competitors and internal dissociative employees or citizens. If their complaints grow, dissatisfied employees and citizens may organize protests that produce different forms of conflict between management and labor or between the wealth upper class and the impoverished, working class.

Wise corporate and national leaders need to recognize that they do not have all the best answers to these dissociative disparities. They need to recognize that workers are not machines or dunces. There is needs to change from a competitive dualistic outlook that pits 2nd level labor and management against each other unto 3rd level, interrelational cooperation. While 2nd level individuals pessimistically highlight dissociative differences with conflicting attitudes, 3rd level individuals optimistically highlight associative possibilities which can produce win/win opportunities. Each side needs to appreciate the other side for the special contributions each makes to the organization or nation. Management and labor, leaders and followers, upper class and lower class need to recognize in an appreciative way that each group has their own special way of contributing to the success of the company or nation. They need to appreciatively recognize that each part is necessary for and contributes to the vitality of the whole. Then, material compensations for managers and workers need to provide at least for the survival, health, education and happiness of all.

David Copperrider at Case Western Reserve University formalized the process of "appreciative inquiry" in which all the leaders and members of an organization or community respectfully and appreciatively work together for their common goal and individual benefit. Sue Annis Hammond in *The Thin Book of Appreciative Inquiry* lists the following assumptions of this process. This is not a quick-fix, silver bullet solution but a systemic process. Note that the assumptions of appreciative inquiry are highly analogous to the eightfold interrelational logistic of all potential energy systems.

1. In every society, organization, or group, something works.
2. What we focus on becomes our reality.
3. Reality is created in the moment, and there are multiple realities.
4. The act of asking questions of an organization or group influences the group in some way.
5. People have more confidence and comfort to journey to the future (the unknown) when they carry forward parts of the past (the known).
6. If we carry parts of the past forward, they should be what is best about the past.
7. It is important to value differences.
8. The language we use creates our reality.

A more concise, immediate, problem-solving method highlights "brainstorming."

1. A problem is recognized and defined.
2. Then in a brainstorming session, everyone is allowed to suggest different ideas pertaining the solution of the problem, regardless how crazy. All ideas are simply recorded and not evaluated at this point.
3. When no different ideas are forthcoming, the most promising solutions are considered in terms of the intermediate steps needed for each proposal to solve the problem.
4. The most likely possibility is then determined, and an executive decision is made as to whether to implement that possibility or not.

Principle of Subsidiarity

The principle of subsidiarity applies to groups, organizations, and nations in a graduated way. Subsidiarity as a principle of social organization was first described and developed by the Roman Catholic Church following the First Vatican Council at the end of the nineteenth century. Some people associate it with the idea of decentralization; however, subsidiarity is not a negative idea but a positive principle. It recognizes that national empowerment is first found in lower, earlier levels of international development. Consequently, for the sake of national and international development, priority and respect should first be given to lower levels of societal formation before given to higher levels of international dominion. This is in accord with the process of energetic logistical development and logical evolution from the simplest systems unto the more complex systems and societies. The principle of subsidiarity advocates that higher levels of authority should take control only when lower levels of authority are not able to solve a problem at hand.

PART III: DEVELOPMENTS

The *Oxford English Dictionary* defines subsidiarity as the idea that a central authority should have a subsidiary function, performing only those tasks that cannot be performed effectively at a more immediate or local level. Subsidiarity is a general principle of law in the European Union. According to this principle, the EU may only act (i.e., make laws) where the action of individual countries is insufficient. This principle was established in the 1992 Treaty of Maastricht. Historically, it was already a key element at the local level of the European Charter of Local Self-Government, an instrument of the Council of Europe promulgated in 1985.

This raises an important, fundamental question: Does authority impose itself from the top down or arise up from the bottom? In the Bible it is said that "all authority comes from God," implying that it comes from the top down. This is characteristic of the 2nd level, authoritarian leaders and governors of that time. Some were malevolent dictators or tyrants, and some were caring kings. Yet as society advanced into modern times, the 2nd level claim of the divine right of kings has been challenged and overthrown. In England this began with the forced signing of the Magna Carta by King John in 1214. Today, people throughout the world strive in a logical evolutionary way to change 2nd level autocratic governments into 3rd level democratic governments, which inversely are built from the bottom up. So, which is right? Top down from a Kingly God above, or bottom up from God's Spirit moving individuals and communities from within?

IPhil recognizes that in the Primal Absolute Interrelational System (PAIS), there is first a movement from the all-giver to the all-receiver. Then, in the second major phase, the given energetic Spirit empowers the all-receiver to become the all-receiver-giver, which then takes the initiative for closing the primal absolute interrelational energetic cycle. Traditional, religious essentialists view all energy as only going from the Almighty to finite individuals. In contrast, IPhil recognizes that subsequently empowered finite receiving individuals are empowered and logistically directed to take the initiative in actively reciprocating the giving. This reciprocal response advances the active phase of the interrelational logistic from the all-giver to the all-receiver-giver.

Thus, IPhil recognizes that in the course of history, all energy and responsibility first went from top down. This first occurs in the patriarchal leadership within each families, extended families, clans, communities, and national governments. Then, once the people are empowered and educated, the reciprocal aspect of the interrelational logistic occurs, and the empowered receivers start to exert their role from the "bottom up" as the new leaders in society. In this way, IPhil recognizes that reality is neither top down nor bottom up, but progressively together, first top down from the all-giver, and then bottom up from the all-receiver-giver. This reversal promotes a reciprocal interrelational closure that is logically capable of evolving a higher corporate interrelational state of life. Thus, all power and authority does come from Almighty God first, transcendentally from above or from the outside. Then, power arises imminently from the received potential energy arising from within each individual receiver-giver.

When children are young, parents naturally exert authority over them. But as the parents grow older, leadership and care of the family shifts to the adult children, who then bear responsibility for the older parents and the assets of the family. Many long-established leaders resist this inverse reordering of things. However, the reciprocal interrelational logistic of energy legitimizes the subsequent governing initiatives of the empowered lower members in societies. Likewise, in the wild, after young males have been watched over and protected by a patriarchal male or matriarchal female, there comes a time when growing young males and females will challenge and unseat alpha-males and matriarchal females in their species. This can be done peacefully or violently. In the absolute order, the transition of initiative from the all-giver to the all-receiver-giver was well-ordered. In this process, it was important that the all-giver "let go" of ownership, dominion, and control in the transfer of absolute potential energy. Only with the "letting go" of its claim to absolute potential energy does the all-giver first empower and then allow the all-receiver to act as a self-initiating all-receiver-giver. In an analogous way in the finite order, the interrelational logistic of potential energy pushes the generating giver to empower generated receivers. Then, pushed by forward-leaning potential energy, giver-receivers are ordered to "let go" of dominion so that the empowered receivers can become finite receiver-givers, who are now responsible for closing and maintaining social order after the retirement of the former leader.

It is in the reciprocal return phase of interrelational exchange that the principle of subsidiarity acknowledges that there is an appropriate time to recognize the empowerment of lesser nations and individuals. Here, long-standing hierarchical powers need to start "letting go" of their overriding control, so that lesser individual nations and political groups are able to grow stronger and take control over what they are logistically destined to control. Wise parents need to do that with their children as they grow, first encouraging them to do as much as they can do, and then gradually, carefully stepping aside. When the transition occurs too quickly, as happens politically in revolutions or in families after a parent dies or divorces, inexperienced children frequently make painful, dissociative, organizational mistakes. In this conversion process, wise parents need to remain close at hand as children grow up and take over by being their tutors and advisors—as in the medieval guild process of masters and apprentices. Unfortunately, it is common for people with power and wealth to want to keep things intrarelationally just as they were in the past. They find security in keeping things as they were in good times. However, the progressive and cyclic character of energy within enduring systems always pushes forward in an inverse, reciprocal way. The principle of subsidiarity is a format in which advancement of lesser members is encouraged and supported.

As the members of a group or society mature and become educated in the operations of a corporation or a nation, the role of the progenitor and leader within society becomes less directive and more supportive in the operations of the group and society. Writing about establishing good order, Confucius recognized that within a well-functioning society, the role of the servant-leader is to be less directive and more symbolic of societal harmony.

The antithesis of a good organizational leader is an insecure "micromanager," who has little confidence that subordinates will do what is right or best for a company. A "micromanager" needs to "let go" of self-centered ownership of a business and let other members of a company show their talents and ability to take responsibility for whatever task or office they hold. Caesar Chavez organized the migrant workers in California and improved their living conditions. But as the movement advanced, his role changed from being a leading protestor to becoming a living symbol of that movement. As the people became stronger, he let them take increasing responsibility of the local operations of the movement, while he became the inspirational bearer of the movement's "vision." Likewise, even though Dr. Martin Luther King Jr. was assassinated many years ago, through his "dream" he continues to support and lead black people and the oppressed. Similarly, founders of religious movements soon die, but in a subsidiary way their personal legacy becomes the spirit that energizes their followers in a united way.

Since the Magna Carta, the role of the monarchy in England has increasingly become a symbolic one. In line with the principle of subsidiarity, the British people have taken control of the governance of their nation on different levels. So why keep the monarchy? By the carryover characteristic of the logical evolutionary theory, the role of dominant-leader is not eliminated but transformed into that of a 4th level servant-leader. In times of trouble, however, when a unified organization of a corporation or nation is threatened militarily or begins to fragment, it is appropriate for the responsible servant-leader to step back into the role of administrative leader, synthetically bringing unity back to the ranks. Then, once peace is reestablished, the servant-leader again needs to fade into the background and let the followers once again take the lead.

In summary, the principle of subsidiarity is *not* a universal, unchanging, essentialistic principle. Rather, the principle is to be followed when it is logistically appropriate—when the people can take increasing responsibility for the governance of society for the common good. Just as it is unwise to make children and adolescents responsible for the maintenance, insurance, taxes, and property rights of one's home, so too it is premature to promote democratic forms of government in a country where the citizens are still immature in their interrelational education and ability to establish enduring good order in a pluralistic society. In the history of the United States of America, responsibility for governance was first grasped by property owners. Then, with the gradual universalizing of education and experience in governance, the voting franchise was demanded and claimed by other groups. Subsidiarity is not a right passively received. With rights come responsibilities, and the citizenry of a country must be educated and claim the right to be part of the well-ordered 3rd level governance of a society. If an individual group is immature and self-centered when they are given adult, interrelational responsibilities, they will tend to strive to impose their immature, self-centered, 2nd level ideological ideas on the entire community. They need to have both a sense of personal strength and a respect for the community's welfare as a whole. They need to "let go" of some of their adolescent ideals and strive to work with others who think differently than them, toward the broad, 3rd level, common good of society.

Resolving Conservative vs. Liberal Stand-Off

The modern conservative deadlock in many democratic countries arises regularly from a stand-off between those with a 2nd level self-centered, traditional worldview versus those with a 3rd level community-centered, progressive worldview. IPhil proposes a general solution to this dilemma, which is currently tearing many democracies apart from within—just as it did in the Senate of historic Rome and the democratic forum of ancient Greece. Are we going to repeat the mistakes of the past, or are we going to learn from them and find a way past them?

All organizations, institutions, and political parties are interrelational within a given country and the world, so it is right that they be licensed to operate within the nations they serve. According to the reciprocal interrelational logistic, societal logistical rights come with societal logistical responsibilities. If institutions are operating within a state for their own benefit, they are also obligated in justice to contribute to the welfare of the state.

All public persons or organizations in a state need to receive a license to exist and operate in that state This is nothing new, for all organizations today need to be legally chartered by the state and give annual reports of their financial earnings and their tax deductions for contribution to charities. To maintain their corporate licenses within a state, every organization needs to demonstrate publicly that their workers and members are receiving sufficient reimbursement for their labors to live a healthy individual and family life so as not to be a burden to society. They need to pay their fair share of taxes to the state(s) they are operating in. They also need to contribute to cultural, service, educational and value-centered institutions for the betterment of the society as a whole. If organizations and political parties fail to demonstrate their financial organizational contributions to the common good of society, they should lose their licenses and not be allowed to operate in that society. If they do so only partially, they should be correspondingly restricted in their operations within that state. More can be proposed on this great societal concern.

Concluding Thoughts

Reflecting on the universal analogous character of the fourfold interrelational logistic of energy, I am reminded of the classical musical form called "Theme and Variations."

Here, the composer begins with a theme. It can be very long and elaborate or short and concise. The composer then takes this theme as a challenge. The composer with great skill produces all kinds of variations of that theme: playing it fast, playing it slow, playing it high and soft, playing it low and loud, adding notes here or there, taking out a note here or there, changing the rhythm and the key, etc. Each new variation brings delight while holding on to the enduring theme, showing how it can be cleverly altered in new ways. Finally, all those variations are knit together in a variegated, developmental way to produce a single piece of deeply moving music.

When orchestra members were first given the music for Beethoven's Fifth Symphony, they laughed at the simplicity of the four-note theme of its first movement: ta-ta-ta-TAH. How could anyone produce a quality piece of music with the simple theme of just four notes—three short and one long? But then they played it, and played it, and played it again—and it has become one of the most memorable pieces of symphonic literature. The four-note theme of the interrelational logistic is like that. It is so simple: receiving, integrating, interrelating, and surrendering. Yet that theme has myriads of variations, each with its own unique tone and delight, whether it is played as a solo, a duet, a quartet, or by a huge orchestra with many different instruments. Recognizing the fourfold interrelational logistic of energy as the theme of everything in our energized world, we suddenly discover that physical reality is a wondrous, pulsating, overwhelming, transcending piece of music.

We are oriented toward acting in a progressive fourfold way not because we are told to do so by others or because we rationally discover this to be the best way. Rather, we feel most comfortable acting in 4/4 time because our inner potential energy is the foundation of our existence and interrelational development. Our potential energy pushes us to acting according to the progressive fourfold metaphysical logistic, and its fourfold physical nature exists and operates in every fiber and atom of our bodies.

This is true in the physical order and in the metaphysical order as well, in material things and spiritual things, in their own logistical order, and in evolved orders. This is how humans and all creatures have evolved from and can interrelate with the Almighty. By logical evolution, the fourfold character of our foundational energy becomes the well-ordering logistical guide and inner push in everything in our universe.

Chapter 18

Who Is a Person? Is the Almighty a Person?

> Common Notions of Person. Historic Origin of the Term *Person*. Expanding the Notion of Person. All My Relatives. Eating in a Broadly Personal World. Different Judgments about God. Rational Arguments for God as Person. Impersonal and Personal Aspects of God. Prayer in IPhil.

Ethically, politically, and individually, there is today a huge range of different ideas, attitudes, and beliefs about the notion of *person*. Who is a person? How and why can one individual treat another individual personally, yet treat a nearby individual impersonally? An individual's responses to these questions signal the individual's views and attitudes toward *everything*. At one end, materialists view nothing as personal; they consider everything, including God, to be only an impersonal, energetic, material reality. At the other end, animists and naturalists view everything as equally personal. They strive to establish interpersonal, spiritual interrelationships with everything, including God. Yet, when two people are standing side-by-side, an individual can be very warm and personal with one individual, and very cold and impersonal to the other. When asked whether the "other party" is a person, the individual out of convention will say, "Of course." However, the physical ways a human actually interacts with two neighboring individuals—personally versus impersonally—can display two different patterns of thoughts and actions. So, asking, "Who is a person?" is socially and philosophically a highly nuanced question.

Common Notions of Person

There is a commonly recognized, historic, legal definition for *person*. Boethius, a sixth-century Roman jurist, gave the earliest legal definition of *person* as an individual substance of a rational nature.[1] This definition advanced Aristotle's definition of a human as a rational animal. Subsequently, following their intellectual bias, rational scholastic philosophers and theologians elaborated on this idea and said that a *person* is an individual whose essence includes intellect and free will. They maintained that only humans, angels,

---

1. Boethius, *Theological Tractates and the Consolation of Philosophy*.

and God are persons because, they said, only these are rational and have free will. In their day, they did not recognize the presence of limited knowledge, reasoning, and free self-initiatives within animals, plants, and things. Because they judged animals, plants, and minerals to be non-rational and lacking free self-initiative, they were not considered to be persons, and they could be used and killed at will.

At the beginning of the European conquest of the Americas, Pope Alexander VI in 1493 wrote several bulls, which became the Doctrine of Conquest.[2] With these documents as justification, since Native Americans could not produce a logical syllogism, European invaders considered them not to be rational, and they were treated as non-persons or wild animals having no property rights. After the pleading of some missionaries, the church soon reversed that judgment. Yet, the Doctrine of Conquest remains, and it was recently used as part by a judgment of the US Supreme Court to settle a case involving long-standing property rights. Similarly, colonialists claimed the right to buy and sell slaves from Africa. "Possession is nine-tenths of the law."

It took many years and much complaining by Christians and humanitarians before groups of British started a movement to override large economic interests and to slowly end slavery as a legal institution. Still, in the original US Constitution, slavery of humans was recognized, and slaves were counted as only partial individuals and thereby only as partial legal human persons. It took the very bloody American Civil War and modern protests on a national scale to end slavery and discrimination legally in federally supported institutions. Today, some people consider human fetuses, non-rational infants, brain-dead patients, and senile elderly as non-rational beings, incapable of rational, deliberate actions, and therefore not full human persons, not deserving full human and civil rights, and open to euthanasia.

Throughout our world today, there are many groups of humans that are discriminated against and treated in grossly impersonal ways. Regularly, Jews, unbelievers, racial groups, and enemies in war have been reduced to the status of detestable dogs, snakes, pigs, and scum. So, the question "Who is a person?" has wide scope. Only after determining what and who is a human person, are we ready to ask whether God is a person.

Historic Origin of the Term *Person*

People's idea of a *person* is generally derived not from a philosophical or legal definition but from their experiences in daily life. Consequently, rather than begin this discussion with a pre-emptive *coup d'état* by philosophically defining a "person" *a priori as* an intelligent, rational, volitional individual, it is important to look at the ways people in everyday situations speak of things in personal versus impersonal ways. If an individual's worldview is strictly materialistic, that individual cannot honestly recognize anyone as a person, even themselves, much less God, for the

---

2. Pope Alexander VI, *Inter caetera*.

notion of "person" is not in their philosophic worldview. Too many modern biological scientists have experimented on humans in ways that treat them not as persons but only as biological specimens.

Historically, the origin of the term "person" is in Roman theater. Here the actors on stage wore masks, which had mouth-holes carved like small megaphones. This enabled the actors to project their voices so that the audience could hear them better. Those masks were carved to look like the different people in the play. They were called "personae" because the actors spoke through a mask (*per + sonare = through + to sound*). Each actor on the stage had one (or more) persona-masks during the play on the stage. Now here is a very important point: *The only "persons" in Roman theater were the actors on the stage. All the members of the audience were not "persons."* Even though members of the audience made loud side comments, lacking a mask fitted to that drama, members of the audience had no recognized *persona*. So, the members of the audience did not count as real persons in the context of the play, even though members of the audience might be intelligent, rational, and volitional. That is, in its beginning, the term *person* was not applied to all humans but only to those individuals who were important to the action at hand.

In times of war and other conflicts, individuals regularly use derogatory names to demonize their opponents. They call their enemies dogs, pigs, snakes, and the scum of the earth to dehumanize them. We don't kill "our own," so by name-calling, we brainwash ourselves to see our enemies as not "our own," not of our human society, but as demonized expressions of dissociated evil. Thinking and acting in a 2nd antithetical way, we debase anyone or anything which threatens to hurt us or our projects in any way. By doing this, opponents lose their social status as persons, and we thereby justify treating them in impersonal and inhuman ways.

It is only in recent centuries that international laws, like the Geneva Conventions, have legally required that in war prisoners be treated as human persons with limited but important rights. Until very recently, women did not have standing as full human persons. In ancient times, even philosophers and theologians judged women to be less than men. They were not fully respected as rational persons because they lacked a penis, they lacked muscle strength, they lacked strength of voice, and they lacked the power to stand up to another man in claiming ownership of property. Property ownership and property-based leadership roles were extremely limited for women, and "possession is nine-tenths of the law." So, women were treated as chattel or the property of men. Even today in wedding ceremonies, the bride's father regularly walks his daughter down the aisle, a sign of his ownership of her. Then the father hands his daughter over to the groom, symbolic of the transfer of ownership of his daughter to the groom. In many cultures, children have had the legal status of slaves. In ancient times, it was permitted in some cultures for fathers to kill unwanted infants with impunity.

Even in our modern day, in schools and in neighborhoods, adolescents frequently name-call, discriminate against, put down, bully, and abuse classmates who are unusual or "different" from the way they deem socially right. Outsiders are judged by narrow-minded cliques to be unfit to be considered peers and to be treated respectfully in an equitable manner as persons. Clearly, the division of "us" from "them" doesn't have anything to do with the other individual's ability to think or choose. Consequently, the essentialistic, legal definition of a person as a "rational substance" has little meaning in the public lives of people today. Within human societies and in groups through history, the crucial dividing point is whether another individual is recognized and treated as a *fellow* human being, acceptable as an active member of *our* group, socially on a par with the other members of *our* group. Rather than recognizing Boethius's 2nd level *a priori* legal definition, all these examples provide ample evidence that the *operative* understanding of "*person*" is interrelationally and socially 3rd level. Evidence indicates that *person* is not a 2nd level, essentialistic category but a 3rd level interrelational category.

The Bible gives the rule: "Love your neighbor as yourself."[3] A Jewish lawyer asked Jesus the very meaningful question: "Who is my neighbor?"[4] In the Jewish culture at that time, the essentialistic mindset of orthodox Jews was that only Jews were "neighbors" to other Jews. Everyone else was deemed an outside foreigner and religiously unclean sinners, and they were socially excluded by orthodox Jews. Samaritans were looked down upon as inferior mixed-bloods, and devout 2nd level orthodox Jews dissociated themselves from them. Jesus broke that cultural boundary by putting forward the 3rd level, revolutionary idea that by his caring actions toward a needy Jew, the Good Samaritan should be recognized as a "neighbor" to that Jew and all Jews. That is, half-blood Samaritans, when they are associative and helpful, should be considered persons worthy of equal respect and social association. For him, it was not the Jewish or racial birth of an individual but the interrelational behavior of an individual that established a *neighbor* as an interrelational *person*. The Samaritan was an interactive player and thus a *person* in the story-drama of that rescue. The priest and the Levite who passed by the injured man did not act like neighbors to the injured Jew. If an individual, even a Jew, was ritualistically "impure" in any way, they would have become unclean by touching him, so they passed him by as a non-person or an animal with whom they did not want to interact.

From the beginning of the United States, many individuals were disenfranchised. Initially, if men did not own property, they could not speak or vote as full citizens. For decades, women and slaves were not considered equal citizens before the law. There first had to be many political protests and fights initiated by groups who demanded equal civil rights for individuals in an excluded group as full human persons under the Constitution. The modern push to be non-discriminating toward all adult biological humans

---

3. Lev 19:18; Deut 6:5.
4. Luke 10:29–37.

comes from the intuition first expressed in the US Declaration of Independence that "all men [humans] are created equal." However, the subsequently ratified US Constitution did not recognize that, and this disparity has been slow to be changed.

Different groups use different criteria for determining the beginning of a metaphysical, spiritual human person—as distinguished from a physical, material human life. Those who associate one's soul with one's breath recognize the beginning of the human fetus at its first breath outside the womb. Using Aristotelean categories, Thomas Aquinas maintained that a fetus receives a human soul when it *looks* like a human. Others maintained that a fetus is a human person at its first independent movement or "kick" in the womb at its "quickening." Others maintain that a human fetus should be recognized as a person at its first heartbeat. Others maintain that an egg that is fertilized at conception is a spiritual human person because it has physical human DNA. However, the corpses of dead humans have human DNA, so does the presence of human DNA in a corpse make that corpse a human person? Science has shown that a biological eggs, even before they are fertilized have the potential to become a mature human by the parthenogenetic process of cloning. Through history, courts' decisions as to who is a fully protected person under civil law has progressively changed, for their decisions reflect the current level of human interrelationality in their society. So, if a person doesn't like the decisions of the court, it is necessary to change the level of interrelational, interactive acceptance of certain groups in one's society.

From an interrelational point of view, the acceptance or rejection of human fetuses as persons depends upon whether a person sees a fetus in the womb as already an important, associative part of a family or of one's society, or whether it is a dissociative part of the woman, family, and society. That is, some fetuses are seen as meaningfully interactive in the womb and in the lives of some women, while some women see them as unwanted intrusions in their lives. Many human cultures currently do not embrace confined fetuses as active, interrelational, and meaningful parts of their human societies—unlike human babies living independently outside of the womb. Is a pregnant woman to be counted as one or two citizens? When fetuses are judged to be interrelationally dissociative and meaningless in the life of the mother and society, they are open to the mother physically rejecting them as a person. Currently, in our individualistic culture, there are strong individualistic 2nd level public opinions in some groups, who do not want to accept human fetuses as 3rd level interrelational, free, independent members of our human society.

While a 2nd level essentialistic focus upon an integrating, knowing, self-initiating human individual recognizes a fetus as an individual person, a 3rd level interrelational focus upon a fetus still enclosed in a womb questions whether a fetus has matured enough to be embraced as a societal person in civil society. Societal rejection of a human fetus as a human person is interrelationally 3rd level, which is the domain of civil law. Some parts of society strongly say the human fetus is not a human person to be treated as other humans, while other parts of society strongly claim that human fetuses are not full human persons and should not be treated legally as if they were. Today, most courts and governments defer from making that public decision, because there is no pubic consensus on this matter. So, this question will be resolved not by a philosophic definition but by communal, interrelational, societal consensus. Since positive laws follow the will of the people, it is important to change the hearts of the people toward interrelationally embracing each human fetus as "one of our own."

A lesson may be drawn from history. In the Greek, Hellenistic, and Roman worlds, infanticide was common. The life of a child was officially the exclusive responsibility of the father of the household, and the state officially recognized that it was *not* its legal place to monitor the behavior of the heads of the households, and fathers could kill their infants with impunity. This is similar to the way modern courts have ruled that the life of a fetus is totally under the dominion of the mother, and the mother has the legal right to kill her fetus with impunity.

In ancient times, the head of a family could and sometimes did kill an infant—sometimes with tragic, long-term, social consequences. For example, to appease the wrath of the goddess Artemis, King Agamemnon killed his daughter, Iphigenia, so he would have favorable winds to sail and attack the city of Troy. This act came to haunt him later, for his wife and the mother of Iphigenia had him killed in revenge when he returned victorious from Troy. In another case, to thwart a negative prophecy, King Laius of Thebes left his son, Oedipus, to die on a mountainside. Saved by a shepherd, he grew up to kill his father and become King of Thebes. In the great Greek tragedies, killing or abandoning a child was recognized as the right of a father, but that dissociative act had long-term negative effects in the lives of all the people involved. Regardless, infanticide remained common and legally sanctioned in the Greek, Hellenistic, and Roman worlds. Then, through the quiet influence of Christian values, by the fourth century, infanticide had all but disappeared—not by any change of laws by legislation or by courts, but by a change in people's loving hearts.

Today, even if banned by law and the courts, experience shows that some people will continue to practice abortion, often with unhealthy, dangerous results. In my mind, the most important thing to do is promote a positive attitude toward all human life from conception, even if there are differences of opinion when a human fetus becomes a human person. The notion of *person* is interrelational and societal, and our potential energy pushes us to be more interrelational toward the other than more intrarelational toward ourselves. A socially mature individual will advance beyond individualistic, self-centered, 2nd level values to greater 3rd level interrelational values in the hope of some type of 4th level transcendence. I believe the best way to reduce abortions is to promote a metanoia, or change of mind and heart, in individuals and in society, embracing and supporting the full potentiality in *all* of human life, starting in the fetus. I believe in a cultural metanoia regarding protecting from abortion the life of a human fetus as a person is possible—just as happened in the fourth century regarding infanticide. Here, the woman's right to choose and respect for human personal life is not opposed but affirmed in choosing the more interrelational outcome. In this way, the solution to this matter is not in law but in interrelational love. This position is both pro-life toward the fetus and pro-choice toward the mother.

Can the reader see the connection between recognizing a human fetus to be a human person—because of its potential to be an interactive part in the life of society—with recognizing that the potential Almighty is a person because of the Almighty's potential to be an interactive part in the life of human society?

Expanding the Notion of Person

Humans regularly "adopt" pets and give them personal names. Many adult humans today describe their pets as their children and as members of their family. In obituaries, both surviving and deceased pets are often listed by name among the deceased's relatives. If an outsider kicks or beats a pet, everyone in that family wages a minor war upon the attacker, who has injured "one of their own." Regularly, when a beloved pet dies, members of its human family grieve and take special care with the pet's burial. Giving pets personal, humanlike names is a sign of our increased thinking and treatment of pets as *persons*. Nonetheless, while these pets may be "animal persons," they lack the communal logistic of human DNA. Their differences in DNA, however, do not stop humans from loving them and treating pets as persons. This is another indication that the notion of "person" in human society is not essentialistically 2nd level but interrelationally and interactively 3rd level.

In recent years, more people recognize the personal characteristics of chimps, dogs, and pets, and they increasingly respect how Native Americans recognize and respect the personal qualities of land, lakes, and animals. Legal cases have been brought before the American courts claiming that before the law, land-persons should be treated equitably with human persons. The courts have resisted these claims out of custom and pragmatism. IPhil provides solid philosophical reasons for legally distinguishing the logistics of these other species, because they are less evolved and therefore have some but not all natural, legal rights as humans, but they do have some. Similarly, it has long been recognized that corporations have personal characteristics analogous to individual persons. However, the logistics of corporate persons are diversified and not as integrated as the logistics of individual persons. Consequently, corporate persons have only some of the natural rights of individual human persons, but not all. In general, there is a tendency for 2nd level intrarelational individuals to equivocate similar things, while 3rd level interrelational individuals philosophically respects the logistical differences between species and between integrated individual persons and diversified corporate persons.

This leads to two questions. How can an atheist treat a pet like a person but vehemently claim that God is not a person? Inversely, how can an individual deal with God in a respectful, personal way and then treat other humans and creatures panentheistically of and in God in a disrespectful, impersonal way? The answers to these questions lie not in the extent that an individual is exclusively, intrarelationally on the 2nd level in restricting their understanding of *person*, but in the 3rd level expansion of our appreciation of their interrelationships with humans, other animals, and with the Almighty.

In western South Dakota there are many large cattle ranches. I had a friend who was a foreman on one of those ranches. I liked to go horseback riding with him as he checked out the herds. As we loped along, he would point out different cows, identifying each with its *personal name*, family history or pedigree, medical history, and individual personality traits. In so doing, he indicated his personal knowledge, care, and affection for each cow. Each animal was a meaningful, personal part of his life. From this experience, it became clear to me why humans use *personal names* for animals and creatures in a logical evolutionary way beyond their biological and material order. This affection breaks the closed boundary of essentialists who call only humans, angels, and God by personal names. Humans establish personal bonds with individuals who are deeply meaningful to them in their lives, regardless of their logistical nature or the level of their intelligence.

Through my participation in the Lakota culture and religion, I learned that every animate or inanimate thing should be regarded as a personal individual. Regardless of their evolutionary order, every individual has personal characteristics which can interrelate with me in deeply meaningful ways. Reciprocally, these characteristics invite me to interrelate with them in interpersonal ways. That is not easy. Just as it takes a long time to get to know and interact with another human in a personal way, so too, it takes much time to grow in personal knowledge of nonhumans through prolong interacting with different animals, plants, or rocks.

While civilians view a ship as a metal machine used for transportation in war, sailors who have lived in its belly call their ship a beloved "she." Landlubbers may not take that feeling and that identification seriously, but IPhil recognizes that the intuited identification of a ship with a loving, caring woman is analogically true. To a sailor, that ship is a real, dear person, and sailors will fight and struggle to keep her alive when her life is threatened—just as they depend on her for their preservation. A ship goes through all four phases of energy's interrelational logistic—conception, assembly, operation, and retirement—just as human persons do.

During the last century, biologists and conservationists have increasingly recognized how animals and floral species play a significant interactive role in the ecology of our world. As a result, young students and the public at large are showing increased personal interest in and care for the physical things in our environment. Humane Societies are socially 3rd level, and they are associatively pushing for expanded appropriate, personal respect toward more and more animals. In contrast, hard-headed and hard-hearted 2nd level rationalists and conceptual essentialists strongly resist recognizing the interrelational and personal aspects of the animals and things in their world. In their black/white, antithetical view of our physical world, rationalists strongly promote their own personal dignity and rights and the dignity of rights of those who are like them, but they become highly antithetical toward recognizing any meaningful interrelationship with the other things in our world. In their academic professions today, scientists, philosophers, and theologians cannot publicly recognize the personal character of anything other than humans. If they did, they would immediately be publicly censured and expelled from their logistically closed professions. Yet, on the sly or in private, many who are closely involved with the plants and animals in our world develop deep personal and spiritual connections with their subjects.

All My Relatives

I had the great fortune of living and working for twenty years with the Lakota Indians in southwest South Dakota. During that time, I led a seven-year dialogue with the Lakota medicine men and Catholic pastors of the Rosebud Reservation.[5] We became close, interpersonal friends. During my participation in Lakota cultural and religious activities, one repeated statement stood out. At the beginning and end of every public speech, prayer, and ritual, the Lakota participants always say, *Mitakuyas'in*, "All my relatives." This expression includes not only their immediate blood relatives but all Lakota. In addition, depending on the spiritual maturity and vision of the individual, the prayer and acclamation, *Mitakuyas'in*, includes humans of every race, all animals, all plants, all rocks, *Tunkasila WakanTanka* (Grandfather Almighty) above, and *Unci Maka* (Grandmother Earth) below. In this way, the Lakota spiritually recognize that we are related to everything in our world not only in a materialistic way, but in a familial and social way. The Lakota recognize that everything and everyone has the capacity and the spiritual calling to interrelate positively and freely with every other thing in a respectful, spiritual, and personal way—as my relatives. Of course, one's closest relationships are with one's blood relatives. Yet, the Lakota recognize each reality as personally, spiritually, and meaningfully interconnected in the life of the Lakota people.

What does interrelational philosophy (IPhil) think about this? IPhil recognizes that through the Logical Tree of Evolution, all species and all individuals are energetically and logistically related, both in material, energetic content, and in spiritual, logistical form. The logistic of every evolved individual inherited from the Almighty has the fourfold metaphysical characteristics of enduring existence, intrarelational individuality, interrelational sociability, and true self-initiative. However, for a receiver to know experientially the personal potentialities of anything, the receiver must be open and recognize the capacity of every species to interrelate with us in an associative 3rd level way, and in a progressive, caring 4th level way. Unfortunately, too many urban humans are so busy with their own self-centered affairs that they do not take the time simply to sit and receive other species in a personally caring way.

5. Stolzman, *The Pipe and Christ*, 206–10.

Just as it takes a human a long time to trust another human enough to be open and interpersonal, so too it takes nonhuman individuals a long time to trust, be open, and be interpersonal with a human. Exceptions are dogs and horses, for their close associations with humans go back centuries.

In contrast, some enthusiastic people want to embrace everything in an equal personal way all at once. That is an adolescent fantasy. This idea usually is a rational head-trip. To be a good relative to all, one must be respectful and physically helpful to the members of one's immediate family first, then to one's extended family, then to one's friends, then to one's neighbors, then to one's community, then to all humans, then in a gradual way to all living animals and plants, and then to rocks and waters. No wonder it takes a lifetime to reach the level where the prayer and acclamation, *Mitakuyas'in*, "All my relatives," embraces everything in our world. People can start small by having in their homes some animals and plants which they can care for deeply. Others can start small by taking the time to sit and enjoy the outdoors in an open, receiving, interpersonal way, where communications can go both ways. Quietly walking through the woods give many people a very deep sense of union with all of nature in one's soul.

In primitive societies, people live a simpler life, and they can afford to have extended periods of time very close to the things of nature. Unfortunately, modern, harried humans in professions and in businesses regularly put their possessions into numbered, impersonal groups to keep them well-ordered and controlled in a 2nd level way. As we landscape around our houses, we often act as taskmasters, forcing all plants to grow only in the way we demand, paying little attention to the individuality and personality of each plant. The Japanese practice of *feng shui* is sensitive to and respects the impact of all items in a room or a garden on each other, thereby producing an environment that is interrelationally harmonious and peaceful.

By contrast, studies in physics, psychology, and sociology often regularly disregard variations from the norm. Most scientists gather individual things in samples large enough that the results can be shown to be "statistically significant." Once experimenters have a statistically significant finding that is not random, they are satisfied with an empirical, mathematical, statistical description of some general pattern of behavior in the group. They simply accept the statistically significant results, and they do not explore the possibility of personal, free self-initiatives of the individual participants apart from that normal distribution. They regularly ignore all evidence and any thought that some departures from the norm may be evidence of some participants' personal self-initiatives.

Darwinian evolution is founded on the application of statistical ideas to large samples. IPhil in the logical evolutionary theory inversely recognizes the special role of a single progenitor, capable of binding with a few members on the antecedent system and leading hem in a new way toward possible evolution. Here, a progenitor strives singularly to better himself and then to better a small group of others in an advanced way from the antecedent logistic and norm. History shows that some of the greatest discoveries come from paying attention to unique, individual differences from the norm. Those variants could be from a progenitor whose inner potential is pushing toward the systemic, interrelational benefit of a truly new, logical evolved system. Is that free, self-initiated, advancing, beneficial, interrelational behavior personal?

Putting on a scientifically "professional" impersonal mindset is thought to be objective. But IPhil finds the statistically significant mindset in statistical studies to be materially prejudiced against allowing self-initiating members within a group to express themselves freely for the sake of new interrelational and logical evolutionary development. When 2nd level tight boundary conditions are imposed upon an experiment to keep the sample well-defined and controlled, all that is 3rd level personal is suppressed, and what is uniquely personal cannot be observed. IPhil recognizes that there is a right time and place for well-defined, impersonal, 2nd level, objective intrarelational investigations of individuals. However, by not getting involved personally in an interaction, the observer suppresses the emergence of new possibilities that arise from the interpersonal, community-oriented side of nature.

IPhil strongly advocates that there is a right time and place for open, interactive, 3rd level, interpersonal discoveries. Fortunately, some experimenters have been willing to break the boundaries of their profession's accepted logistic to view abnormal exceptions as possible logical evolutionary progenitors to new discoveries and perhaps an advanced, evolved system. When that happens, the new discovery, invention, or mutation becomes "their baby," and they treasure it and nurture it, often secretly, in a deeply personal way.

Fortunately, modern life does have areas in which we are willing to break from the professional, dictatorial constraints of the status quo. In the privacy of one's home and with confidants in professional "bull sessions," we are encouraged to share what is dissociative and extraordinary because we are surrounded by other accepting, open, and supporting individuals. There, we let our friends express their individual personalities by their own free self-initiatives and by their spontaneous personal responses. It is interesting that in this situation many people start talking to their pets and plants. Even though pet owners speak a logistically different language, they claim that at times they are really communicating with their pets in nonverbal, interpersonal ways. The popularity of various "animal whisperer" television programs displays the abilities of certain humans to have in-depth understanding of and communications with different animal species. Caregivers soon no longer consider themselves merely the animals' "owners," for they incorporate them in personal ways into the network of their families. Even though materialistic scientists, academicians, and insensitive individuals may mock such supposed interpersonal relationships, pet caregivers know there is something deeply personal, loving, and spiritual going on between them and their pets.

PART III: DEVELOPMENTS

St. Francis of Assisi is embraced worldwide because of his love of animals. But most people do not recognize the great depth of his personal respect for animals and all creatures. St. Francis loved animals and creatures not only in themselves but also because they image God. His famous Canticle of Brother Sun and Sister Moon begins:

> Be praised, my Lord, for all your creatures.
>
> First for brother sun, who makes the day bright and luminous.
>
> He is beautiful and radiant with great splendor.
>
> He is the image of you, Most High.[6]

For Francis, being an image of the Most High is not a passive aspect of the sun but an active and advancing one. He recognized that the sun is analogous to God. It is not God but the sun "who makes the day bright and luminous." The lines that follow express what is called "mirror spirituality." In it, not only humans but all creatures are analogically images of their prime image: their Creator. St. Francis and many others recognize that all the animals and creatures—Brother Sun, Sister Moon, Brother Fire, Mother Earth, and so forth—are created and evolved from the Almighty. They thereby inherit and manifest interrelational characteristics of the Almighty in their own special ways. St. Francis's designations of them as persons is more than metaphorical. By the evolutionary creative process, their designation as active persons comes from the Almighty and is analogously true—albeit in a partial way. So, our respect for the things in this world should be more than materialistic; it should progressively become interpersonal and ecological.

In IPhil, the Sun and every creature are images of the Most High, because the logically evolved logistics of PAER, PAIS, and the panentheistic all-receiver (Logos) are at the logistical core of each creature, motivated existentially and interrelationally by the potential energy of the forward pushing, all-communicator Spirit. Still, each creature is individually distinct in its evolved interrelationally logistic, and by the "letting go" mercy of the Almighty is capable of acting freely within that logistic. Within that logistic every creature can love its own and others in increasingly personal ways toward enlarging one's peaceful, mutually caring family.

The root of the word *ecological*, "eco-," is from the Greek *oikos*, meaning house or home. We care about and support ecological concerns because earth is our *home*—not only for fellow humans but also for all related, fellow creatures in our world.

By the evolutionary process, there are differences and similarities between a human person, an animal person, a material person, and an absolute Person. However, while each has a *different* interrelational logistic, by logical evolution each is activated by the *same* interrelational potential energy, which originates in the Almighty through the big bang. IPhil recognizes that the notion of *person* is not a static, idealized, univocal concept. IPhil recognizes that the logistic of the Primal Absolute Potential Energy

---

6. Francis of Assisi, *Canticle of the Sun*.

is at the core of all evolved finite individuals, climaxing in humans. The potential energy of graced, receiving individuals provides each with the metaphysical opportunity (1) to physically exist, (2) to be a self-aware, integrated reality, (3) which can have mutually supportive interrelationships with others, and (4) *can* self-initiate either automatically or to some degree deliberately toward others in interrelational, progressive, and even personal ways. The more interrelational an individual is with another reality, the more interpersonal that relationship becomes, and the more the other reality is recognized as a person in their own way.

Eating in a Broadly Personal World

This leaves us with a problem. We have no difficulty eating some*thing*, but we find it repugnant to eat some*one*. Youngsters in 4-H often face the dilemma of seeing an animal that they have raised in a caring way, being sold after competition to a slaughterhouse for a meal on someone's table. Nonetheless, as they grow up, they are gradually able to deal with this dissociative experience as an associative, understanding, and accepting adult.

In Jainism, Hinduism, and Buddhism, however, followers are strict vegetarians for several reasons: They believe in the possible reincarnation of relatives into a nonhuman animal. They follow the principle of *ahimsa*, which is nonviolence toward every living thing. For these reasons, they eat no meat.

In contrast, IPhil has an evolutionary understanding of the sacredness of eating anything. IPhil recognizes that in the logical evolutionary process, every interrelational evolved system needs to take in material elements from antecedent systems to endure and grow. By doing this, the elements taken from lower orders become consumed participants in the higher, more interrelational order of the consumer. In this way, logical and physical consumption of lower species advances the interrelationality of what is consumed in the advanced, logically evolved consumer.

Biologists call the consumption of individuals in a lower biological class by individuals in a higher biological class, the "food chain." That is what is observed, and that is what IPhil's logical evolutionary theory explains and supports. In the consumption of lower species by higher species, the deaths of individuals lower on the food chain is indeed dissociative and therefore difficult and painful to watch. Nonetheless, IPhil recognizes that the painful deaths of creatures lower on the food chain is only temporarily dissociative. In its immediate consumption, what is consumed or used quickly leads to integration and advancement of the lower individuals into another individual or thing which is more interrelational. Unfortunately, many people just focus on the dissociative deaths of things in the food chain, and they fail to see how a dissociative death can be part of an associative advancement to a higher form of life. Self-centered, 2nd level essentialists have no tolerance or acceptance of any form of physical evil, especially death. They hold onto the dissociative, evil images associated with killing and eating, and they cannot get past that. However, on the 3rd level, IPhil tolerates some dissociative physical evils—as long as they are temporary and directed toward a higher level of interrelational existence. We not only learn intellectually from our dissociative mistakes; we also grow from our associative consumption of the plants and animals we kill. This is how progress takes place in our evolving, energetic world.

However, some deaths and destruction of creatures are *not* directed toward what is more interrelational. Useless killings and destructions of resources are contrary to the interrelational logistic of the Almighty. They are immoral and sinful before the logistic of the individual as well as the overarching logistic of the Almighty. Like Inuit people who eat and utilize every part of the whales and seals they kill, we need to treat all animals and matter as sacred, as they bestow on us their blessings. Consequently, humans are inwardly pushed by the interrelational Spirit of potential energy to find wastefulness repugnant and distressing. But we feel joy when we find ways to use all by-products in some advanced interrelational way. Conversion of waste and trash into useful, beneficial products is logically evolutionary and beneficial to our world. We need to advance interrelationally everything that is mined, cut down, or killed, in thoughtful, efficient, and personally caring ways.

IPhil recognizes the sacrificial character of this process in which there is a material loss at a lower level for a spiritual gain on a higher level. Many people object to the notion of "sacrifice," usually because they are shortsighted. They focus on the dissociative pain of the initial loss and don't let that go. They fail to push through that dissociation to rejoice in the higher associative gain. In baseball there is a "sacrificial bunt." When there is a player on first base, the batter can hit a grounder in the infield, usually as a bunt. That grounder gives the player on first base enough time to advance to second base, but not enough time for the batter to reach first base before being thrown out. Here, the batter experiences the loss of getting on base, while the other player is now on second base and in position for getting home and possibly scoring a run if a following batter hits only a single. In this way, in a "sacrifice bunt," the individual batter "takes one" for the future good of the team.

The word *sacrifice* is from "sacra + facere" (holy + to make) = to make whole/holy. A sacrifice advances a 2nd level dissociation into a 3rd level association and possibly a 4th level transcendence. In eating something, what dies on a lower level is logistically incorporated into the higher logistical level of the receiver. In IPhil, the sacrificial process first occurred in the primal absolute order when the all-giver "lets go" of all interrelational energy to the all-receiver, reducing the all-giver's ownership of interrelational energy to zero. In that process, the all-giver energetically "dies" so that the all-receiver may energetically "live." Furthermore, the all-giver then becomes open to receive all interrelational energy back from the all-receiver-giver and thus becomes an all-giver-receiver. By the subsequent completion of the absolute energetic interchange, the reciprocal Absolute Potential Interrelational System (PAIS) becomes a logically consistent system, capable of logically evolving our finite world. So, in the

2nd stages of the absolute interrelational logistic, in dissociatively surrendering of all potential energy to the all-receiver, the all-giver and then the all-receiver-giver "takes one" for the PAIS team. By the successful reciprocal sacrifices of the all-giver-receiver and the all-receiver-giver, both surrendering participants in PAIS become systemic winners, and the subsequently evolved finite world becomes a receiving player in the universal interrelational game of finite existence and life.

The universal practice of prayer before meals recognizes that there is something deeply sacred about a meal. Different people and cultures understand this sacredness in their own way. Beyond thanking God as the primal source of the food we eat, IPhil recognizes that it is fitting to recognize and thank all the creatures who have sacrificed their life that we may live. American Indians still make tobacco offerings toward a red willow plant they are going to cut for making Indian tobacco, or toward a deer before the hunter kills it as food for the hunter's family. These food prayers and offerings are reciprocal 3rd level gifts for what we are about to receive. In 2nd level societies, however, only God matters, and all intermediate, instrumental individuals are ignored because of their comparative lack of significance. Going beyond that, on the 3rd level, the place and importance of all finite and intermediate terms are recognized. The quiet, prayerful time before eating allows for deeper recognition and thankfulness toward all the creatures whose deaths foster our life. Their materiality and sacrifice are raised up, recognized, and made holy by our consuming them for our (and their) higher interrelational life. Recognizing that, thoughtful consumers will not waste any food or resources and will be motivated to make the most of our activities, which are for our personal welfare, the welfare of others, and the glory of those who gave their lives that we may live.

Different Judgments about God

Regarding disbelief and belief in God, most people tend to fit into one of four categories, reflective of the fourfold characteristic of the interrelational logistic.

1. *Agnostics* say they are unsure whether there is or is not a primal absolute who is a person because they lack physical evidence. All human physical experiences and conscious knowledge are limited and finite, while God is absolute. Furthermore, since the Almighty is Primal Absolute Potential Energy (PAPE), and potential energy cannot be directly observed, physical experience of God is impossible. Consequently, scientifically it is impossible to establish directly from our limited and physical experiences that God as PAPE exists or is a person. So, from a strict physical, materialistic point of view, it is proper to say, "By physical evidence, I don't know whether God exists or not, much less whether God is a person or not."

2. *Atheists* firmly assert that there is no God. All things they experientially know are material and finite. Consequently, they proclaim with scientific certainty that there is no all-powerful, all-encompassing, spiritual, metaphysical God. Atheists

do recognize evidence for the big bang, but their understanding of the energy behind the big bang is framed as a mathematical algorithm, which expresses the postulated unified field theory. They dismiss the question as to where this primal algorithm and this primal energy came from in the first place. For them, physical laws operate only mechanically and materially, and not personally or spiritually. So, the notions of "person" and "spirit" are judged to be imaginative creations of naive humans. Striving to maintain strong boundaries around their highly logical, materialistic world, they maintain that physical and material evidence is the way to true knowledge and the *only* way. Everything outside of the finite, material boundaries of their physical world, like God, is a fiction and folly.

3. *Deists* recognize there was an Almighty God from whom came the energy with its algorithm that caused our world to develop. Deists further believe that once our physical world was created, its physical development was not under the aegis of the Almighty but strictly under the control of the local physical entities. These are described by the equations of the physical sciences. Their primary focus is on discovering the interrelational physical laws of our universe, rather than on determining metaphysically how they came to be, or how they are still energetically supported by the Almighty. In this mindset, Deists judge all miraculous actions attributed to God to be contrary to the laws of nature. For them, our world is a closed mathematical system, and its originating God is efficiently outside of it.

4. *Theists* hold that the Almighty is a personal reality who can act deliberately in a well-ordered, reasonable way. Furthermore, God can and occasionally does communicate with humans, and these personal revelations give reassurance that God is personal. IPhil further recognizes that the potential energy of PAPE freely pushes everything in our world systemically forward, starting with the big bang, thereby establishing a well-organized corporate universe, according to its logically evolved logistics. These come by logical evolution from God who as Absolute potential energy—like an encouraging, guiding parent behind the scenes—remains as the potential energy foundation of all that is kinetic in our world. The logistic of the all-receiver and the *Logos* of the all-receiver-giver guides all finite receivers, who are pushed by the energizing communication Spirit, guide, develop, and evolve according to the interrelational logistic of the potential energy of the all-giving Almighty.

Rational Arguments for God as Person

Here is a quick rational argument for establishing that God is a person. I recognize myself to be a person with respect to myself and my peers. The thing that makes me a subjective, spiritual person more than an objective, material individual is that I care about myself and others in a potentially benevolent way. Because of the causative and panentheistic character of the Primal Absolute Potential Energy (PAPE), the Almighty contains all energetic potentials. If I recognize myself to be a person with respect to myself and my peers, and if I am a person panentheistically *in* God and *of* God, then God who contains all my potentialities must be to some extent personal. Then, since being a person is not just a partial characteristic of me but a characteristic of me as an integral individual, in an integral way God must be a person. That is, if I am a person and I am panentheistically *of* God, then God is, to some extent, personal and therefore a person.

Inversely, if the Almighty is the potential energy source of all things, and if the Almighty is not a person, then there is no possibility for Primal Absolute Potential Energy to be personal. If the Almighty and PAPE are not personal in whole or part, then by logical evolution I would not receive the potential to be a person. However, I know from my experiences that I am a free, caring person. Consequently, the proposition that the Almighty does not have the potential to act as or be a person must be false. As the ancient adage says: You cannot give what you do not have.

Another argument: Looking at matter from a metaphysical point of view, potential energy pushes to be more interrelational. To do that, interrelational energy must have the four metaphysical characteristics of: existence, self-identity, interrelational logistic, and self-initiative. One's self-initiative occurs and operates within one's interrelational logistic and is directed toward greater interrelationships, which are beneficial to all with whom I interrelate. If existence, self-identity, knowledge, and free will are the metaphysical prerequisites of being a person, then by the principles of the logical evolutionary theory, not only is the primal absolute personal, but every evolved and existing reality will have the fourfold metaphysical logistic and physical capacity to interact in a personal way, according to their respective received, evolved logistics.

In humility, we must recognize that our philosophic knowledge of God, or of myself, or of anyone else as a person is not complete but only partial. Also, because the 4th transcendental faculty is the ability to go beyond one's current logistical state, the internal movements of love or hate of another individual cannot be directly known. They can only be secondarily inferred from observing the external, physical actions of another individual. A couple may be married for years, but at times a spouse will be surprised by some new behavior and says, "I never knew he/she had that in him/her." Or a spouse can turn to the spouse sleeping alongside and ask, "Who is this person really?" Similarly, IPhil recognizes that in lieu of the dissociative aspect of all free interactions, it is impossible to know as philosophically true the proposition that the Almighty is really all-might and personal eternally.

The nature of space, the game of catch, and a human person are metaphysically spiritual realities beyond the material terms of their intrarelational constitution. When a person reflects on the free will that keeps a game of catch going, that interactive system is beyond our physical observation and an awesome spiritual mystery. It is mystery because the metaphysical aspects of its constitution are only secondarily known. Because of an individual's lack of total experience of one's potential, every physical individual does not have full conscious knowledge of one's own individual logistic. Because an individual does not know their own full logistical potential or the full potential of any individual; every individual and a free, caring person is a *mystery*—including the Almighty and oneself.

Materialists say that the events in our world are not from God but happen only randomly by chance. Let's pause and think about this a bit. All free, random kinetic events are empowered by some form of kinetic energy. An individual's free self-initiatives arise from the individual's potential energy. An individual's potential energy is received, and it is a part of the potential energy that is the existential foundation of our world from the kinetic big bang. Phil recognizes there is more than this. By conservation of energy the big bang was not a solitary action but a singularity. The logical evolution of the antecedent Primal Absolute Potential Energy produces our finite world in a free, altruistic way. The big bang, then, is evidence that the primal Almighty is a personal reality that is progressively developmental and altruistically supportive of our universe and everything in it.

IPhil pushes people to several things. (1) Be open to the Almighty's potential goodness that through the big bang is the foundation of all kinetic events in our world. (2) Recognize the Almighty's individualizing goodness in our world. (3) Be thankful for the Almighty's push to make goodness interrelational in our world and in our lives. (4) Be confident that there will be more and better goodness from the Almighty in the interrelational, evolutionary advancement of ourselves and our material-spiritual world.

Impersonal and Personal Aspects of God

The Lakota refer to the Almighty as *Tunkaśila Wakan Tanka* (Grandfather-Most Powerful). *Wakan Tanka* is often translated as "Great Spirit." In Lakota, the substantive adjective *wakan* means awesomely powerful, and *tanka* means "great." So, *Wakan Tanka* is better translated as the "Great Awesome Power," which is coldly impersonal. The first term, *Tunkaśila*, however, is translated "grandfather," who is a warm, endearing, close relative who loves us greatly and whom we deeply revere in return. Thus, the traditional Lakota description of the Almighty has both a powerful and a caring aspect, like the two aspects of potential energy.

It is common for religions to describe the Almighty with two terms, one impersonal and the other personal. In the Hebrew tradition, the Jews refer to the "Lord God." In the Jewish religion, the special font for Lord refers to the Tetragrammaton, which out of respect is never spoken and is translated as "I AM." The title "I Am Who I Am" is a duplication and emphasizes to the Jews the real existence of their God, who with power led them out of Egypt. The title "Lord" in Hebrew also is a common name, referring to anyone with responsibility over many things, like a farmer who overlooks all in a benevolent way. In English, the title "Lord God" refers to the distinguished power and caring oversight of the Almighty.

Christians speak and pray to "God the Father." The title "God" indicates God's dominion and power over all things. More intimately, Jesus addressed and talked of God as his *Abba*, the affectionate name given by a child to his caring "Father." Some have said that there are two different Gods in the Bible—a wrathful God in the Hebrew Scriptures and a compassionate God in the Christian Scriptures. However, the impersonal and personal aspects of God are found in both sets of Scriptures.

In the Qur'an, Allah usually speaks to Mohammed using the personal, intimate pronoun, *I*. In giving decrees to the people, however, Allah spoke using the "imperial plural," *We*. In Daoism, the yang/yin expresses the strength and tenderness complementarity found in the Dao. Similarly, in IPhil, the Primal Absolute Potential Energy or PAPE before the big bang has *Existential Being* as its founding gentle aspect, and its potential has *Forward-Leaning Becoming* in the dissociative physical formation of the things in this world.

The image of God as an imperial, judgmental, demanding deity is strongly emphasized by antithetical youth in the 2nd level phase of their development. The image of God as loving and caring comes from an adult, 3rd level interrelational view of the absolute. Which is true? They both are, not in a diametrically opposite way but in a dialogically progressive way. The associative phase of Christian love does not replace the dissociative phase of Hebrew separated holiness. In a logical evolutionary way, the Christian logistic carries over the aspects of the Jewish law and their understanding of the Lord God in a way that is said to "fulfill" them.

One of the big questions a person asks is: Who am I? The verb "am" in this question indicates my energetic existence. Being of energy, my identity has fourfold aspects. (1) Potentially, metaphysically my logistical soul panentheistically has been a "part" of the universal potentiality of the Almighty from all eternity. (2) Individually, my intrarelational integrity establishes my enduring self-identity, which I know intuitively and consciously as "I" or ego. (3) Socially, I am a member in many different organizations, of which my family is the first. In these communal settings I am addressed with different public names and identifications. (4) By logical evolution, I am a relative of all things. So, I am an "interrelational center" with all finite things, and logistically analogous to the transcending and unifying Almighty. The more interrelationally I know who I am, the more I experientially know who God is.

Prayer in IPhil

IPhil recognizes that the Almighty and humans are energetically interrelational. Consequently, there must be some form of communication between them. Usually humans experience communications from the Almighty via the metaphysical Spirit of the Primal Absolute Energetic Reality (PAER). The Spirit often in a nonverbal way urges the individual toward what is more interrelational. In a complementary way, the hoping individual communicates responds nonverbally and verbally to and through in the Spirit toward some greater potential good. The forms of interior prayers and efficient sacrifices a human can make must be according to the logistic received from the Almighty. In prayers expressed in faith, believing free individuals and communities open themselves to receive higher level of energy from the potentialities of the free Almighty, as well as expressing hope for the actualization of some energetic potentials. IPhil recognizes that prayers expressed orally and physically, especially in a communal setting, are interrelationally more advanced than prayers said mentally and from the heart in private. Yet, each is valuable and has its own time and place

Strongly influenced by the Age of Enlightenment, George Washington and political contemporaries like Thomas Jefferson were deists who believed in the impersonal God of Nature. They nonetheless prayed for the improvements of the providence that comes from the Almighty. Norman Vincent Peale wrote, "Personally, I believe that prayer is a sending out of vibrations from one person to another and to God. All the universe is in vibration. When you send out a prayer to another person, you employ the force inherent in the spiritual universe."

In a multi-sectarian society, it is more appropriate that public prayers not be sectarian but general and universal, so that each participant in that communal event can participant fully in that group prayer. These prayers would be based not on the tenets of particular religions but on the universally shared energy that is found in nature and in the Almighty. Some may wish to use the founding tenets and conclusions of the interrelational philosophy as the foundation of a trans-sectarian, universal religion. Nevertheless, just as individuals and groups can advance from impersonal communications to interpersonal communications with other humans, so too, communications between the potential, spiritual Almighty and the potential, metaphysical aspects of humans and the community can advance from impersonal nonsectarian communications, to personalized intercommunications, and even to logically evolve into some kinetic, physical transformation or event.

Chapter 19

Religions

Homo sapiens. Homo religiosus. Faith and Reason. Religion Goes Beyond Physical Science. How Spirit-Focused Beliefs Arose. Religions in History. Characteristics of Religion. Religions in the East and in the West. Theology and Nones. Dealing with Different Religions. Images of God.

Homo Sapiens

Of all evolved physical things and biological species, humans show themselves to be the most interrelational and most intelligent. While birds and insects ingeniously create nests in the most precarious of places, humans can construct much larger homes and buildings. Animals can distinguish an enemy from a friend, and they vocally and nonverbally warn their peers of mutual enemies and guide their peers to food and beneficial resources. These behaviors show they are capable of some level of abstraction and generalization. Nonetheless, their conceptualizations and their corresponding communications have a vocabulary that is much smaller than that of humans. Bears claim their territory with claw marks upon trees, and dogs and wolves mark their territories with urine. These are forms of writing, but the range of their ownership is small compared to humans, who claim not only physical property but also intellectual property. Rats, squirrels, and other animals can find their way through complex mazes and solve simple problems, but not at the level of human beings. Some knowledge, reasoning skills, and communication skills must be present and active in all physical species. So, physical evidence shows that being *sapiens* (wise) is not the *essentialistic* distinction between animals and humans. What is?

Homo religiosus

Neanderthals lived 100,000 to 70,000 years ago. While some animals are recognized to grieve the death of their companions and their children, those expressions of grief last only a short time. In contrast, Neanderthals at times showed that their grief was long-term and transcendental. They often marked their buried deceased with red ochre or vermillion. They often surrounded them with flowers. The quality of these gifts points toward some type of metaphysical belief in an enduring afterlife state that is better for the deceased.

Roughly 11,000 years ago, at Göbekli Tepe in southeastern Turkey, a temple was constructed.[1] This temple complex has large circular structures. There are numerous T-shaped monolithic pillars of limestone that stand up to ten feet high. A pair of bigger pillars was placed in the center of the structures. Carved into the pillars are reliefs of foxes, lions, cattle, hyenas, wild boars, herons, ducks, scorpions, ants, spiders, snakes, and only a small number of humanlike figures. The dates of this and other unearthed worship sites indicate that in many places around the world, the building of worship complexes preceded the building of cities. Astronomical characteristics built into these worship centers indicate at least annual gatherings of large numbers of nomadic people. The size of these worship centers indicates the coordinated efforts of many clans. Such physical structures indicate some awareness of spiritual realities by large groups of nomadic humans. The physical remains of many worship sites show that the establishment of human religion is part of how our world evolved from God.

The worldwide evidence of these mega-worship sites indicates that humans are uniquely different from all animal species in their religiosity. Physical evidence of the religious orientations of humans from ancient times is found in Stonehenge in England, Notre Dame Cathedral in Paris, St. Peter's Basilica in Rome, the National Cathedrals in Washington, DC, Mayan temples, the Aztec temples outside of Mexico City, Shinto Shrines in Japan, Buddhist and Hindu temples in India, the Kaaba in Mecca, and various centers of prayer and worship in every population center on earth. In contrast, there is no evidence of animals creating such sites. Today 80–85 percent of the world's population believes in some form of spiritual realities, especially some type of energetic absolute or Almighty reality. However, in today's anti-religious, secular, rationalistic professional world, many scientists and intellectuals deliberately dismiss or reject that spirituality and religion are an essential aspect of human life—often because they are not in theirs. Nonetheless, religious art, literature, and rituals throughout history and throughout the world are parts of how our world evolved from God. That is why religion is a part of this book.

Belief in some kind of "Higher Power" is worldwide. To believers, their faith is not a figment of their imaginations but very real. But are their faith experiences true? Truth is the conformity of internal knowledge with external events. But *finite material* reality cannot provide evidence of the *absolute spiritual* reality of God. Verbal and written statements about one's inner experiences and beliefs are secondary. Texts written by spiritual sages and prophets speak of things that cannot be physically verified after the fact. Yet sacred scriptures have guided the faith and devotion of millions of followers. Although secondary and tertiary in their testimony, they play a significant part of human history and culture, and they need to be studied seriously because

1. Curry, "Gobekli Tepe."

their enduring logistics indicate some level of logical consistency and some level of conformity to our physical world. Their endurance is found not in physical details but in metaphysical ideas and processes.

There are many variations and disagreements in religion, and critics use these difference and dissociations as evidence of the falsehoods of these religions. However, through the centuries there have been many erroneous scientific hypotheses that have been put forward as true, until demonstrated otherwise. Why is it that erroneous ideas have been soundly and repeatedly condemned in religion, but erroneous ideas in the sciences has been ignored as part of the scientific process? That is a common behavior in closed 2nd level thinking and communications. Professional people who studied religions, religious statements, and rituals from around the world recognize common themes and characteristics, which indicate a high probability of truth in these statements.

Comparing the lives of atheists or believers, evidence shows that believers are more generous and promote in organized ways more interrelational goodness and peace in this world. Individually, many atheistic humanists are caring and generous, but communally they lack logistical coherence, and they do not have as great, positive, or enduring impact on our world as do institutional religions. The modern study of comparative religions has facilitated greater understanding of religious matters. If one applies the evidentiary criteria, "By their fruits you shall know them," the transcending aspirations of believers in religions stands out, and they deserve recognition and serious study.

In summary, scientific evidence shows that animals have some ordered knowledge of the physical world they live in, but there is no evidence that animals have any knowledge, behavior, or physical structures oriented toward something beyond their immediate physical world. While the scope of human physical science is truly great, it is religion that frequently pushes humans beyond their current physical state toward a better, interrelationally caring life, and toward some type of union and interaction with what is transcendent to our physical order. Attempts of humans to interrelate with some type of spiritual realities are truly unique to humans. The physical evidence of worship sites and rituals shows that humans are more than *Homo sapiens*. Worldwide historical evidence shows that uniquely, humans are *Homo religiosus*.

In atomic physics, not all interactions are directly observed. Neutrinos have no electric charge, and they cannot be electromagnetically observed. Yet physicists recognize their existence secondarily by the impacts they are able to observe on other particles. In a similar way, potential energy is always recognized secondarily. Here the criteria for truth advances from 3rd level direct conformity of external phenomena to interior ideas, to the 4th level ancient maxim: "By their fruits you shall know them." It is impossible to know that the absolute Almighty caused something because a finite knower can only know in a finite mode. Nonetheless, the fruits of spiritual actions by the Almighty in extraordinary events in the physical world and in people's lives do point in that direction—not definitively but indicatively.

Faith and Reason

It is common in our materialistic society to antithetically oppose religion and science, faith and reason. In contrast, IPhil synthetically interrelates and progressively orders the two. IPhil sees the relationship between faith and reason, as well as the spiritual and material orders, as analogous to the relationship of potential and kinetic energy. On the 1st level, an individual's sequencing of perceptions depends upon believing the accuracy of those perceptions. On the 2nd level, prior to an individual's connective self-knowledge, the individual must trust the intrarelational character of that knowledge. On the 3rd level, an individual's conscious logical reasoning is founded upon one's faith in induced premises. On the 4th level, belief in the probability of an advanced afterlife is the foundation of logically evolved theology about a transcendental afterlife. Thus, in every stage and level of cognitive development, faith in the accuracy of some insight precedes the subsequent ordering process of reasoning with those insights. Those insights can be obtained through logistic carryover from one's ancestors, from personal experience, by communal instructions, and/or by transcending inspirations.

Nonetheless, in the physical order, 4th order transcending inspirations will regularly and justifiably be opposed as logically inverse to 3rd order institutional and scientific norms or law (*jus*). However, rather than establish a wall of separation between church and state, IPhil advocates the application of the principle of subsidiarity, where their distinctive interrelational domains and competency are respected and logistically ordered.

Religion Goes Beyond Physical Science

In the logical evolutionary process, a logically consistent system built upon a given set of axioms is closed, but it is also incomplete. Only by breaking the boundary of a logically consistent system can a new, advanced system be evolved. IPhil recognizes that religion is a new step on the path of evolution beyond strictly material reality. That is why it is impossible to extend deductively what scientists have established from the physical evidence of kinetic energy in lower material orders unto the metaphysical, transcendental aspirations found in spiritualities and religions throughout the world. IPhil admits that human discoveries and understandings of the religious and spiritual evolution of the human logistic are often chaotic. So too, children's 1st level view of their world and their early exploration of this world are quite chaotic. But that does not stop the child's pursuit to know more truth about our world in an organizing way. Similarly, the advancement of scientific knowledge has not been linear but quite chaotic, with many proposed theories proven wrong. We need accept that human advances in the sciences and in religions has been slow. Despite past intellectual failures in the physical sciences, medicine, social theory, religion, and spirituality, we need to respect their efforts and be slow and careful to accept their findings as true to exterior reality both in the physical and metaphysical manifestations of kinetic and potential energy.

Unfortunately, idealistic, essentialistic, antithetical, 2nd level skeptics expect that religions and what they say pertaining to the Almighty to be perfect and unchangeable. Antithetically, materialists also regularly discredit and disregard religions as unreal because spiritualities and religions are at times destructive because of their closed-mindedness toward other religions and other intellectual ideas. These critics ignore the ways that physical sciences at times have been significantly destructive of human life. In contrast, synthetic 3rd level individuals recognize and tolerate the material imperfections of our world. Also, they break through past closed paradigms to recognize in an evolutionary way that there are things in our world that are not materialistic, but they are transcendentally spiritual. The 3rd level academic discipline of comparative religions is finding many inner, interrelational similarities in different religions, whose rituals and beliefs at first external glance appear to be very different.

Some religious adherents are strongly 2nd level. Faced with the presence of many different religions in our cosmopolitan world, they dualistically assume that their own belief system has all truth and the others do not. They further insist that their moral standards are right, and others are wrong.

It is easy for adolescents to conclude that when people do not practice a religion, those individuals cannot possibly be moral, righteous, and even spiritual. Yet, Buddhists do not believe in God, and who would say the Dalai Lama, a Tibetan Buddhist, does not live a highly moral and spiritual life? In the West, virtuous atheistic humanists have claimed moral principles developed by Christianity, while they do not believe in the Christian theology of God and Christ. Scientific observers of animals recognize and have recorded the rules of social behavior within groups of animals. Rules of conduct among animals support the successful operations of their groups and the promulgation of their species. How can wolves, who do not know God, form a well-functioning, mutually sensitive, caring, and social wolf pack? Adult wolves, lions, and apes show affection and concerns for their young and their mates. Where does the moral behavior of these animals come from if they don't know God or have religion?

Energy is inherently interrelational, and Primal Absolute Potential Energy (PAPE) pushes every species to maximize enduring, well-ordered, reciprocal interrelationships on the 3rd level. By the principles of logical evolution, the well-ordered, reciprocally respectful interrelational logistic of PAIS is carried over into all subsequently evolved species in their received metaphysical logistic. In this process, the well-ordered, enduring, 3rd level communal logistic is given to all species, including humans. Nonetheless, being religious goes beyond being morally good. The spiritual and religious concerns of human are 4th level and transcendent of material concerns. Morality and religiosity are not disjoint but related in a logical evolutionary way, where 3rd level morality toward other humans and things is carried over into 4th level religious beliefs in spiritual realities and even the Almighty. Nonetheless, spiritual and religion experiences and knowledge need not be grasped by those on the 3rd level of social engagement. By the principles of logical evolution, it is expected that a spiritual person is morally good, but a morally good person may not be religious.

How Spirit-Focused Beliefs Arose

How did humans logically evolve from 3rd level moral societies, as found among animals, to transcendental 4th level spiritual religions? Archeological and historical evidence indicates that worship and religion did not begin with a belief in one, absolute, humanlike God but in many local animal spirits. As indicated above, at the worship center at Göbekli Tepe in southeastern Turkey roughly 11,000 years ago, its columns had carvings of foxes, lions, cattle, hyenas, wild boars, herons, ducks, scorpions, ants, spiders, snakes, and only a small number of humanlike figures. This raises the question: *How* did worship of animal spirits arise among early humans?

Before answering that question, I need to cover a few preliminaries. First, essentialists define religion as the worship of one God or many gods. For them, religion is this way and only this way. They view religion idealistically. Then, when evidence indicates that a religion changes, they consider that religion to be imperfect and superstitious. However, truth is the conformity of internal ideas with external reality. The physical record shows that as humans have progressively evolved physically and intellectually, so too their worship and religions have also evolved and continue to evolve interrelationally. Consequently, the *a priori* view that the essence of a true religion does not change does not match the physical evidence of history. In contrast, IPhil is developmental and evolutionary in its philosophical outlook, and the question becomes: What progressive model of spirituality matches historical religious evidence?

I begin by looking at this question from today's experiences, and then I will analogously move backwards in time. An inverse corollary of the logical evolutionary theory allows me to do this. The term "worship" comes from *worth-ship*. We worship the things that are worthwhile to us. We interrelationally attach ourselves to and worship the things that are personally valuable to us. If money has the greatest meaning in an individual's life, that individual will think and plan things around money. In this way, they worth-ship or worship money. If individuals are dedicated to sports, their minds are filled with statistics and the prospect of their favorite team winning. They worth-ship or worship their star players and consider their favorite team #1 in their lives. Watching a sports event on television is more meaningful to them than going to a religious worship service. Youth often idolize movie stars as the heroes in their lives, and "they worship the ground they walk on." There are many idols that are worth-shipped, worshipped in today's society.

Whenever a teenager is in a group of adults, that teenager is socially pushed to say and do things in accordance with the expectations of that adult group. However, bring in another teenager, and soon they will "play off each other." Through nonverbal and verbal communications, they start to interact in adolescent ways. If there are more teenagers, they soon organize into a peer group, which operates with increasing independence of the surrounding adults. Here there are many physical, material individuals, but their playful and competitive interactions soon go beyond expressed physical interactions to a metaphysical awareness of the group as a whole, more than a collection of a number of individuals. Those interactions soon take on the metaphysical, interrelational *spirit* of the group. Their 1st level actions, words, and totems represent not only the 2nd level individuals in the group but also the 3rd level interactive group as a whole, which regularly "pushes the envelope" in 4th level innovative, creative ways that are beyond standing conventions. Here, not individuals but the *spirit* of the group as a whole, pushes members to do clever, creative, and sometime stupid and dangerous things.

The bond of friendship, the bond of marriage, the esprit de corps of a military unit, of an athletic team, of a school, of a corporation, or of a nation are more than pluralistic realities. They have interrelational, spiritual qualities that cannot be physically identified but can only intuitively be sensed in a metaphysical, spiritual way. Those bonded interrelationships "chain" individuals together spiritually, and that logistical bond pushes and pulls the participating free, dedicated individuals toward actualizing their group's communal identity. That is, their communal logistic pushes and pulls upon the members of a group, just as the communal logistic of the proto-spatial system pushes and pulls on spacepoints when various forces attempt to move some spacepoints apart from their logistically specified quantum distances. Once an interrelational institution, organization, or club is established in a reciprocal, enduring, logically consistent way, it is possible for that union to produce something new—such as a child in a marriage, an invention within a company's Research & Development department, or a trophy for a spirit-filled winning team.

A charismatic person can attract many followers by his or her message into what is called a "movement." When that person dies, dedicated followers are moved to continue that movement and organization according to the *spirit* of the founder. The antithetical spirit of our materialistic age fosters individualism and rejects organized religious and social conventions. In contrast, the synthetical spirit of the logistic of interrelational philosophy (IPhil) is real and dynamic. The communal, interrelational, metaphysical spirit of a close group guides participant individuals to a fuller, positive life by interrelating disciples toward their metaphysically, spiritual communal pursuits. As potential energy guides and stimulates kinetic actions, the spirit of a group dynamically expresses itself in the beliefs and practices of its disciples. This is not imaginary, for its inner dynamic can not only be perceived mentally, it also can be felt emotionally and expressed physically. Our experience of group spirit today prepares us to understand IPhil's proposed explanation of early human experience of animal spirits and their development of spirit-centered rituals and religions. Their worth-ship rituals, which are directed toward spirits of nature are like the spirited rituals of pep rallies and like the pep talks of spirited coaches today. IPhil proposes that because of the carryover characteristic of logical evolution, the experience of "spirit" in groups today evolved from the spirit-rituals, beliefs, and commitments of groups in primitive times.

When a person looks at the archaeological evidence of the earliest worship centers, like Göbekli Tepe and Stonehenge, there is little or no physical evidence of recognition of anything absolute in them. Rather, the dominant markings in the earliest worship centers and in caves are those of the animals they see around them. This evidence indicates that the earliest humans were animists. The depictions of animal figures in cave painting and stone carvings of worship centers indicate that the central figures in their worth-ship were not the Almighty but toward animal spirits. Daily concerns of early clans, tribes, and nations were immediate within their environments. They pursued specific animals, *and* they also pursued some type of beneficial, spiritual, communal interrelationships with these animals spiritually. Even today, Lakota medicine men primarily invoke animal spirits—the white tail deer spirit, the eagle spirit, the spider spirit, etc. How did the physical interrelational experiences of humans with animals advance from individual physical encounters with material individuals to metaphysical belief in the interrelational spirits of those animals?

Just as a pep rally recognizes and augments an esprit de corps that can corporately influence a team in their group activities and competitions, so too native groups can pursue not just individual animals but the spiritual, communal logistic of those animals. An animal spirit of many is communally one. The worth-ship of pursuing many individual animals easily generalizes and transcends from a metaphysically guided pursuit or avoidance of the communal logistic of the individual animals they wish to positively interrelate with. Their ritualized pursuit of individual, 2nd level prey then is cast in terms of the 3rd level communal logistic of a species, such as a bear-spirit, rather than toward specific individual bears. From this perspective, the animals painted on the walls of primitive caves are projective of idealized, dynamic, spiritual logistics of an entire species rather than portraits of just particular individual members of that species. What they worth-ship is the metaphysical, communal, logistical spirit of that animal, beyond particular material individuals. Here is a profound logical evolutionary shift from 2nd level to 3rd level relationship of individual animals, which are metaphysically *of* the communal spirit of those animals. Communal logistics are not only intellectually known, they are externally real. That is, there is a spiritual prototype of communal logistic that is not only imaginary in the mind but also real within nature.

Here is a crucial question for the reader: Can you make a paradigm shift from seeing reality only on the 1st and 2nd level material individualism but also on the 3rd and 4th level communal realities—as occurs in sports teams and rallies, in political parties and rallies, in the esprit de corps of military units, in the unifying spirit of a nation, especially in time of war? Is the communal spirit of each of these realities just in your mind or imagination, or is the distinct communal spirit experienced in each of these communal realities truly real and existent in our world? These communal spirits cannot be physically measured, but evidence points toward the real existence and forceful dynamics of the metaphysical, communal logistic and spirit of different groups. Furthermore, these spiritual communal realities (spirits) are not eternal. Rather, like every finite reality—here as potential energetic, spiritual realities—they pass through the four stages of logical evolutionary birth, empowering integration, extended interrelationships, and quiet demise.

IPhil maintains that communal logistics really do exist spiritually, metaphysically, and they exert a real, free dynamic force on the individual members of that logistically bonded group as well on interrelating with other groups. In the primal, absolute order of the Almighty, beyond the distinct, interdependent, individual elements of the Primal Absolute Interrelational System (PAIS) is a true, free corporate reality, not just an intellectual construct. In the human order, your physical body is more that an interactive aggregation of its parts, it is a real, free, corporate reality in itself. It is also a part of the human communal logistic, which spiritually pushes and pulls me according to the contemporary state of the spirit of the human communal logistic. Corporate persons are as real as individual persons are real. In primitive societies, prototypical communal animal spirits were seen to be spiritually as real as individual animals were materially real.

In a bank loan, the head of a corporation and the head of a bank ritualistically signs a document that metaphysically influences all the members of that corporation and bank—according to and within the communal logistic of that bank. Similarly, in primitive spirit religion, a medicine man communicates with the spirit of an animal, and that ritual act spiritually impacts the family members of the medicine man and the other members of his group. IPhil proposes that this is the way early humans came to the notion of the prototypical, metaphysical spirit of the animals in a given species.

There are not only individual deer, there is also a communal deer logistical spirit that internally forms and guides all individual deer. Similarly, a team's spirit drives the activities and successes of all the individual members of a team collectively. Recognizing and promoting a team's spirit will make that team prosper as a whole and as individual members. Recognizing and promoting an animal herd's spirit makes all the animals of that species and associated tribes of people prosper. The devotional promotion of the communal spirit of all deer is linked to beneficial interrelations with all individual deer. When a group of humans gathers in faith and hope to recognize the worth of the communal spirit of the logistic of an animal species, they believe that their own spiritual, potential, communal logistic can interrelate spiritually with the spiritual, potential communal logistic of that animal to their benefit. IPhil proposes that this is way the worth-ship and worship of the spirits of the animals, plants, and material formations surrounding them developed within primitive, tightly bonded clans and communities. In this way, IPhil breaks the closed boundary of materialistic individualism, which only recognizes material, physical individuals in the domain of kinetic energy, breaking into recognizing spiritual, metaphysical species as real, potentially dynamic elements in the domain of potential energy.

The Bible recognizes that every name signifies a dynamic spiritual reality that has power and interrelational influence (Eph 1:21). In modern society, a name on the façade of a building, on a letterhead, or the side of a plane has a public, spiritual dynamic that transcends the bricks of the building or the wings of a plane. We need to break out of the modern individualistic, materialistic mindset that limits dynamism to individual physical things. We need recognize also that all communal names represent spiritual realities that have their own special interrelational dynamisms that influence many greatly. It is frequently said, "You can fight city hall." Regardless whether that statement is true or not, "city hall" is a real corporate reality that is difficult to deal with. When a person says, "Nicky is a dog," there is a cascade of metaphysical, spiritual interrelationships and powerful dynamics that are associated with and are beyond the individual Nicky because Nicky is a member of the dynamic dog species. If you hurt Nicky, many humans will condemn you not only for hurting that individual dog but also for hurting a member of the dog species. Looking only at the individual dog, a person also spiritually, metaphysically sees the interrelational dynamic of the communal logistic of the dog's species.

Whenever two or more individuals of the same species get together, an advanced, spiritual, communal dynamic kicks in. The communal existence of spirits is recognized in advanced interrelational relationships and activities when two or more of the same kind are together and interact together in union. In this way, IPhil understands and explains the existence and the advance potential activities of "logistical spirits." They are not figments of human imaginations but philosophically are very real and significant in our world. Spiritual and religious people know that, and they interact within that truth for their own spiritual benefit and the communal benefit of their own.

Finally, Duns Scotus philosophically recognized real natures. Real natures are not just in people's minds.[2] Likewise, IPhil also recognizes advanced, spiritual, corporate, and generic realities. For example, the concept of gravitation is known in our minds, but that does not mean that gravitation is a figment of our imaginations. Gravitation is not directly observed, but it is secondarily known through observations. Gravitation is not material, but it is more than imaginary; it is logistically *real*. Gravitation involves real energetic communications between material terms. Therefore, strictly speaking, gravitation is a spirit—one of the "Powers" of the panentheistic Almighty.

---

2. Bates, *Duns Scotus and the Problem of Universals*.

Religions in History

IPhil recognizes that religion historically began (1) in the Spirit Age, in which humans in small groups sought to worth-ship the natural communal spirits of the animal, plant, and rocks around their dwellings and their local ritual sites. (2) This logically evolved in the Mythical Age into worth-ship of many self-centered, humanlike spiritual gods and heroes, who defended and interacted with city-states, cultures, and empires. Unfortunately, poets took descriptions of communal spirits from earlier smaller communities, combined them in a synchronistic way and used extravagant hyperbole to described Zeus, Psyche, Venus, Athena, etc., in myths that were entertaining but distant from real human experiences. (3) Then, this polytheism inversely evolved in the philosophic Axial Age to worth-ship of one God, the Almighty God as an absolute, spiritual, and rationally consistent. (4) In the Modern Age, humans are moving toward a personal, transcendental spirituality and union of all in the Almighty all.

The fourfold advancement of the Jewish religion is well-documented in the Hebrew Bible: (1) Prehistory in Genesis 1–11 depicts events that had pre-Abrahamic relationship with the Almighty. (2) This led to the establishment of a covenant from God with Abraham and his descendants, and the establishment of the law and religious discipline under Moses, and then to (3) the establishment of the Jewish Kingdom and temple in Jerusalem around the time of King David. (4) Then the Jewish state became diminished and scattered during and after the Babylonian Captivity. (5) They returned to the country or Israel, even as they were absorbed into Hellenistic and Roman culture. (6) Revolting against Roman domination, the destruction of the temple and government center was destroyed. (7) After much success and struggle in many countries in the world, (8) the Jewish people are returning to the promised land with its capital in Jerusalem.

Christian history likewise can be divided into four main segments: (1) The founding of the Christian religion through Jesus and the missionary work of his apostles in the Roman Empire. (2) Subsequently, there were many internal struggles in the church from exterior persecutions by the Romans and by various internal heretical group. (3) After the Dark Ages, the institutional church flourished in the Middle Ages. (4) The diffusion of Christian churches has occurred worldwide since the Reformation. (5) There is an expanded openness to work with diverse religions and cultures. Each world religion has its own historical development, which in many ways shows waves of fourfold development. But these need to be laid out in greater detail in another place.

Characteristics of Religion

The term *religion* comes from the Latin (*re* + *ligare* = to bind again). Scholars regularly note that religion has three classical marks: creed, code, and cult. IPhil expands and reorders that list to match the fourfold interrelational logistic of energy:

1. *Tradition.* Religion is first passed on (*L., traditus*) in a personal way within one's extended family and clan. Young and uneducated individuals are open to inspired, charismatic leaders, who share their spiritual experiences and their understanding of the Almighty. This is initially done through sacred stories and myths about the past. Out of respect and seeking security, lesser members hold tightly to these teachings, rituals, and traditions in a childlike, unquestioning way.

2. *Code.* As youth and religions mature, they draw lessons from those myths and sacred stories. Approved moral and behavioral expectations within the group are drawn out by teachers from the revered sayings and practices of the founders and authorized leaders of a group. A clear, legalized code of conduct is established by the leaders of that religion, speaking in the name of the Almighty and the spirit of the founder(s). Fear of the consequences of breaking these laws keeps adolescent members in line and strengthens their religious identity. Here, "fear of God is the beginning of wisdom." Following the rules promotes unity, peace, and eternal security within the religious group, making it interiorly strong and exteriorly highly defensive.

3. *Creed.* Advanced rational reflections on the myths, stories, and ritual practices of a religion lead to extensive commentaries and perhaps canonized religious statements. These credal statements promote consistency and harmony in words and deeds among all the membership of the group, and they soon become institutionalized. Traditional writings and rituals are consolidated, formalized, and canonized by the religious leaders. Using leader-approved sacred texts and traditions, they increasingly assert their guardianship and their authority in the domain of their religious beliefs and practices for the sake of keeping their religious community internally bound together and externally distinct.

4. *Cult.* The religious community periodically gathers to worship. Here, members of the gathered group publicly make their commitment to the spirituality and beliefs shared by the group. Major myths and stories are retold and even reenacted ritualistically, often annually. When the leading beliefs and practices are publicly affirmed in faith, the worshiping participants become open to receive blessings and guidance from the Almighty and from spirit(s) of the founder(s) of the religion. In these rituals, the members become spiritually uplifted in their surrender in faith to the transcendental aspirations and hope of the founder.

Just as everything energetic has an enduring metaphysical side and a transient physical side, so too each religion has enduring faith tenets, memories, and variable moral standards. Since the faith tenets tend to become logically consistent in form, they can logically evolve, carrying over past beliefs and going beyond them in different directions. The leaders of a religion try to keep this advancement in line in various authoritarian ways. Moral exhortations are pushed by the Spirit of the potential energy and logistic of a religious institution, as the religion faces temporal and environmental changes.

Religions in the East and in the West

The geography of the earth for millennia separated the cultures of the East and the West, causing the historic developments of their religions to become significantly different. Nonetheless, using the analytic tools of IPhil, it can be shown how both developed in fourfold ways that are harmonious with the fourfold interrelational logistic. The religions in the East emphasize 2nd level development of the individual. In contrast, the religions in the West emphasize 3rd level institutional authority. In this way, the religions of the East emphasize *yin* and the progressive, potential energy side of reality, while the religions of the West emphasize *yang* and authoritarian, communal, logistical structure.

In the East, each major religion/philosophy emphasizes one aspect of human development.

- Hinduism is the oldest religion and is polytheistic, although sages point out that all the images of and stories about the gods are only facets of the One God. In childlike and familial ways, they joyfully celebrate holidays dedicated to the god(s) that watch over their local communities like guardian angels.

- Buddhism inversely focuses inwardly on the individual. This profound psychology seeks to bring associative peace inwardly to distressed, dissociative individuals through the Four Noble Truths, and then externally to the individual within the community through the Eightfold Path.

- Confucianism promotes the virtue of mutual respect. It seeks to establish good order within the government and in the home by following many detailed protocols. The ideal ruler operates not with 2nd level strength and authority but operates best as a 3rd level centralizing figurehead serving and guiding society, uniting distinctive segments of society in a well-ordered, respectful way. Enduring order comes not from the top down but from socially responsible individuals, starting in the family and working up.

- Daoism is regularly practiced by individuals who in their retirement are no longer bound by the formal rules of government and Confucianism, and instead go their own way, "letting go" of general rules to perform "effortless action" according to the inner guidance and peace they draw from the imminent, transcending Dao.

Because these Eastern religions/philosophies express different aspects of an individual's interrelational logistic and life, it is common for some individuals to practice two or more of these religions/philosophies at once, acting according to one or the other as appropriate in a given situation. Note that political leaders have rarely had an enduring influence in the development and implementation of these religions/philosophies.

Nonetheless, in contemporary China, two major religions from the West, Christianity and Islam, are currently being variously suppressed by a Communist regime, whose atheistic political philosophy originated in the West.

Patriarchal and political authorities have strongly impacted the practice of Western religions—which emphasize obedience and submission to authority figures. More than acting horizontally toward others, the primary concern of religions in the West is directed vertically toward God and laws canonized in scriptures, subordinating all the Faithful hierarchically to authorized religious leaders.

- In Babylon, Egypt, and in Rome, the official polytheistic religions of the state were supported and influenced by the leader of their country. These countries regularly experience internal conflict over matters of secular and religious authority and power. Each country had its national gods or religious expectations. Their rituals were directed toward their gods as protectors, and tenets of the faith were more the concerns of leaders than of ordinary people. Refusal to worship in the official, approved way was considered antithetical not only to the national gods but also to the national leader and to the welfare of the state. Moral norms between citizens were not part of this early form of institutional religion.

- Islam is a strongly monotheistic religion. It rose inversely from tribal polytheism, under the guidance of Mohammed. The Qur'an contains the revelations that God told Mohammed through the archangel Gabriel. When different groups or individuals did not accept the tenets of Islam, religious wars and social exclusions were often waged in the name of Allah. Even today, the conflicts between Sunnis and Shiites are over who holds authority in that religion. Fundamentalists are antithetically quick to condemn or even kill anyone who dishonors Allah, Mohammed, or the Qur'an.

- Judaism advanced in monotheism. It first recognized their God as the best God, and then recognized their God as the only God. Its written guide is the Hebrew Bible, especially its first five books, called the Torah. These contain God's revelations to Moses as he militarily led the Israelites out of Egypt toward the conquest of the nations that occupied the land promised them in the covenant with Abraham. Subsequently, prophets authoritatively spoke "the word of God," insisting that the people return and remain loyal to the LORD and his law. The current flag of Israel bears the Star of David, who expanded the territory of the Israelites to its maximum, covenantal extension.

- Christianity from the beginning had to deal with matters of authority. This began with Jesus' challenge of traditional Jewish practices and his proclamation of the coming of the kingdom of heaven. His followers recognized Jesus as the anointed messiah who from his heavenly throne would put all nations under the kingdom of God. Jesus identified himself with the Son of Man, especially as expressed in Daniel 7. Before the High Priest, Jesus claimed that his elevation to the right hand of the Almighty established him as divine Son of God. Christian apocalyptic writings indicate that all nations and spiritual realities will have Christ as their Lord and ruler. In the course of history the authority of the bishop of Rome had a major part in associations and dissociations within the church.

In summary: in the East, diversity of religions is tied to the four different stages in the lives of individual humans. In the West, diversity of religions is tied to the fourfold, progressive, authoritarian history of organized religion. The antitheses of Western religions can become synthetically harmonious when they transform their view of authority from that of 3rd level corporate authoritarian dominance to 4th level servant leadership.

Theology and Nones

The word *theology* comes from two Greek words *theo* + *logia*—God-words or God-logic. Usually, theology focuses upon words and text from the sacred traditions of a given religion, seeking to understand them in an essentialistic way and interrelate them in a logically consistent way. This adds understanding to faith and gives the believer greater security. This is important, because many texts appear to be contradictory, and theology seeks to resolve those apparent inconsistent, theological conclusions—whether by Hebrew or Greek mode of argument so that they become a major part of 3rd level orthodox teachings of institutional religions.

A modern tendency is the rejection of institutional religions by individuals who have been labeled "Nones." This name arose from multiple-choice surveys, which asked a person's current religious affiliation. The survey lists many traditional religions, followed by the option: "None of the above." There are two major groups of "Nones." One large group is 2nd level, antithetical. They reject participation in institutional religions because they disagree with some of the teachings, or they are turned off by past institutional actions they consider offensive. Another group of Nones rejects institutional religion because they desire to have personal experiences with God, but they do not have the spiritual God experiences they desire from the formal teachings and rituals of traditional institutional religions. In doing this, they are aspiringly 4th level insofar as they are seeking some type of advanced and advancing spiritual experience of God. Too often they earnestly seek to discover by their individual effort a spirituality where they have personal experience of God. Unfortunately, they soon tire and eventually, become disappointed, and abandon the search, even though the hearts these None still desire some type of personal experience of the Almighty.

IPhil recognizes that in advancing and evolving to a higher level, it is a most difficult task if a person tries to start from scratch building an advanced interrelationship with God. In a logical evolutionary way, it is easier when the individual uses the practices and teachings of some antecedent, traditional religion as a base for an advanced and advancing spirituality. People in what are called religious orders have been doing that for centuries. Respecting the religious writings and traditions from past, they go beyond them in various advanced, transcendental ways. Support from the spirituality of a group assists them in discovering in a logistically advanced manner, an intense spiritual, interpersonal, unitive interrelationship with the Almighty.

Dealing with Different Religions

At one time, everyone in small, local communities practiced the same religion. Today, in our increasingly pluralistic society, different religious traditions rub elbows—sometimes affectionately or at least without incident, and sometimes divisively. Depending on their spiritual maturity, different people have different attitudes toward the religions and spiritualities of other people. On the Lakota reservations in southern South Dakota, I led a seven-year dialogue between the Lakota medicine men and the Catholic pastors. In that bicultural, bilingual, bi-religious setting, four very different ways of relating these two religions emerged.

1. *Welcoming*. When lay people are physically hurting or in need, they are open to whatever religious tradition and rituals may work for their material or spiritual benefit. In the local hospital, the doctors were open to whatever religious practitioners the patient requested. Often leaders from many different religions and spiritual practices came and were accepted, provided that the practitioners worked for and prayed for the health and welfare of the patient.

2. *Opposing*. Indian militants and Fundamentalist Christians strongly opposed every religion but their own. They insisted that their followers be pure and total in their commitment to the one, right, religious way—*their* way. They condemned all other religious traditions, and they pushed to expel an unwanted religion from their community, even when many did not want that.

3. *Accommodating*. Peace-loving religious people look at different religions with rose-colored glasses, emphasizing similarities and ignoring differences. These peace-focused accommodators tended to be syncretistic in their rituals, mixing elements from the different religious traditions, showing respect and warm friendship toward people of different religions. They usually focused on the external aspects of rituals and cultural principles, rather than the interior, spiritual foundations and deeper meanings of the different religions' beliefs and rituals.

4. *Respecting*. Recognizing that different religions are in some ways similar and in some ways different, respectful individuals bring religions together in ways that respect both their similarities and their differences. Rather than be syncretistic and ignore the differences between religions, these respectful individuals keep differing elements apart. They are able to combine those elements that are clearly similar and harmonious, *and* they keep separate those elements that are unique to each religion. They examine and talk about religious and spiritual differences carefully, seeking respectful understanding. They do not base their discussions on the writings of outside, professional "experts," but upon their own personal spiritual experiences. While similarities are spiritually reassuring and comforting, the differences provide opportunities to appreciate the unique spiritual contributions of each religion. After much discussion and meditation, it becomes clear how their differences are not diametrically opposite but enrichingly complementary. In this way, individuals, who grow up in one religious tradition find fundamental harmony with the other religion in some matters, while experiencing new, deeper spiritual experiences through their participation in the unique elements of the other religious tradition.

From personal experience, I realized that the Lakota religion focuses primarily on the horizontal axis and connects the user of the Pipe with the many beneficial spirits of our physical world. In contrast, Christianity's primary focus is on the vertical axis, connecting earth and heaven through Christ. I found that these two different spiritual experiences complemented each other and expanded my spirituality and faith without any theological inconsistencies.

Sometimes theological differences in religious beliefs were difficult to reconcile. Nonetheless, the participants held a 4th level hope for finding some deeper bridge between the spiritual elements of these two religions. We found that it was not through rational analyses or arguments that these differences were resolved. Rather, recognizing that these are matters of God and the spirits, we turned to them for answers. Through personal spiritual experiences in prayer and traditional rituals, we resolved the most difficult questions we had.

Images of God

In the earliest years of human history, people in different parts of the world had different needs, and they recognized that certain aspects of our world played an important role in their survival and their prosperity. For this reason, different cultural groups fostered devotions toward energy-communicating animal spirits and deities that were of benefit to them. Nonetheless, as groups grew and migrated, they met other groups, and out of respect they showed devotion to all their benevolent, ancestral, spiritual realities too. As many different groups blended together, however, poets created mythological stories about spirits and gods in which hierarchies of natural deities acted in ways similar to local political leaders. Consequently, such imaginative syntheses were filled with many inconsistencies and humanlike problems, which the poets heightened for dramatic effect.

In the sixth century BCE,[3] the shrewd and witty pre-Socratic philosopher Xenophanes observed that Ethiopians say their gods are flat-nosed and black, while the gods of the Thracians have blue eyes and red hair. He speculated that if cattle or horses or lions had hands and could draw and sculpt like men, the horses would draw their gods like horses, and cattle like cattle—each god in the artist's respective likeness. IPhil agrees with this observation and asks what is wrong with that. As a 2nd level idealistic essentialist, Xenophanes assumed that there was only one right depiction for all gods. Anything less than that, is imperfect and should be ridiculed, condemned, and dismissed. IPhil's 3rd level point of view, however, is more tolerant. In logical evolution, conceptions and images of spirits and the Almighty are always partial and analogously true. *Quidquid recipitur in modo recipientis recipitur* (What is received is received in the mode of the receiver). In each culture and religious belief system, it is intellectually legitimate to describe the Almighty as the best and most interrelational reality possible within the logistic of one's own system or species. Thus, the best way for humans to describe the Almighty is not abstractly or negatively but anthropomorphically. This presentation is not totally true, but it is as true as humans are logistically capable of knowing it or describing it. When that happens, IPhil finds the difference between our subjective image of God and the real objective image of God to be interrelationally insignificant in our communal, societal logistic.

Intellectually immature individuals claim their idealized conceptions or images of the Almighty and anything else is the only way to describe something. They idealistically claim that their ideas and their images are absolutely true. In contrast, intellectually mature individuals know that our physical experiences are only partially true. Still, by expanded experiences, human knowledge is becoming more comprehensive and true in religion—just as it is in science.

That reminds me of how the image of Santa Claus is humorously perceived in the four ages of life.

1. A child knows there definitely *is* a Santa Claus.

2. A youth knows there definitely is *no* Santa Claus.

3. An adult *is* Santa Claus.

4. An elder *looks* like Santa Claus.

---

3. As quoted by Clement of Alexandria, *Miscellanies*, V, 110, and VII, 22. Butterworth, *Clement of Alexandria*.

As the editor of the *Sun Times* famously wrote: "Yes, Virginia, there is a Santa Claus. . . . He is as real as love is real."[4] IPhil agrees. Love is more than a feeling. Mature love is an enduring, metaphysical, reality that bonds two or more individuals together—as every communicating spirit does. Like love, images of Santa Claus as a 3rd level metaphysical reality manifests itself in formally different ways in different cultural settings. Unfortunately, in our modern world, scientists, philosophers, and adolescents keep trying to reduce 3rd level communal, spiritual, metaphysical realities into 2nd level individual, material, physical realities.

Going beyond the humor of the last line, IPhil recognized in the 4th phase that everyone who is a gift-giver is a Santa Claus, and Christmas can be celebrated every day of the year. IPhil recognizes that Santa Claus is a real communal spirit, the personification of 3rd level "associative love." In different environmental and cultural contexts, "Santa Claus" goes by many names and analogous descriptions—with St. Nicholas as their historical progenitor. Similarly, Satan is a real communal spirit, the personification of 2nd level "dissociative evil." Around the world, "Satan" goes by many names and analogous descriptions, with different historic villains as local progenitors. Mother Earth also is a real communal spirit, analogous to mothers worldwide.

---

4. September 21, 1897.

Part IV
End

Chapter 20

Soul and Afterlife

Early Views of the Soul. Other Descriptions of the Soul and Afterlife. Modern Views of Souls. IPhil's View of Soul and Afterlife. Peer Pressure in the Afterlife.

A term that I have used only occasionally in this book is "soul." Like the expression, "free will," it is laden with many different meanings acquired through the centuries. In this chapter, I give many different descriptions of the soul, each rising from some particular worldview. Finally, IPhil's description of the "logistic-soul" arises from the logistical nature of potential energy.

Early Views of the Soul

The earliest signs of the belief in the continued life of an individual's soul and its probable afterlife are found in graves of Neanderthals. They often decorated the physical remains of their departed with flowers and ochre. These indicate signs of devotion to their relatives and a belief in some type of better afterlife. Throughout the world, folk cultures and modern cultures have extensive burial practices. These are physical signs of their subjective, metaphysical, spiritual beliefs about the human soul in this life and into some type of projected better afterlife. Whether there actually is an afterlife is another matter to be discussed later in the chapter.

For twenty years I lived and worked among the Lakota in South Dakota. They had a fourfold mythological understanding of the stages of the life of souls, analogous to the four-stage logistic of energy.

- A newborn child is called *wakanheja* < wakan + heca = "it is holy," or, "it is an awesome mystery." The family looks with awe upon the child as a gift from God, *Wakan Tanka*, "the Great Holy," "the Great Awesome Mystery." During life, the individual participates in the seven rituals of the Sacred Pipe, beginning with the Name-Giving Ceremony. A Lakota's name is a sign of a particular spirit interrelationship or quality, which will be with the person throughout life. By partaking in a ritual of the Sacred Pipe, the individual in a traditional way begins to go down the Pipe's spiritual Red Path, going from north to south.

- When a person dies, it is believed that his soul remains near to his relatives for a time. On the one-year anniversary of the death, the extended family puts on a large give-away ceremony, in which what is sacrificially given away is thought to be given to the deceased in a spiritual way to assist the soul on its way to the south. They believe there is a spiritual old woman in the south who stands before a fallen log which goes across a river to *Tahca Makoce*, "Deer Country," where all deceased relatives gather and feast. Non-Lakota call this the "Happy Hunting Ground." However, if the old woman judges the deceased person not to be worthy, that soul must walk alone on earth for a long time. To this day, traditional Lakota put out ritual food at the beginning of each meal for these pitiful, lonely, wandering souls.

- If a soul is allowed to cross the log bridge to the south, *itokagata*, literally, "the direction toward which we face," the happy soul is able to celebrate with its relatives and with all animals in a prosperous, joyful afterlife.

- Lastly, the souls of great medicine men and heroes are believed to go to the mountains in the west, from which these caring, powerful, *wakan* individuals look down upon and help the people remaining on earth.

This scenario is similar to the views of souls and the afterlife in many cultures throughout the world. Is there some truth to this belief? I find it is presumptuous for materialists summary to dismiss these highly analogous mythologies without any in-depth metaphysical analysis. But then materialists only believe in what is physical, and their eyes lack the ability to see what is metaphysical and spiritual in our world. They have physical eyes, but they cannot see the metaphysical, spiritual realities behind those physical realities—much like most people in Socrates's cave. Yet, somehow, they do have eyes that can see what is metaphysically possible in the projective equations materialists have drawn from physical experiences. But that's a start.

Other Descriptions of Souls and Afterlife

In the Hellenistic world, a common belief was that the souls of ordinary deceased persons went down into subterranean Hades or Shades. Many Hellenists also believed in reincarnation after an individual was purified from one's wrongdoing in the River of Forgetfulness. Socrates and Plato believed the soul in a body was like a bird trapped in a cage. At death the soul, like a bird, could fly up to the realm of the Good with its ideals. They also believed in a reincarnation in which some memories from a former life could be drawn upon to make advanced rational conclusions. Also, a few heroes and emperors could be inversely inducted into the heavenly realm of the gods.

In reading these mythologies, do not look literally at these 2nd level physical descriptions, but like people of those days, *see* 3rd level analogous, metaphysical interrelationships within them.

The Hindus believe that most souls of deceased individuals will be reincarnated in an elevated or reduced form according to the law of Karma, which rewards or punishes an individual according to the individual's past good or bad deeds in this life. For most of them, reincarnation is an eternal recycling process, similar to the cycle of preserving, replanting, growing, and harvesting of crops in their agrarian society. Only select virtuous individuals and contemplatives could advance to a transcendent life within Brahman. Inversely, Buddhists hold to the doctrine of *no-soul*, and they maintain that an individual at death will dissipate like the smoke of an extinguished candle into the One, primal, universal Buddha.[1]

In ancient Jewish psalms, it is said that after death, individuals go into a dark, inert state, called *Sheol*.[2] There was no spiritual interaction or happiness in this state. Here, the soul is not separated from but remains close to the remains of the body after death. Even today Jewish teachers maintain a holistic view of the human person rather than a separable body-soul dualistic view. In this life and in the afterlife, Jews maintain that an individual remains an integral whole. Still, in the biblical account of God forming the first human, God first formed Adam's body from earth and then breathed into it a spirit.[3] So, initially, the two were distinct realities. Also, when Saul visited the witch of Ender, the ghost of Samuel comes as "a preternatural being rising from the earth."[4] Thus, Hebrew Scriptures do indicate in a few places that body and soul can exist and act somewhat separately. Most Jews today, however, hold a united, holistic, body-soul view of humans because God promised that a messiah will come, and he will reestablish for the Jewish people of God the kingdom of David with Jerusalem as its capital. At that time, deceased, buried Jews will experience a *resuscitation* of their body as it was in history. For a time they will joyfully share in the fulfillment of God's promise to Abraham and the Jewish people.

The earliest Christian scenario of the afterlife is found in Paul's Letter to the Thessalonians.[5] There, he described the traditional Jewish scenario, in which an individual sleeps or rests in the grave until awakened. Then, like the risen body of Christ, the physical bodies of faithful will be evolved into spiritual bodies on the Great and Final Day of the coming of the eternal kingdom in which the promised Messiah is Lord of all. Nonetheless, a few centuries later, Christians adopted the Hellenistic, dualistic scenario of the afterlife where the body remains in the tomb until the resurrection at the end, while the separated soul could go to heaven before the end after a personal judgment and/or purgation process immediately after death.

Thomas Aquinas, following the Aristotelian way of analyzing the different metaphysical aspects of reality, identified the human soul with the "substantial form" of

1. Batchelor, *Buddhism without Beliefs*.
2. Ps 28:1; 30:4; 88:5–13; 143:7.
3. Gen 2:7.
4. 1 Sam 28:13.
5. 1 Thess 4:13–18.

the living body.[6] Unfortunately, people commonly view the soul as a pseudo-physical substance. Because of the possible split of body and soul after death, most Christians maintain the Neoplatonic view that the human soul could exist and act outside of the body after death. This view goes beyond Christian Scriptures and claims that before the end, the souls of ordinary people—not just martyrs—could be judged worthy to come in the presence of God and have a Beatific Vision of God.[7]

In summary, throughout the ancient world there were many different descriptions of the soul and the afterlife. Many of these descriptions are quite similar. The reason for the similarities may simply be cross-cultural communications. However, the worldwide belief in an afterlife, even in highly isolated cultures, indicates that there probably is some truth here.

Modern Views of Souls

In ancient times, the soul was identified with one's physical breath, for death and cessation of breath were seen to go together. With the rise of rational Greek philosophy, their concept of the human soul advanced to the seat of conscious thinking. To Plato and his Neoplatonic followers, because of the enduring conceptualizing ability of the human mind, unlike animal souls, the human soul would continue to exist after death. They argued that if the mind is capable of knowing eternal ideas, the mind-soul must be eternal. Such philosophers held that during life the conscious-soul is like a bird in the cage of one's body. When an individual's body dies, one's conscious, knowing metaphysical/spiritual soul escapes the cage of its material body and flies upward, like a bird, to the place of all eternal forms, ideas, where the all-perfect Good resides. Here, the knowing soul was joyfully united with all perfect, eternal, ideals/forms.

While Plato saw the human soul as immortal and destined for union with the idealistic Good, with the possibility of reincarnation, Aristotle argued to the contrary. He identified the soul or *psyche* of an individual with the "first actuality" of a naturally organized body, and he argued against the soul having a separate existence from the physical body after death.[8]

In the modern West, René Descartes dualistically described humans as consisting of a corporeal body and an incorporeal soul. He maintained that the soul and the body were connected to each other through the brain's pineal gland. That explanation, however, was never generally accepted, for when the body dies, so does the pineal gland. In this model, the spiritual soul continues to exist after death, but there is no means by which the soul can learn anything more.

During the Age of Enlightenment, the Age of Reason, the Scientific Revolution, and into modern time, only the mechanical side of Cartesian dualism was accepted

6. Eardley and Still, *Aquinas*.
7. Rev 5:11–13.
8. Sachs, *Aristotle's On the Soul and On Memory and Recollection*.

in academic circles. Because of the materialistic and empirical foundation of physical experiences and scientific knowledge, modern thought dismisses the existence of a distinct, enduring spiritual side to life. Modern philosophers argue instead for the nonexistence of the soul and the nonexistence of anything spiritual. All explanations and arguments that soul is the "ghost-in-the-machine" operating inside the human body have been found to be intellectually questionable and wanting.[9]

Although the word "psychology" literally means "study of the soul," modern psychology focuses upon the physical operations of the brain. Today, the dominant view of psychology is the brain/mind identity theory. Many correlatives between observed physical changes in the brain and psychological experiences in the mind have been found. Since the mind is commonly recognized to be the central, conscious component of the soul, accumulated physiological and psychological correlatives provide experimental foundations for the brain/mind identity theory. This theory does not argue against the belief in the existence of the soul; rather, it simply renders the notion of "soul" archaic, unfounded, and unneeded, removed by Occam's razor.

Nonetheless, the psychologist James Hillman, the founder of archetypal psychology, has worked to restore the notion of soul to contemporary discussion.[10] There are many unanswered questions here which challenge the currently popular mind/brain identity theory. However, Gilbert Ryle's ghost-in-the-machine rejection of Descartes's mind/body dualism still stands academically. Furthermore, advances in neuroscience steadily undermine the validity of a mindful soul that is independent from the brained body. Richard Swinburne's argument for the existence of the soul is considered only tenuously unassailable.[11] Consequently, many modern scientists and philosophers do not believe in the notion of *soul*. These materialists consider the idea of a soul a tradition or figment of our imagination, something like Santa Claus, which is comforting and enjoyable but unreal. Empirical scientists measure all things physically, and then tease out metaphysical characteristics secondarily. However, physical evidence points to the conclusion that human souls really do not exist metaphysically in our pervasively materialistic, physical world.

IPhil's View of Soul and Afterlife

IPhil has its own description of "logistic-soul," based upon the fourfold interrelational nature of energy, which interrelational philosophy holds as the foundation of everything energetic in our world. IPhil objects to the identification of soul with the *physical* aspects of 1st level bodily breathing, with 2nd level self-awareness, or with 3rd level rational thinking. Rather, IPhil identifies the soul of an individual with the individual's *individual logistic*. That is why I have replaced the term "soul" with "logistic-soul."

9. Ryle, *Concept of Mind*.
10. Hillman, *Blue Fire*.
11. Swinburne, *Evolution of the Soul*.

Traditional views of soul are generally materialistic as a separable substance. Rather, IPhil recognizes the logistic-soul to be foundationally metaphysical and spiritual, capable of developing a physical body and brain when material and energy are presented to the logistic-soul in proper historic order. At death, the physical body terminates but the metaphysical logistic soul remains. As potential energy is the antecedent foundation of kinetic energy and physical, material individuality, so too the logistic-soul is of the foundation of one's metaphysical and physical individuality.

An animal's DNA physically has material components, but the foundational unique character of an individual's DNA is found in its metaphysical individual logistic found in its enduring DNA *code*. One's DNA code remains basically the same throughout one's lifespan as the person grows and changes physically. It is to this enduring metaphysical logistic that an individual's name is applied during life and afterwards. Consistency of the application of a person's name indicates that an individual logistically remains the same through many physical changes during the individual's life. That name is also applied to that individual remembered after death.

The individual's personal metaphysical logistic is much like an equation that retains the same mental, metaphysical form, even before and after physical values are assigned to the variables in that equation. IPhil's identification of an individual's logistic-soul with the individual's fourfold metaphysical logistic is very close to the Thomistic identification of one's soul with the metaphysical "substantial form" of an individual. Like an enduring "equation" through time—whether activated by variables or not—IPhil recognizes that the logistic-soul of an individual metaphysically exists before, during, and after a person's physical life.

An individual's logistic is a well-ordered logical reality with fourfold metaphysical characteristics. It is panentheistically analogous to the fourfold Logistic-Soul of the all-receiver-giver of PAPE. In this way, it can be said that all enduring individual logistic-souls are in and of the eternal all-inclusive Logistic-Soul of the Almighty. Unlike the classical argument that the human soul is eternal because its rational concepts are unchanging and eternal, IPhil goes beyond that in a logical evolutionary way by saying the human soul—and every soul—is eternal because its individual logistic is logically evolved from the fourfold Logistic-Soul of the Almighty. Like an equation with open variables, the individual's logistic-soul is an open, potential receiver, capable of being actualized by the input of energy and matter at a time determined by the giver of that energy and matter, such as one's parents or a system's progenitor. Consequently, an individual's open logistic-soul first exists in a waiting metaphysical state before it is freely actualized by the reception of energy/matter from one's immediate parents and one's environment. Also, when a logistic-soul "lets go" of its energy/matter at physical death, the physical body disintegrates, but the metaphysical logistic-soul remains. Sometimes, when an individual strongly resists fully "letting go" of its materiality, the logistical-soul of an individual can appear as a ghost, until such time as the individual's soul trustingly "lets go" and is at rest.

Every individual's logistic-soul expresses itself progressively in a fourfold way in both the metaphysical and physical orders.

- In its 1st metaphysical phase of existence, the individual's logically evolved logistic-soul is within the panentheistic logistic of the all-receiver-giver. It is not yet animated but simply open to receive potential and physical energy from another.

- Then, when physical conditions are right from the received matter and energy, the 2nd phase of the logistic-soul takes place. Here, the open, receiving logistic-soul physically receives matter and energy from one's parents and one's physical world. At this point the open, receiving logistic-soul takes on a growing physical body. The logistic-soul and the name of the individual basically remains the same through the material life span of the individual.

- In the 3rd metaphysical phase of existence, the energized logistic-soul interrelates physically with others to form enduring logically consistent interrelational systems according to its received individual logistic, advancing the interrelationships in our physical universe,

- In the 4th metaphysical phase of existence, the individual's logistic-soul "lets go" of all its energy/matter in physical death. Here, the logistic-soul of the individual again becomes metaphysically open but inactive.

- Finally, the logistic-soul of all individuals *can* again be freely empowered by the Almighty in a transcendental spirit/material 5th metaphysical phase of its existence in a transcendental communal afterlife.

IPhil recognizes that one's physical body energetically emerges as a partial, energized, kinetic expression of the enduring, holistic metaphysical logistic of the individual. Thus, the soul is *not* in the body; rather, the body is *in* the soul. This is analogous to kinetic energy emerging from one's potential energy. That is, an individual's physical, material body is within the individual's metaphysical, spiritual logistic-soul. If a person loses an arm in a workshop accident, the entire holistic logistic-soul of the individual remains the same, even though the body is diminished. If a highly qualified surgeon is nearby, that doctor can reattach that arm—not in a random way but according to the enduring logistic of that individual. One's logistic "form" and destiny potentially surround a person's current physical state before the person acts physically in our material, kinetic world. The metaphysical, logistic-soul of an individual contains all one's energetic potentialities, and the current physical body of an individual is only a partial expression of those possibilities. We tend to look at things externally and fail to realize that in well-operating systems, the system acts and changes in an enduring way only according to its received, enduring logistic and toward its operational destiny.

More than identifying an individual's soul with physical breath, more than identifying an individual's soul with bodily animation, more than identifying a human's soul with conscious knowledge, IPhil identifies an individual's enduring logistic-soul and name with one's individual metaphysical logistic. Consequently, to repeat, one's metaphysical logistic-soul is not in one's physical body, but one's kinetic, physical body is in one's potential, metaphysical logistic-soul. Conscious rationality is only one operation out of many in the human logistic-soul. As temporary kinetic energy is from and of eternal potential energy, so too every temporary, material body is a partial physical manifestation of a receiver's enduring, metaphysical, logistic-soul.

IPhil recognizes that while an individual's DNA is material and will terminate, the logistical *code* of one's DNA is enduring. In fact, the logistical code of an individual's DNA can exist potentially *before* a child is conceived. Even today, knowing the DNA code of parents, it is medically possible to know whether a child will likely have a congenital disease before a child is conceived. Likewise, the logistical code of one's DNA will metaphysically endure *after* a person is dead. The military takes blood samples of all its soldiers, and their DNA code is kept on file. That is why the military can name the remains of a dead soldier many years after a soldier's death. Because the DNA code of those remains mark the identity of the dead soldier, those remains are respected and given an honored burial. Because of the dissipation of the molecules from a physical body after death into the lives of other material individuals, IPhil provides a way that the body of every dead person can be raised as a distinct individual in the resuscitation or the resurrection of the dead. IPhil recognizes that the metaphysical logistic of an individual is not dissipated, and the kinetic matter/body of a deceased person can be accurately reconstituted or resurrected via the potential energy of an individual's enduring logistic-soul. Also, by Einstein's famous equation, we know the equivalence of enduring potential energy with fleeting matter.

By identifying one's soul with one's individual logistic, one's logistic-soul is not physically material but metaphysically spiritual. As such, it exists interrelationally beyond the current temporal and spatial limitations of one's body, which is but a temporary, changing, physical expression of one's enduring, logistical identity. Recognizing that an individual's logistic-soul is greater than one's current material body gives foundation for the existence of an energetic aura around an individual's body. This also explains how an individual can sense if there is someone else nearby.

In the history of religion and spirituality, perfectionists held and promoted the dualistic, 2nd level belief that materiality was imperfect, and that materiality lowered an individual's spiritual union with the ideal, perfect God. In contrast, IPhil inversely recognizes that physical materiality and kinetic energy are advanced, albeit partial, interrelational forms of the metaphysical potential energy received from the Almighty. That is, the metaphysical, potential, spiritual order is not reduced but is advanced interrelationally through the incarnational energizing and materialization of one's eternal, spiritual logistic-soul. IPhil recognizes that both the notions of imperfection and perfection are found in the primal interrelational process of PAIS. That is, PAIS is imperfect until it reaches its four-stage perfection in the closure of the energetic logistic of PAIS. With the logical evolution of our world and all the things in it, the panentheistic world through its partialities advances until their final interrelational associations and happiness takes place within the Lord and with all-in-All, according to the received individual logistic of one's logistic-soul. Physicality and materiality do not diminish the Lord but increase the interrelationality of the Almighty. Our responsibility is to maximize physicality and materiality according to one's logistic logically evolved from PAIS, to maximize the interrelational, associative joy and sense of fulfillment of all-in-All.

Furthermore, since an individual's logistic is logically consistent, the individual's logistical-soul *can* advance toward a possible, logically-evolved, interrelational state. Just as the parameter of time can only go forward, so too the logistic or the logistic-soul can only become more interrelational. That is why it is legitimate to greet a person with an honorific title, like president, general or mayor, even after the person has retired from that office. While the human intellect in its normal operations seeks harmonious operations with everything in our finite world, the logical evolution of the human's logistic-soul inversely leans toward some type of advanced interrelationship that is beyond the logistic of our current finite world, and unto a greater, conscious interrelationship with the Almighty. Thus, in an evolutionarily possible 5th stage, the logistic-soul of a deceased individual *can* logically evolve toward a more advanced union of all-in-All.

Peer Pressure in the Afterlife

A perplexing theological question is: Would free individuals in heaven ever sin? If not, why not? A common essentialistic answer is that in heaven, knowledge of God as the supreme good would be so strong, individuals would always choose to remain loving and never sin in the transcendent state of heaven. However, the greatest pain of sinners in hell is said to come from the knowledge of their separation from the all-loving Almighty. That would mean the doomed sinners had to have knowledge of their separation from the absolute good, and that would require knowledge of the absolute Good as the opposing term of that separation. These two conclusions are logically inconsistent. That argument requires the damned to have knowledge of God as absolute Good, but of itself, that knowledge should be so strong that no one—not even the sinner—could ever remain a sinner in the afterlife. This observation raises serious doubts about this line of argument.

The problem with this line of argument, I find, is that it is 2nd level individualistic. I believe the answer to this conundrum is 3rd level communal. I find peer pressure and social sanctions very strongly keep people from acting individually, like mavericks, even though they have the freedom to do so. Beyond individual stubbornness arising from self-righteousness, peer pressure and group loyalty are regularly stronger and harder to brake from. Only on the rarest occasions do individuals go against the pressures they experience from their communal society. Where did peer pressure come from? IPhil locates the earliest example of free individuals submitting to their communal logistic in the Primal Absolute Interrelational System (PAIS). There, the interdependent elements/persons of that trinitarian system are individually indeed free, but they are pushed by the energy in their logistical PAIS to always give absolutely to the other terms in that absolute system. Analogously, by logical evolution, very strong peer pressure is found within individuals in every evolved interrelational system. Peer pressure can be found in the free members of 2nd level subgroups, which act in a 2nd level, self-centered way or in a 3rd level other-centered way.

In this setting, the crucial, transformational, turning point is one's advancement from unwanted, dissociative, temporary peer pressure away from outsiders, to wanted, associative communal, eternal love of others. Here, there continues to be the possibility of metanoia from self-centered love to other-centered love. However, experience indicates that the free inverse activation of that possibility is highly unlikely in this world and in the next.

Chapter 21

The Finite-Absolute Progenitor

Logical Evolution of the Grand Interrelational Logistic (GIL). Finite-Absolute Progenitor. Identity of Finite-Absolute Progenitor.

Energy does not operate in a haphazard way but according to a definite four-stage interrelational logistic (receiving, integrating, interrelating, and giving). Not only individuals but also systems have their own logistic, each with their own fourfold process. This indicates that our finite physical universe as a whole, which began with the big bang, has its own cosmic interrelational logistic, which I call the "Grand Interrelational Logistic" (GIL). It contains all finite systems that have or can logically evolve from the proto-spatial system generated in the big bang.

Logical Evolution of the Grand Interrelational Logistic (GIL)

By its absolute reciprocity, the antecedent Primal Absolute Interrelational System (PAIS) is logically consistent and therefore is capable of logistically evolving our finite world through an inverse absolute-finite progenitor. This became the first, most highly energized finite receiver. By acting according to its received, logically evolved logistic, the absolute-finite progenitor became the first spacepoint. Initially, it had no spatial dimensions, but in operating according to the interrelational logistic in a finite way, it become the progenitor spacepoint of the finite, logically consistent, quantized proto-spatial system. It was through the quantized character of this proto-spatial system that the four forces of physics and all finite material things were logically and physically evolved.

In addition, the cyclic character of the logistic of all energy leads to the conclusion that there is a cyclic Grand Interrelational Logistic (GIL) guiding the progressive development of the entire history of our finite world. Random freedom is not unrestrained but are self-initiated energetic expressions that are always within the boundaries of the individual's logistic. All logically evolved interrelational logistics are cyclic. For our physical world that means what physically has come from the Almighty is metaphysical oriented to logistically return to the Almighty, as in a grand game of catch.

Because of the logical consistency of PAIS, the logically evolved GIL is also logically consistent. So, GIL itself is capable of logical evolution through a logically inverse progenitor, empowered by the potential energy received from the antecedent all-giver.

There are two critical points in the logical evolutionary process of our whole finite physical world; they are at the beginning and at the end of GIL. At the beginning of our finite physical world, there must be an absolute-finite progenitor spacepoint whose inverse finite character gives way to the energetic explosion of protospace in what is called the big bang. At the end and possible evolution of our finite physical world, there *can* be a complementary finite-absolute progenitor who, in a total "letting go" of its received energy, metaphysically reconnects the finite order to the absolute order by a logical evolutionary process through a logically inverse progenitor.

How could that happen? Won't the evolutionary process just keep going, producing more and more complex systems? No, because of the finitude of our physical world. Just as the energy of the participants of the logistic of PAIS cannot be infinite but must logically be absolute, that is, of a definite amount, likewise, the energy of the logically evolved big bang must be a definite amount. Because the energy in the big bang was finite, there is only a finite amount of energy available to advance interrelational systems. In our limited and quantized energetic world, as the logical evolution of our world becomes increasingly interrelational, the possible number of more inclusive interrelational systems will begin to reduce in number.

In the early 2nd phase of our world's Grand Interrelational Logistic (GIL), material dissociations dominated, and like a growing teenager, the Tree of Life expanded rapidly. Contrariwise, as in the 3rd or mature adult phase of interrelational expansion, things and systems will become increasingly consolidated interrelationally and move toward becoming more and more interrelationally one in the Grand Interrelational Logistic (GIL). With expanding human communication and domination in our modern world, the earth is becoming increasingly integrated, and systemically the newly evolved systems will numerically become interrelationally integrated and numerically smaller. This is similar to the way in which many small businesses are consumed by a few big businesses. As businesses grow by gobbling up other businesses, the number of big businesses becomes smaller. Our knowledge of all material things in our universe is becoming intellectually unified, and things are becoming interrelationally, spiritually consolidated, toward systemically becoming interrelationally one.

We recognize how evolved species became increasingly widespread by increasingly dominating the things in their environment with few ecological concerns. In modern times, however, humans are increasingly recognizing how all things in our world are dependent on one another, and people are striving to bond things together in new and lasting systemic ways. With limited resources, this consolidation process is taking over the diversification process of the past, as everything in our world increasingly advances from individual independence to communal interdependence. This process need not externally suppress individual freedom, but it can in a respectful and cooperative way expand the range of everyone's and everything's interrelational freedom. In this way, they happily participate in the unification of our world with more and more tightly bonded things and information, freely operating within the nested boundaries of their respective logistics.

The initial progenitor of the big bang was able to begin the evolution of our finite world through an absolute-finite inverse, followed by a carryover of the absolute interrelational logistic of PAIS into the finite order. In a complementary way, logically consistent GIL is ordered toward advancing the evolutionary process of our entire finite universe through a finite-absolute progenitor toward logistically bridging our finite world with the integrated absolute PAIS. Following the breaking of the boundary of the finite logistic of our cosmos, in this all-inclusive evolutionary process, the finite-absolute progenitor will advance into the absolute order in a logically inverse way, while carrying over finite members of antecedent systems into the new corporate union of all-in-All.

The minds of philosophers strive to order all things in a logically consistent way into one comprehensive intellectual model. Unfortunately, by the deductive logic of the 2nd level, essentialistic philosophers do not allow themselves to break the boundaries of their current rational systems. Physical and environmental sciences likewise seek to unite things in a logically consistent, mutually beneficial way in our world, but only within the current paradigms of our physical world. In all these efforts, future hopes are not transcendent beyond current logistics of existence, because they do not know of a way to advance from the current material and human order into anything more logistically advanced.

IPhil recognizes that through the logical evolution of the Grand Interrelational Logistic (GIL) of our finite world, there *can* be a progenitor that *can* establish an advanced interrelational system via a finite-absolute progenitor, which can become a part of a new, more interrelational order that synthesizes the finite order into the absolute order. That is, through the logical evolution of the Grand Interrelational Logistic (GIL) of our cosmos, a logically inverse progenitor can break the boundary of our finite world so that it can be united with the absolute world, carrying over members of our current finite world in a new, transcendent, interrelational way into the Almighty.

By conservation of energy the new order will *energetically* be the same as the current one, but this new order will be *more interrelational* for both the absolute members of PAIS and the spiritually evolved members of the finite systems of our physical world. In this way, the panentheistic process that began with the logical evolution of the first spacepoint finds its completion through a finite-absolute progenitor, which establishes a new, evolved world that interrelationally advances both the Almighty and every subsequently energized finite individual in our current world. Unlike the essentialistic belief that God is always perfect and never changes, IPhil recognizes that while the Almighty never changes in terms of having absolute energy, the Almighty can advance interrelationally. Since potential energy pushes to be more interrelational, IPhil recognizes that the Almighty PAPE is oriented toward the interrelational advancement of all-in-All. Within the interrelational logistic of energy, that interrelational advancement is first actualized dissociatively, establishing many new things in an outgoing, dissociative manner. Then, in the second half of the Grand Interrelational Logistical cycle, it is oriented toward bringing all things together associatively, so that the Almighty has all-in-All. The logical evolutionary process shows that the final unification requires the existence of a finite-absolute progenitor. But what would be the characteristics of this finite-absolute progenitor?

Finite-Absolute Progenitor

Advancement into any logically and physically evolved system requires a progenitor that has the actual characteristics of the antecedent system and potentially the characteristics of the subsequent, evolved system. Logical evolutionary advancements are possible only because of the inner push of the foundational potential energy of the eternal Primal Absolute Potential Energy (PAPE) channeled through the Primal Absolute Interrelational System (PAIS) into subsequently evolved systems. It was through received potential energy from PAER, that the Almighty, through the all-communicator spirit of potential energy in the all-receiver-giver that the first space-point was pushed to go from a projected, virtual reality into an actual finite reality. Subsequently, through the logistical progenitors of subsequent finite systems, subsequent, different systems and species and systems have logistically and then physically evolved in our finite world. In this way, it can be said that the Almighty is the primal potential energetic source and the carryover existential support of all the evolved systems and species in our world.

In breaking the boundary of a given system, a logically potential progenitor must have at least an intuitive knowledge of the advanced, inverse state which its received energy is pushing toward. Consequently, if the logical evolution is oriented toward reuniting the finite order with the absolute order, the finite-absolute progenitor would have at least an intuitive awareness of the absolute aspect of its finite-absolute logistic. History shows that pre-human individuals do not have some level of intuitive knowledge of the Almighty, as displayed by their lack of being *religiosus*. However, physical evidence shows that the human species does have knowledge of the transcendental absolute order. That is, the finite-absolute progenitor had to have the logistic of a *homo religiosus* and therefore initially be human. Also, unlike other physical species, only modern humans have the intellectual and volitional drive to understand and unite all things in our universe. Thus, the finite-absolute progenitor could not have been born immediately after the big bang. Rather, the finite-absolute progenitor has to have been born in the "fullness of time," when the finite-absolute progenitor as a human would be capable of receiving energy that is originated from the Almighty to be capable of grasping potential and intuitive knowledge of the Almighty, and be freely responsive to the push of that potential energy toward some type of associative union of all-in-All.

THE FINITE-ABSOLUTE PROGENITOR

Initially, the potential energy received by the finite-absolute progenitor would push in a diffused way. As a child, knowledge of its finite-absolute logistic and calling would be at most semiconscious, diffuse and intuitive. Then, as the logistic of the individual advances, the finite-absolute progenitor would advance from unconscious acceptance to conscious pursuit of fulfilling the evolutionarily received finite-absolute logistic. Then in a 3rd level way, the received potential energy pushes the progenitor toward building increased interrelational associations in an advancing, logically consistent, systemic way. Finally, the received energy pushes the finite-absolute progenitor toward 4th level sacrificial surrender of materiality on the antecedent level to "let go," potentially to be energized and glorified in an evolved way, as the finite-absolute progenitor establishes a bridge between the finite order to the absolute order. Only by gradual physical, kinetic experience of the received absolute potential energy would the finite-absolute progenitor gradually become conscious of what was his finite-absolute evolutionary logistic and destiny. That is, prior to experiential knowledge of the union of all-in-All, the finite-Absolute progenitor had to pursue that finite-absolute goal in faith and hope.

In addition, the interrelational logistic of potential energy has both individual and communal aspects. Consequently, the evolutionary finite-absolute progenitor will be pushed to gather around himself a support group of humans, characteristic of the antecedent system, so as to carry over and establish the new finite-absolute system logically consistent system in the absolute order in the all-receiver-giver.

Contrariwise, traditional leaders and thinkers of contemporary human societies would strive to maintain the integrity of their current communal, closed, religious systems, and they would strongly oppose the breaking of their enduring system. Consequently, they would oppose the breaking of their system's logical boundary by the finite-absolute progenitor and the evolutionary support group of his disciples. Breaking the provincialism of human societies, the finite-absolute progenitor would push to broaden interrelationships worldwide, not only with humans of his race but also all races, extending the domain of this transcendental society to include all humans and all the species of the earth, and the earth as well in a universal, logistically described, systemic dominion. Orthodox and conservative leaders would strive to silence every rogue innovator by imprisonment or death. Meanwhile, the logistical process of establishing an evolutionary organization would move forward in its fourfold development of establishing (1) an initial progenitor, (2) a small support group of dedicated followers, (3) then an internationally expanding systemic institution, (4) the prophetic vision of establishing a universal logistical interrelationship of all in the finite order, and finally (5) the physical transcendental transformation of all-in-All through and in the finite-absolute energy potentially found panentheistically in the all-receiver-giver in the absolute order.

The 4th stage of the progenitor's logistic requires a total "letting go" of his control of all his finite material and energetic possessions. In other words, in its 4th phase, the life of this finite-absolute progenitor needs to be ordered toward the total sacrifice of his energetic resources for the sake of bringing his received finite-absolute logistic to fulfillment. To do this, he must be confident that the universal push of his received potential energy and his logistically evolved logistic are truly ordered and destined to build the bridge from our imperfect, finite world to interrelational union with the perfect absolute Almighty. In other words, the logistic of the finite, material phase of this finite-Absolute progenitor's life requires that the progenitor have total faith in the potential, prophetic interrelational process of the enduring Spirit of the enduring finite-absolute logistic. This is done by totally "letting go" of his finite materiality, and by being "open" to the possibility of subsequent actualization of receiving potential energy onto a transcendent life beyond the current finite, physical order. Exactly how this finite-absolute progenitor "lets go" of his physical totality will happen will depend upon the physical and community situation of the progenitor at that time.

In this way, the finite-absolute progenitor is the advancing *logos* or logistical "seed" to which others can attach, thereby advancing not only the progenitor but also evolutionarily advancing all humans and all things that become intrarelationally attached to him—while others in the world remain only humanly associated with him. Together, his associated supporters and followers *can* establish a more evolved system and order, which draws together all the energized resources and realities of the cosmos to form a new, advanced, integrated interrelationship in and with the Almighty. In this way, the finite-absolute progenitor becomes the cosmic corporate head of a new, transcendental, absolute organization of all-in-All. Through him, associated members will corporately inherit the new, evolved, Absolute characteristics of the evolutionary progenitor.

Looking at the big picture from the standpoint of the panentheistic all-receiver, what is the relationship of the above finite-absolute progenitor to the all-receiver-giver? The finite-absolute receiving-giving progenitor is the initiating "seed" that strives to interrelate physically all finite things in our world within its universal communal logistic unto an evolutionary non-finite, absolute end. In addition, the all-receiver-giver panentheistically contains all finite receivers metaphysically from the beginning to the end of the cosmos, from which they express themselves physically in generated and evolved finite ways at their appropriate times. From this standpoint, the individual logistic of the finite-absolute progenitor and the panentheistic individual logistic of the all-receiver-giver are logistically the inverse of each other. Since it is the individual's logistic that establishes the individual as a person, the finite-absolute progenitor and the all-receiver-giver are logically distinct persons. Nonetheless, the finite-absolute progenitor's communal logic physically ends in an absolute state, while the all-receiver-giver begins and remains absolute. This is as the beginning stage of the original absolute potential energy of our world going from the absolute order unto the finite order.

In addition, recall that the physical end of a reality is metaphysically, potentially present in its beginning. There can be only one absolute, so the absolute receiver-giver as a panentheistic reality and the all-inclusive finite-Absolute progenitor and head of a potentially universal corporation, while interrelationally distinct, must be the same energetic reality. This brings us back to the Arian controversy. Just as energy has two faces—enduring existential potential energy and interrelationally advancing kinetic energy—it can be argued that even though the all-receiver-giver and the finite-absolute progenitor have two different faces, they are the same potential/actual absolute energetic reality.

Identity of Finite-Absolute Progenitor

IPhil maintains that the beginning of our universe and the Grand Interrelational Logistic (GIL) from PAIS took place in a logical evolutionary way through the absolute-finite progenitor of the first spacepoint. In a symmetrical way, at the end of the Grand Interrelational Logistic (GIL). the logical evolution of our finite world into the absolute world will take place through a finite-absolute progenitor. IPhil recognizes that just as the beginning spacepoint was a real, physical, individual energized term, so too IPhil recognizes that in the evolutionary bridging of our finite world with the absolute order, there likewise has to be a real, physical, historical individual that is the progenitor of that evolution.

Because potential energy has many kinetic possibilities, philosophically it is impossible to determine exactly when or who this finite-absolute progenitor would be historically. Still, as argued above, this finite-absolute progenitor must arise from the God-knowing human population. There are several possibilities of who might be considered the finite-absolute progenitor.

- In the *Bhagavad Gita*, Krishna is an avatar of the Vishnu, one of the dimensions of Brahma (God).[1] He presents himself as the charioteer of Arjuna, who is struggling with the moral dilemma of whether to fight against one's relatives or not. During the discussions in the story, Arjuna manifests himself as the divine Krishna. The account of this theophany is awesome, profound, and very encouraging. However, in this story, Arjuna-Krishna does not go through the phases of human life, so he is not fully human. IPhil, on the contrary, recognizes that the progenitor must be fully human, going through all four stages of life to be the fully human catalyst for the material-spiritual evolution of our limited, finite, dissociative world into a universal, associative, finite-absolute, communal reality.

- Siddhartha Gautama was a historic figure of the sixth century BCE. He is called the Buddha, or the Enlightened, for reaching the state of enlightenment and for teaching the Four Noble Truths and the Eightfold Path, telling people how to reach the state of universal enlightenment. Buddhism originally was not a 4th level religion or a 3rd level metaphysics but a 2nd level psychology. That path focuses upon detaching or "letting go" of conscious concerns about the

---

1. Chidbhavananda, *Bhadavad Gita*.

dissociations found in our physical world, preferring the peace of the non-rational knowledge found in simple awareness. Because 2nd level peaceful psychology is the central focus of Buddhism, it strives to detach from, rather than embrace, the 1st level material side of our world, which Buddhists described as an illusion. In contrast, IPhil recognizes that the progenitor must be physically material, and in an evolutionary way the progenitor must use the partial, finite materialities of our world as an intermediate and enduring means for raising not only oneself but the whole world corporately to the level of absolute interrelationships. Buddhists say that Buddha is present in all places, in all beings, in all things, in all land, not just in the monastery. In the course of time, Buddhism has elevated Siddhartha Gautama to a divine status, in whom all find peace here and spiritual union eternally. Buddha cannot be the finite-absolute progenitor, for Buddhism does not recognize the carryover characteristic of our energetic and materially progressive and evolving world.

- Muhammad lived in the seventh century CE. The Qur'an, which he dictated, came from revelations received from God through the archangel Gabriel. The teachings of Muhammad and the Qur'an are cornerstones of Islamic surrender to God. Belief in only One God is a pillar of Islam, and Muslims recognize Muhammad as the last and greatest human prophet. However, the historic finite-absolute progenitor would have to have both a human and a divine nature. This is explicitly contrary to Muhammad's own words and Islamic beliefs.

- Jesus of Nazareth, born during the reign of King Herod ca. 6 BCE, most closely fits the profile of the finite-absolute, human-divine progenitor described by IPhil. His conception was described as being from God the Father through the Spirit in a human virgin, Mary. He lived a full human life of birth, adolescence, adulthood, and early death. He called himself "Son of Man." The word for "man" is adam, which is from the Hebrew word for "earth," adama. So, the title "Son of Man/Adam/Earth" can be analogously expanded to include the earth and all material things. In this way, Jesus analogously identified himself as one from the earth and one who was descendant from the first human. By referring to himself as "Son of Man," he also identified himself as a prophet in line with other ancient prophets who called themselves "son of man." Finally, at his trial before the high priest (Mark 14:60–64), Jesus identified himself with the corporate, eschatologically glorified Son of Man, as described in Daniel 7, seated at the right hand of the Ancient of Ages. Here, the title "Son of Man" refers to both an individual and the corporate head of the people of God. In Daniel's prophesy, the "son of man" individually and corporately was prophesied first to suffer at the hands of evil men, and then be elevated to the throne of the Ancient of Ages in heaven. When he quoted Daniel, Jesus extended the meaning of that passage to affirm before the high priest that as Son of Man, he would be *seated* at the right hand of the

Almighty and be given dominion over all nations and all material creation (Ps 110:1). Because of how Jesus identified God as his Father, and especially because of how he described himself as seated at the right hand of the Almighty, which indicated equality in divinity, Jesus was condemned to death for blasphemy by the High Priest and the Sanhedrin.

However, Jesus never recorded any of his statements. All of Jesus' sayings and all accounts of his deeds were recorded later by his followers, who would have been selective in their memories. Because of this, it can be argued that the words and actions of Jesus were not recorded objectively, for his followers would have been prejudicial and prone to exaggerate legends of their founder. Consequently, it is easy to argue that the divine side of Jesus Christ's existence was not real but the fabrication of his followers. Historian Bart D. Ehrman in *How Jesus Became God* described quite accurately how leaders of the early Christian church stretched and logically evolved Jesus' claim to be Son of Man into the church's claim that he was the resurrected and ascended Son of God (e.g., Mark 1:1). I find most of Ehrman's historical conclusions to be quite accurate, except those found in the chapter on Jesus' self-knowledge, preaching, and actions. It can be argued that Jesus himself displayed an evolutionary style of thinking and preaching that respected and went beyond the Mosaic laws, beyond the predictions of recognized Jewish prophets, and beyond contemporary Jewish thought.

In addition, extended christological claims are not only grounded in physical, historic facts but on some extraordinary spiritual experiences of Jesus after his resurrection. These statements arise from experiences and scriptural conclusions of Jesus' early disciples, who became the societal base for the formation of the enduring, spiritually evolved Christian community or church. Not verifiable with material evidence, these statements must be held in faith.

- Another possibility is still maintained by the Jewish community: that the promised Messiah has yet to come. However, the Messiah of Jewish tradition is not a universal but a nationalistic figure. Yet, Jews strongly hold to the covenant God made with Abraham, in which the Lord said, "All the communities of the earth shall find blessing in you" (Gen 12:3). According to this prophesy, the promised Messiah will be a descendant of Abraham, King David, and will bring God's blessings and peace both to the Jewish people and to all nations in the world. They continue to wait for his coming.

Even if there is no historic figure who currently fills the profile of the finite-absolute progenitor of the evolutionary return of the finite universe into the absolute order, the interrelational logistic of IPhil requires and projects that sometime in the future there will be such a historic figure. During his lifetime, Jesus' disciples asked him: "Are you the promised Messiah or are we to look for another?" Jesus never answered that question directly.

PART IV: END

Our intuitive understanding of a transcendent future is built upon what has already gone before us. So, our faith, hope, and loving life are metaphysically well-grounded, thanks to the guidance of spiritual sages and religious leaders, who have great, though partial, knowledge of the spiritual side of energetic existence. More importantly, our energy-driven intuition tells us that we are headed in the right direction and that sometime, somehow, we are destined to be united all-in-All eternally.

Chapter 22

The Big Collapse & New Beginning

Modern Views of the End of Our Physical World. The Big Collapse. After the Big Collapse. Hell from an Interrelational Perspective. Afterlife in IPhil.

Modern Views of the End of Our Physical World

In the 1920s, after recognizing that our universe started with a big bang, scientists projected two possible endings for our expanding universe: either fire or ice. Because of the gravitational attraction of the total mass of the cosmos, if the speed of expansion of our material universe after the big bang were less than terminal velocity, the cosmos would eventually stop expanding and collapse upon itself into a fireball similar to the original big bang. In contrast, if the speed of expansion were equal to or greater than terminal velocity, the cosmos would continue to expand, and according to the second law of thermodynamics, all communicable energy would dissipate into space, and everything would get colder and colder until each celestial and atomic body exists in its lowest energy state in icy isolation from all others.

Recent astronomical observations have provided evidence that the cosmos is not expanding at a uniform rate but at an accelerating rate. From this, scientists have concluded that the "ice scenario" will be the final fate of our cosmos. Also, even if parts of the universe collapsed and exploded several times in a cyclic manner, because of the second law of thermodynamics the net usable energy would diminish overall. Thus, even in a cyclic universe, its demise would be that of isolated bodies or "ice." In this scenario, all matter in the cosmos would drop to its lowest atomic energy state, from which it would be incapable of sending any energetic communication to another.

IPhil recognizes that humans, animals, plants, atoms, and all energized realities go through their own four stage process. Similarly, the universe as an energized, corporate reality will go through its own fourfold logistic, which IPhil calls the Grand Interrelational Logistic (GIL). Looking at the human body's rate of metabolism, professionals recognize that there is a maximal number of times that the cells in our body can replace themselves. Scientists currently project that approximately 125 years is the maximum possible human life span. At physical death, the organic structure of the body will disintegrate and return to the earth in diffusing material parts. Similarly, because the energy and matter in our universe is finite, the interrelational logistic of the cosmos is similarly finite in its operations, and it will have a maximal temporal lifespan. IPhil recognizes that because of the partial character of our finite universe and the partiality of the energy exchanges in our universe, the usable interrelational energy will gradually decrease, as predicted by the second law of thermodynamics. Consequently, IPhil recognizes that like the 4th stages of development of every individual, the logistical development of our physical world as a whole will end in physical dissipation. In this way, IPhil agrees with the scientific description of entropic reduction of our spatial universe into total isolation chaos. But there is more.

The Big Collapse

IPhil recognizes that chaotic, cold isolation of matter is *not* the ultimate end of our universe. The above explanation of entropic dissipation makes a big assumption: that after the big bang, once generated space is eternal. By the principle of conservation of energy, IPhil does recognize that potential energy and its forward-leaning parameter of time are eternal. However, the evolved spatial system is logistically progressive, and its interactions are not necessarily conservative and therefore are not eternal. As the spatial system had a quantized beginning and expansion, so too space itself will have a quantized end.

By the premises of logical evolution, IPhil recognizes that the logically evolved quantized proto-spatial system was generated according to its received, logically evolved, fourfold logistic through the communication of energized spaceons between spacepoints at the maximum velocity c. As long as these spacepoints receive, hold, direct, and send communicational energy to neighboring spacepoints totally, the evolved and progressively expanding spatial system will continue to expand.

However, in the communication of energy between spacepoints by spaceons, in the 2nd phase of spaceon's interrelational process, some of the communicated energy of spaceons *can randomly* be intrarelated and substantialized into new spacepoints. When a new spacepoint is randomly generated between two neighboring spacepoints, the quantum boundary condition of the quantized space's logistic causes those neighboring spacepoints to separate. As indicated in the chapter on gravity, this causes space to expand interiorly slightly at that point. This acceleration provides IPhil with an explanation for the logical evolution of the gravitational force, mass, as well as the "dark energy," which is responsible for the accelerative expansion of space. This causes the energy density of space to decrease. Through this evolutionary process, the amount of energy available for the maintenance of the expanding quantum space as a system is reduced. That is, not only the things in space but also the expanding spatial system itself displays an entropic reduction of available, usable, interrelational potential energy.

Furthermore, as in the cooling of a gas in one part of a closed chamber, the depleted exchanges of communicated energy between spacepoints at one place in space will tend toward a uniform, energy distribution. That is, a drop in spatial potential energy density in one part of space will quickly be spread and be equalized throughout space. Because of the reciprocal process within the logistic of the spatial system, as space expands, the commutable potential energy held by quantized spacepoints will slowly decrease throughout space.

Furthermore, the evolved and generated spatial system is not infinitesimally continuous but quantized. Here, maintaining reciprocal interactions between neighboring points requires the interrelational exchange to be of at least one quantum of energy. As a result, because the amount of potential energy in the space system is finite, and the energy density in the spatial system is finite, its energy density is constantly reducing. Because of the transformation of communicating spaceons into enduring spacepoints, there will come a critical time when there will be insufficient potential energy in the cosmic spatial system to sustain quantized intercommunication between the spacepoints in the interactive network of quantized space. After that point, the quantized fabric of space would no longer be able to sustain itself interrelationally.

When there is insufficient energy to keep spacepoints apart within its 3rd level spatial system, the 2nd level intrarelational character of energy will inversely cause that space system to revert from its physical, interrelational, kinetic state back into its former 1st level potential state. That is, when the spatial energy density drops locally to a level at which it can no longer sustain the spatial system in a quantized manner, a tear will occur in the quantized fabric of space. Since all space by then had been communally dissipated to its lowest, quantized energy state, that tear will travel at the rate of c throughout the fabric of space, and the whole expanding, cosmic spatial system will collapse like a deflated balloon, producing what I call the "big collapse."

This big collapse will *not* take place within physical space but will occur in the metaphysical *logistic* of the spatial system itself. Furthermore, since all logically evolved physical forces and material forms have the proto-spatial system as their metaphysical, logistical base, all finite physical forces and material forms will collapse, regardless of their respective states. Because space simply collapses, there is no motion. There is no observable kinetic motion where there is no space. Also, because they are established through logically advanced logistics, all materialized forms of energy in our finite universe will cease to exist physically. In other words, just as IPhil proposed that the big bang occurred because of the inner interrelational push of an amount of potential energy that was quantitatively equal to the absolute, so too IPhil proposes that the big collapse will occur when the amount of energy exchanged between spacepoints is less than one energy quantum, returning all kinetic energy in our universe to the Primal Absolute Potential Energy of the Almighty.

Nonetheless, since information is not of itself energetic but metaphysically interrelational, since potential energy logistically retains interrelational memory of all antecedent states, the potential energy of in the final state of potential energy (FAPE) will retain remembered metaphysical information of all the past events of GIL and carry it over into any subsequently evolved state. That is, if the potential energy of FAPE is used to re-energize a subsequent, logically evolved system beyond the antecedent, finally dissipated, logically consistent kinetic GIL, that potential energy would already be primed toward producing a logically evolved new world that is transcendent to the former system, carrying over information and logistic characteristics of our current world into the next.

It is a common experience that even as people watch the physical decline of a loved one, they are always caught by surprise at the exact moment when death holistically happens in an individual. While physical decay may spread slowly through the body, the death of the metaphysical integration of the individual happens all at once, followed by continued, gradual, physical decay. The exact time of this terminal metaphysical event cannot be physically determined precisely in advance, only approximated, by making projections about the decay of the physical systems of the logistically integrated metaphysical individual. This is IPhil's explanation of why the physical death of an individual as a metaphysical person happens suddenly, even as the logistic-soul of the individual endures. Similarly, the physical decay and the metaphysical collapse of the universe's spatial system will follow the same scenario. Just as the metaphysical collapse of a finite individual is related to but distinct from the physical deterioration of a body and often surprises observers as to exact time, so too the metaphysical big collapse of the spatial system and everything spatial within it will physically happen all at once and catch all participants and observers by surprise.

Sometimes an injury in one part of the human body can be so severe that holistically the physical evil is mortally consequential within the individual logistic, and the body as an integrated system is no longer able to continue vital communications between parts. Here the injured body no longer has the logistical consistency to continue to exist as an enduring system. In like manner, a similar cataclysm could theoretically happen in our universe's spatial system, and this could cause it to lose its logical consistency, prematurely ending the enduring physical existence of the spatial system before the metaphysical Grand Interrelational Logistic (GIL) is fully actualized. As an early death can happen at any point of an individual human's life, IPhil recognizes that a consequential dissociative cosmic event could occur at any point in the history of our universe's spatial system, leading to its early physical demise.

Since we currently do not know the amount of potential energy in the spatial system of our universe, or the magnitude of energy quanta needed to sustain interspatial communication between spacepoints within the spatial system, we do not know what kind of physical event could cause a fatal tear in the fabric of quantized space or when. We therefore need to be careful with dissociative events in the life of our earth and our spatial universe. IPhil recognizes only *that* the end of our physical universe could happen in the future *sometime*. When? We don't know.

Consequently, as the prophets of old warned, we need to live in an associative, good way, rather than in a dissociative, evil way. Even then, we need always to be prepared for the end, for we know not the day or the hour!

Looking at the duration of the Grand Interrelation Logistic (GIL) from the perspective of the all-giver-receiver, the general theory of relativity presents another insight. Relativity describes how a clock that is under the influence of a stronger gravitational field will tick slower than in an outside observer's clock. Consequently, when a human's clock on earth measures a thousand, millions, or billions of years, a clock in the energetic Almighty will measure a much slower passage of time. Thus, to the Almighty, the passage of time on earth will be much faster, relatively speaking. To people in the cosmos, the events of the evolution of our finite world will appear to pass slowly over billions of years. From the perspective of the energetic, panentheistic Almighty who possesses all potential energy in its mass equivalent as a singular reality, general relativity indicates that all the events of our long history would be to the Almighty as a blink of an eye.

After the Big Collapse

We experience now, but not the future. Since truth is the conformity of metaphysical mind to physical reality, there is no observational truth or philosophic truth in any future statement. However, the metaphysical logistic of a potential individual endures through time. Thus, the constancy of the four-phase logistics of the things in our world allows for the projective recognition of *probable logistical truths* in the future. Our experiences of the past give us partial knowledge of the logistic of things, and from this knowledge we are able to make probably true predictions of the future. While 2nd level, essentialistic idealists accept only one absolute, universal truth, IPhil is able to make the following probably true propositions about the future and the afterlife from 3rd level logistical projections:

- The foundation of our world is potential energy, which is conservatively eternal. After the termination of our physical universe in the big collapse, all kinetic energy of the big bang and our physical world will return to a potential state panentheistically in the Almighty.

- The logistic-souls of all individuals are in the metaphysical order, and as such, their logistic-souls are likewise ongoing and eternal. Being logistically consistent, every metaphysical, individual logistic-soul is capable of some logical evolution, even after physical death. The metaphysical evolution of an individual's logistic-soul would involve a carryover of the formal associative and dissociative metaphysical aspects of the antecedent state. These carryovers of the antecedent system will imitate in a secondary way the advanced characteristics acquired by logistical association with the progenitor in that evolutionary advancement.

Being possible, however, does not mean it will be actual. This is not a numerical probability but one that is grounded upon the individual's energized freedom to act or not act.

- The big bang was a systemic event, and it was the kinetic start of a communal logistic of our universe according to the holistic Grand Interrelational Logistic (GIL). The physical evolution of our world began through a logically inverse absolute-finite spacepoint progenitor. In a complementary way, because GIL is logically consistent, the universe is open to logical evolution through an inverse finite-absolute progenitor. This finite-absolute progenitor has to arise from our physical world *and* be capable of breaking the finite logistical boundary of our finite physical universe to establish an advanced logically consistent and enduring afterlife system through the forward push of the absolute, potential energy commuting Spirit of the Almighty. Here, all the members of that advanced system *could* participate with the finite-absolute progenitor in a panentheistic finite-absolute way in an advanced, evolved finite-absolute spiritual-material world.

- Just as an equation in physics can be actualized according to the logistical interrelationships described in that equation, similarly, the evolutionary advancement of metaphysical logistics of GIL after the big collapse can be evolved into an advanced state, provided that the evolved, potential logistic-souls of individuals are again somehow externally energized by the Almighty. Since there are no other givers, only the potential energy freely given by the all-giving-receiving-giving Almighty after the big collapse can actualize the logistic-souls of all things in our antecedent world into an advanced, evolved, finite, "heavenly"-absolute world, through and in union with the inverse, symmetry-breaking, evolutionary action of the finite-absolute progenitor.

Hell from an Interrelational Perspective

In the West, heaven and hell are traditionally described in a 2nd level way as antithetical, diametric opposites. With heaven being up and hell being down, never the two shall meet. Still, it is said that God is everywhere, even in hell. How can these two opposing ideas be reconciled? How can IPhil's 3rd level scenario of the afterlife describe how heaven and a hell can remain as interrelationally contrary realities within the panentheistic Almighty?

Let me answer that question by describing a true incident that happened when I was pastor at St. Ignatius Church in White River, Minnesota. A good family had several sons; unfortunately, one of the sons was an alcoholic who regularly became violent. One day he came home very drunk and started to tear up the house. His brothers tried to stop him, but he turned with greater, uncontrollable violence on them. His mother then came forward telling him to stop and behave himself, for this was their home. He responded by starting to beat up on his mother. With that, his brothers ganged up on him. Seeing he would not win this fight, he dashed out the door, got into his car, and started to drive wildly down the highway. A mile down the road, driving down the wrong side of the highway, he smashed into another car coming this way. The drunken young man and the entire family in the other car were killed. There was great mourning in that community. I tried as best I could to support the families before, during, and after multiple funerals.

About two months later, I heard a soft rap at my door. It was the grandmother of the young man who caused the accident. I invited her in, and she took a seat at my table across from me. Her eyes were downcast, and it was clear that she was deeply troubled. I waited silently as she found the courage to say what had been bothering her. Finally looking up, she said to me, "Father, I love all my grandchildren very much, and I want to be near all of them. If any one of my grandchildren goes to hell, then heaven will be hell to me."

I looked at her with sympathy and love. This was a deeply profound and meaningful matter to her. I opened my mouth to answer, and I was overwhelmed as I heard the wisdom the Spirit was putting into my mouth. "That is a very difficult question, and with difficult questions, they often are best answered with a story. So, let me tell you a story about a teenage girl. She wanted to go to a movie with her friend one afternoon. So, she asked her mother for money to go to that movie with her friend. Her mother responded by reminding her that today everyone was gathering to celebrate grandpa's birthday. So, the mother told her that she couldn't give her that money, for that would mean she was participating in the girl's missing the family birthday party and not properly honoring her grandfather. Rather, she would have to accompany her to that honoring birthday party.

"When they arrived at the grandparent's home, the girl was very angry. She had not said a word in the car all the way there. When they entered the home, everyone was already there, and they greeted them with happiness. The girl, however, kept her head down, went into the living room, plopped into the furthest corner of the couch, folded her arms, stared into the floor, and loudly said, 'Ugh!' While everyone else was glad she was there with the rest of the family, she saw no one at the party. They were in heaven; she was in hell."

The grandmother listened intently to the story. Her eyes said that she understood. She smiled softly in appreciation for what I said. Then quietly she got up, left, and went home with a bittersweet sense of understanding and peace.

In the above incident, the grandmother's irreconcilable pain came from the 2nd level, idealized descriptions of hell and heaven that she had been taught her by religious teachers, who presented the concepts of hell and heaven as antithetical, diametric opposites. For them, heaven and hell were on opposite sides of the fence—totally white-and-black realities. In contrast, my 3rd level response described hell and heaven as dialectic, synthetic contraries. Here, hell is not outside of heaven, but rather, hell is panentheistically within the Almighty and within God's interrelational heaven.

In the end, must all mistakes, sins, and dissociations be materially corrected? This rational expectation is built on the 3rd level principle of justice. However, 4th level metanoia involves "letting go" of all one's dissociative imperfections, which in mercy the Almighty will forgive, while actualizing an advanced material/spiritual state that evolutionarily is built and advances the individual's former logistic. In contrast, a stubborn individual intentionally holds on to some significant materialities of one's past, and in justice the unrepentant individual will suffers the natural consequences of those dissociations from one's Almighty-grounded logistic.

Within the 3rd level perspective of IPhil, because of the carryover characteristic of logical evolution, it is possible for a free individual in their 2nd level self-centeredness to continue to be dissociative and experience the pain and anger that is found in going against the Spirit of one's received interrelational potential energy. A hardhearted individual can deliberately remain stuck in one's self-centered, dissociative, 2nd phase of one's four-phase logistic. In a logically evolved way, my 3rd level image of the advanced afterlife more closely matched the grandmother's own personal experience, and it was understandable to her, albeit it in a bittersweet way. Note that in IPhil, heaven is not an all-happy place—because of the possibility that some individuals might remain at a 2nd level self-centered state within a self-isolating, dissociative hell within heaven.

IPhil recognizes that potential energy is the foundation of kinetic energy, and by the carryover process of logical evolution, Primal Absolute Potential Energy is the imminent, existential foundation of everything in our world. Thus, where there is potential energy, there is the Almighty, potentially supporting and guiding all individuals to be more interrelational. Consequently, it can be said that we are always panentheistically of God, and therefore we are already *of* heaven—in a graced, potential way. Heaven is also traditionally described as "up above the stars." In IPhil, the logistic of the Almighty is absolute and panentheistically transcendentally contains all in All in a surrounding, embracing way. Still, by the "letting go" of the all-giver, progressive finite individuals have their own individual logistics and free self-initiatives. Thus, their self-centered logistics are formally distinct from the logistic of the absolute Almighty. Nonetheless, the finite-absolute progenitor is destined for union with all-in-All within the absolute order, while carrying over the Grand Interrelational Logistic of the antecedent finite earthly order. In addition, an evolutionary human support group will participate in the establishment of the advanced, empowered, communal logistic of the finite-absolute progenitor. By the carryover process of logical evolution, all other finite receiver-givers will be associated members of that new finite-absolute order, continuing their former antecedent logistic/nature *and* advancing in a transcendent way in the expanded, all-embracing PAIS of the Almighty.

Afterlife in IPhil

Interrelational philosophy's understanding of the afterlife is not built upon the cacophonous ideas of different speculative humans but upon the fourfold logistic/nature of the potential and physical energy of our world. Because the afterlife is a metaphysical reality, we cannot now have direct physical experience of it. So, it is impossible to make any *philosophically true* statements about the afterlife. Nonetheless, the ongoing logistic of conservative energy is well known through multiple experiences. Consequently, it is possible to project the *probably true* characteristics of a logically evolved afterlife for all energized reality. Humans intuitively have been making similar projections about the afterlife for millennia. Beyond that, IPhil's projections are experientially more probable because they are based on the conservative, advancing, evolving, interrelational potential energy of our current world.

In dualistic essentialistic thought there are only an idealized heaven for those humans who meet a particular moral standard, and an idealized hell for those humans who don't. Heaven is only for individuals logistically their equal or greater. The rest of creation is ignored and dismissed, as were servants and things in ancient times. In contrast, using the logical evolutionary theory and the logistical principle: *Quidquid recipitur in modo recipientis recipitur*—What is received is received in the mode (logistic) of the receiver—IPhil proposes that the experience of heaven in the re-energized afterlife in the end times will be gradated. As kinetic energy is partially of and in potential energy, so too, hell is of and in heaven, panentheistically in and of the all-receiver-giver in PAIS. Self-centered, nonreciprocal individuals will experience deep interior pains according to their God-given individual and communal logistics from their continued dissociation from the continuous push of the energy-communicating Spirit toward being more interrelational. In contrast, moral humans will be happily interrelated in a logically evolved, *finite* logistic; they will be perfectly happy especially being with their own. Beyond this, those who interrelate with the finite-absolute progenitor will participate in an evolved, transcendent societal order with the divinized progenitor. Moreover, this transcendental order will be happily, lovingly, interrelationally associated with all antecedent, finite, material orders of creation in a Godlike way.

The following is a list of some of the probable attributes and experiences of deceased logistic-souls—if and when they are re-energized by the Almighty into an evolutionary afterlife. Because of their hope filled impatience, humans regularly elide distant, transcendent future events in the end with proximate future events shortly after death. Just as energy pushes to fill a vacuum, so too humans push to fill the vacuous state immediately after a loved one's physical death with an afterlife that fulfills their hopes and dreams for their loved ones. In contrast, IPhil recognizes that the state of a logistic-soul immediately after death is unconscious and uneventful, until logistic-souls are again freely energized by the Almighty. In this way, to the re-energized individual, it would seem that no time had lapsed between one's death and one's re-energization.

1. *Re-energizing of logistic-souls.* Just as one's metaphysical logistic-soul needs to be energized and materialized by one's parents to become physically interactive in this life with increasing consciousness, so too after a person's death, one's metaphysical logistic-soul will remain in a dormant, unconscious state until re-energized by another. As the forward-leaning, characteristic of energy, the absolute potential energy of the Almighty pushed the all-giver to evolve our world physically in the big bang, so too, the Almighty after the big collapse can push energetically to again be an all-giver-receiver-giver that freely self-initiates a more evolved finite transcendental physical world in the all-receiver-giver-receiver. Also, IPhil recognizes that after a person dies, the prayers and longings of loved ones and devotees, as expressions of their interrelational potential energy, can partially stimulate the dormant, open, individual logistic-soul, making a partial spiritual connection with the logistic-souls of saints and loved ones—for a time.

2. *Eternal endurance of logistic-souls.* The logistic of an individual's DNA physically endures unchanged through life. Beyond this, the metaphysical *logistic code* of that DNA in the afterlife is logically evolved from and within the fourfold logistic of the Almighty. That logistic accumulates memories from past experiences, which are carried over into the current situation of every individual. However, in potential energy those memories are known only metaphysically and subconsciously, and they become consciously known only when those potentialities are energized according to the logistic-soul of the individual.

3. *Finite-Absolute Progenitor.* In the beginning of our world, an absolute-finite progenitor was necessary to break the logistic of the logically consistent Primal Absolute Interrelational System (PAIS) to generate the quantized proto-spatial field. This took place at the beginning of the Grand Interrelational Logistic (GIL) of our finite world. At the end of GIL, it is likewise necessary that there be a finite-absolute progenitor to break that logistic of our finite world to establish an evolved metaphysical order that is in union with the absolute order. Also, the logical evolutionary process requires a support group of members of the antecedent (human) system to establish and generate an enduring communal system in the absolute order in union with the finite-absolute progenitor.

4. *Review of Life in Afterlife.* Memory is a carryover metaphysical aspect of forward-leaning logistic through time. Memories are unconsciously present until an individual's logistic is energized for an advancing interrelational purpose. In viewing the memory of one's interrelationally deficient past actions, the average individual will consciously experience some sadness. During life, when an individual is logistically dissociative in doing what is wrong, the interrelational

energy of the person will push the individual to have a change of heart—a metanoia—and to "let go" of ownership of those minor, temporary (sinful) dissociations. When that happens, the interrelational Spirit of energy will "let go" of the demands of 3rd level justice and extend 4th level mercy. In that extended mercy, memories of past misdeeds become inconsequential and are in effect forgotten as the individual moves forward interrelationally. The material aspects of those wrongs forgiven in mercy can in justice be covered by loving human associates. However, if the dissociative misdeeds are major and the individual stubbornly does not "let go" of these memories and intrarelationally holds on to them as part of her individual identity, 3rd level justice demands that the energizing Spirit will continue to press for repentance as long as the sinner remains hardhearted—which could be eternally. This enduring antithesis is possible and right in the energized Almighty because 2nd level interrelational antitheses are within 3rd level interrelational synthesis, which presses to enflame one's hearts to be more interrelational, harmonious, happy, wholly holy, and at peace.

5. *Life in the Parousia.* The Almighty does not interrelate and love everyone equally, but according to their respective, individual logistical ways. That is, the Spirit of the Almighty communicates potential energy interrelationally and lovingly with all individuals, according to their received or logically evolved logistic. In this way, each individual will feel personally fulfilled. There will be no jealously, because *Quidquid recipitur, in modo recipientis recipitur.* "Whatever is received, is received in the mode of the receiver." Individuals, rocks, plants, animals, and some humans with lesser evolved logistics will not feel slighted because they will not recognize the logistic of those logically evolved to a higher level of interrelationships. Each level of interrelationships will be both individually and communally fulfilling, according to their received, evolved, eternal logistic. They will experience heaven according to their present logistical expectations.

IPhil recognizes that the expression "all-in-All" has a fourfold development. (1) Before all ages, all finite individuals are potentially of/in the potentialities of the Absolute. (2) In history, all finite individuals are kinetically are of/in the foundational potential energy of the Almighty. (3) With the coming of the finite-absolute progenitor, all finite individuals are embraced in the universal logistic of/in the transcendentalizing progenitor. (4) In the end, both the Almighty and all finite receivers in the logistic of the all-receiver-giver progenitor logically evolve into an advanced, universal, joyful, associative union of/in PAIS.

In conclusion, the interrelational philosophy (IPhil) in this book is grounded upon the nature of potential energy, which is the ur-stuff of reality from before the big bang to after the big collapse. In this book, I have described how energy is initially in a potential, metaphysical state, and then it is expressed as physically evolved as material and kinetic realities. Energy gives existence, forms intrarelationally material bodies, organizes interrelational communities and species, and empowers free self-initiatives in our world. Energy metaphysically and physically has a fourfold characteristic, which being logically consistent established in a quantized way our logically evolving 4-dimensional spatial world, the four fundamental forces of physics, the fourfold character of everything cyclic, the four phases of human life, and many different fourfold human activities. Thus, in many ways, IPhil shows how our world energetically and interrelationally evolved from God.

Glossary

Philosophic terms are 3rd level, and they have meaning not in themselves but within specific paradigmatic worldviews. The following glossary indicates the special meanings and connotations the following terms have in Interrelational Philosophy (IPhil).

Absolute—IPhil recognizes that the ur-stuff of reality is the potential energy from which came the singular kinetic big bang of our finite world. This amount of energy is all that there is in our world and therefore it is *absolute* in our finite order. This absolute energetic reality in the form of antecedent potential energy is called the Almighty.

Act, Actual, Actuality (as opposed to potency, potentiality)—In IPhil, *act* and *actuality* describe the state of a reality in the process of giving or receiving interrelational energy. The terms "real" and "reality" refer to what is energetically existent, and "act" and "actual" refer to some form of changing or becoming.

Almighty—See Absolute.

Analogy—Two realities or events are analogous if they have some common characteristic(s) among other differing characteristics. IPhil recognizes the foundation of analogies is the common ancestry shared by two entities on different branches of the tree of logical or physical evolution.

Artificial (or Domestic) Evolution—The formation of biologically transformed individuals through selective breeding practices by humans. This was originally done for increasing bodily characteristics valued by humans.

Being & Becoming—Energy, the founding, dynamic principle of our world, is interrelational, having terms and connectives. Thus, in IPhil, *being* is the founding aspect of energy, giving terms enduring existence. The interrelational aspect of energy gives terms some forward-leaning capability of *becoming*. Classical, essentialistic philosophies held that "being" could exist alone in a substantial state as unchanging "absolute Being," which those philosophies identified with the Good or God. In contrast, IPhil holds that the substantial foundation of all that exists is Primal Absolute Potential Energy (PAPE), which, as the Almighty, potentially exists in a state of "being," *and*, as interrelational, it is inherently oriented toward expressing itself kinetically as "becoming." If energy does not have a forward-leaning character, it is not energetic.

Biological Evolution—The transform of biological entities through random mutations of some genes in an ancestor's RNA or DNA. The driving force behind random biological evolution is the push of a thing's potential energy to be more interrelational in a given environment or society.

Communal Logistic—The sum of all possible interrelationships which members in a group have in common with all the other individuals in their group or species. It is inherited through one's parent(s) or advanced through an evolutionary progenitor.

Communicator—An energized metaphysical or physical operator within a system that transfers energy from one term to another term within an interrelational system.

Conscience—Intuited 1st level knowledge drawn from one's potential energy, indicating whether a planned, free action is according to one's received individual logistic—or not.

Consciousness—The advanced form of knowledge found intrarelationally in individuals enabling them to recognize, discern, and actualize energetic possibilities for themselves or with others. Once a progressive goal has been achieved, conscious concerns fade back into unconscious systemic awareness.

Energy—Energy enables things to endure and interrelate. An individual's potential energy sustains enduring existence, pushes for the formation of intrarelational individuals, promotes dynamic interrelational associations, and empowers free self-initiatives involving logistical possibilities toward advanced associations.

Generative Operator—A generative operator can interrelate existing terms and systems to produce more complex terms within a given system.

God—Unlike essentialistic philosophies which maintain a static, imperial, perfectionistic view of God, IPhil recognizes that God is an absolute, energetic, developmental reality, which through logical and physical evolution is the prototype of all finite, evolved, and developmental energetic realities in our world. Since the interrelational logistic of energy is fourfold, God's metaphysical aspects and phases of development are fourfold, namely, PAPE, PAER, PAIS, and our panentheistic world.

Imaginary—Pertaining to the metaphysical quality of intrarelated sense images within the mind, which are not yet energetically actualized in the physical world. In this way, imagined unicorns and geometric figures are real in the mind and can be analogously presented on paper, but they are not now energetically existent in the physical world.

Individual Logistic—The sum of all the intrarelationships that orders all parts into one logically self-consistent, enduring individual. The individual logistic is a particular expression within the communal logistic of one's ancestral group. Being logically consistent, one's individual logistic endures, while kinetic expressions of its received potential energy change the individual within the logical possibilities of the individual's logistic. It is similar to the Aristotelian notion of an individual's substantial form.

Interrelational Logistic—A fourfold, metaphysical algorithm describing how energy progressively advances through the four major kinetic stages of an individual's or a system's development: receiving, integrating, interrelating, giving.

Interrelational Philosophy (IPhil)—This philosophical system presents a worldview whose foundation is interrelational energy, and whose paradigm describes how something becomes energetically existent, capable of internally organizing intrarelational individuals, forming advancing logically-consistent systems, and founding free self-initiatives in a progressive or evolutionary way.

Inverse Operator—An inverse operator reverses the action of a generative operator. Because the operation of an inverse operator is subsequent and opposite that of a generative operator, its operations within a given logically consistent system logically are normally symmetrically bound within that system. Logical evolution *can* occur when the operation of an inverse operator energetically and asymmetrically breaks the reciprocal boundary of a logically consistent system, thereby initiating the formation of a new, evolved system, which contains and exceeds the old.

Intuition—In IPhil, potential energy is the foundation of kinetic energy. This existential potential energy is oriented toward the intrarelational formation of integral individuals. One's potential energy pushes a forming individual intrarelationally according to its individual logistic, which can be vaguely or subconsciously known.

IPhil—Interrelational Philosophy. See above.

Knowledge—A fundamental, inherent ability of interrelational energy of an individual to be aware of potential interrelations within itself or with another. As potential energy pushes toward physical, intrarelational and interrelational goals, one's knowledge correspondingly becomes more conscious and deliberate in the individual. Thus, knowledge of interrelationships initially is unconsciously instinctive, then semiconsciously sensate, then consciously rational, and then speculatively planning.

Logical Evolution—The process of forming an advanced logical or metaphysical system through the extension of the range of operation of an inverse operator, leading individuals from and beyond an antecedent system, resulting in the formation of a new, advanced system that is logically consistent with, contains, and logically goes beyond the antecedent system.

Logistic—Taken from military parlance, the logistic of a system is holistically the sum of all logical and physical interrelationships of elements involved in the accessible interactions within an interrelational system. IPhil differentiates individual logistics from communal logistics insofar as the former deals with the potential operations of an integrated individual, and the latter deals with the interactive operations of a corporate system or group of individuals.

Material (as opposed to spiritual)—A thing is called material insofar as it retains and expresses its received energy kinetically. Potential energy takes on material characteristics starting in the 2nd phase of energy's metaphysical development. Some objects are material even if they are not visible. IPhil considers spacepoints in the proto-spatial system to be material because they are confined within a quantum of space, and they are the energy-receiving dialogical terms separated by and energized by spaceons, which are energy-communicators going between separated spacepoints.

Metaphysical (as opposed to physical)—In IPhil, metaphysics pertains to the characteristics of potential energy, which are unobservable in themselves but are indirectly knowable through interrelational kinetic interactions stimulated by potential energy—hence, after-physical.

Object, Objective (as opposed to subject, subjective)—The term "object" refers to the receiver of a physical, metaphysical, or mental action. The term "objective" refers to a mental idea in the brain that has a physical foundation outside the subjective knower. An idea is said to be objectively true when the internal, subjective idea in the mind of the knower matches the external, objective foundation of the idea outside of the mind of the knower. Because of the quantum characteristics of our energetic world and of our senses and knowledge, all knowledge of things and interrelationships are objectively only approximate.

Operator—A metaphysical connective between terms in an interrelationship, indicating the possibility of transfer of potential energy. Since numbers arise metaphysically from sets of distinct terms, operators can connect numbers.

PAER (Primal Absolute Energetic Reality)—The 2nd stage of development of absolute potential interrelational energy in which absolute energy intrarelates and establishes One, individual, absolute reality. In different religious traditions, conceptualized PAER is called God, the Almighty, the Absolute, the Transcendent, etc.

PAIS (Primal Absolute Interrelational System)—The 3rd stage of ongoing reciprocal exchange of absolute energy between the all-giver-receiver and the all-receiver-giver through the all-communicator produces a logically consistent absolute system. Hindus variously named and described the primal absolute Trimurti as consisting of Brahma, Vishnu, and Shiva. Christians have named and described the absolute Trinity as of the Father, and the Son, and the Spirit. These are interdependent, free, intelligent, absolute individual realities in one absolute systemic whole.

Panentheism—IPhil recognizes that the creation of the world is not a deductive process but a logistical evolutionary process in which all finite receivers have their own finite logistic and their own free self-initiative within the all-receiver of PAIS. This stands in contrast to pantheism in which all finite realities are logically deduced from God and therefore have the same nature as God. In pan-*en*-theism, all finite, energetic individuals are *in* the all-receiver in PAIS, but by the logical evolutionary process, all finite receivers do not have the same absolute nature as the primal all-giver. In IPhil, all finite receivers are logistically distinct, and have a different nature from God through the logical evolutionary process.

PAPE (Primal Absolute Potential Energy)—The 1st enduring stage of absolute potential energy. Potential energy has an initial existential state of "being" as well as energetic, forward-leaning, potential "becoming." These aspects are interdependent because if PAPE existed only as perfect being, nothing new could happen that is formally distinct from the current state of PAPE. Although oriented toward intrarelational unity, the initial state of PAPE is not yet intrarelated and therefore can be said to be unformed or diffused. Consequently, applying a "name" to this reality is intellectually not justified. As diffused potential energy, PAPE is therefore a "no thing." As Lao Tzu wrote long ago, "The Dao which is named is not the real Dao." PAPE is only an intuited, pointing acronym, rather than an integrally founded reality.

Paradigm—A philosophical or theoretical framework within which generalizations, drawn from experience, lead to a set of theories, conclusions, and experiments, which together produce an all-encompassing, logically consistent system or worldview.

Person—Essentialists define a person as an individual who is rational. In contrast, IPhil recognizes persons as those individuals who are recognized as actively associative with others. Applied analogously to other levels of interrelationship, the subjective application of the term "interrelational person" in history is regularly applied exclusively to favored groups. Those outside of this favored class are effectively "non-persons." IPhil is logistically oriented toward recognizing all energized realities on all evolved levels as actively, interrelationally personal.

Physical (as opposed to metaphysical)—IPhil recognizes that potential energy can express itself kinetically by interacting with others in space and time, such that the event can be somehow observed.

GLOSSARY

Progenitor—In a reciprocal, logically consistent system, there *can* be a member that is so potentially energized to be able to break the boundaries of that system in a logically inverse way. This enables this dissociative individual to be the first member of a formally advanced, evolved, and generative system.

Potency, Potential (as opposed to act, actuality)—In IPhil, potency describes the state of a reality that *can* give and receive interrelational energy. The state of energetic potency begins with a state of existential being and leans toward subsequent states of interrelational becoming, toward making some type of change in oneself or another, and into a self-initiated subsequent state of potential energy or kinetic energy. In the transformation of potential energy into kinetic energy, the potency to change does not end but remains in a carryover evolutionary way as the enduring, existential foundation of those dynamic changes. When the capacity to act kinetically ends or is no longer sustained, the kinetic energy of an individual reverts to its foundational potential energy state.

Real—In IPhil a thing is "real" insofar as it is energetically interrelational on some level of metaphysical or physical interrelationship. A unicorn is *metaphysically real* within the human mind and as an illustration in a book. However, unicorns are not *physically real* because their existence has not yet been physically evolved and observed to be a part of our interactive physical world.

Realism—The philosophical affirmation that links what is known subjectively in the knower with what objectively, physically exists outside of the knower.

Self-Initiative—The 4th metaphysical ability of an energized receiver, enabling an individual to transform some or all its interrelational potential energy into some form of kinetic energy. In early evolutionary realities, receiving individuals are not yet intrarelationally well-developed, so their free, energetic initiatives are not individually self-initiated but randomly initiated. In more evolved realities, like humans, individuals can consciously self-initiate potentials that are more interrelational or evolutionary.

Soul—In IPhil a soul is the "individual logistic" of an intrarelated existent. It is like the "substantial form" in Aristotelian and Scholastic metaphysics. Since logistics are metaphysical, logistic-souls precede the formation of a body and remain after an individual person dies physically. One's DNA *code* is an important part of one's logistic-soul. One's physical body is in and of an individual's logistic-soul. Because all logistics are logically evolved panentheistically, they are metaphysical evolved in PAIS before the big bang. Thus, all logistic-souls are eternal, and they *can* progressively guide actions in the physical order, and they *can* find their fulfillment after physical death in an externally energized and evolved afterlife.

Spirit—In IPhil the term "spirit" has two meanings: (1) The unified metaphysical push of potential energy and information in an interrelated group urging the members of the group to communally act in a team-like, corporate way. (2) A communicator of potential energy and information to an individual receiver, urging the individual receiver to act in an advancing interrelational way.

Spiritual (as opposed to material)—In IPhil a thing is called spiritual insofar as it possesses and transfers potential energy and information interrelationally in various forms, especially in energy fields between and in terms. Because these operations are between terms, they are physically invisible to the terms but secondarily known by the terms. So, what is spiritual and metaphysical is invisible.

Subjective (as opposed to objective)—In IPhil the term "subjective" refers to the self-initiator of a physical, metaphysical, or mental action. In common parlance, however, use and application of the word "subjective" can be confusing because sometimes the term "subject" can refer to the object of one's mental inquiries.

Truth—Truth is the conformity of an individual's knowledge of another's reality. It can be found on all four metaphysical levels: metaphysical truth between potentiality and actuality, personal truth between self-knowledge and experience, communal truth between proposition and systemic practices, and corresponding truth connecting metaphysical awareness and efficient, physical expressions. Because physical observations are partial, increased external and internal confirmation moves knowledge toward comprehensive *certitude*.

Bibliography

Alpha Institute for Advanced Studies. "Beyond Einstein." *Nexus*, August/September 2018.
Al-Rawi, F. N. H., and J. A. Black. "A New Manuscript of Enūma Eliš, Tablet VI." *Journal of Cuneiform Studies* 46 (1994) 131–39.
Aristotle. *Metaphysics*. Translated by Hugh Tredennick. Cambridge: Harvard University Press, 1933.
Aristotle. *The Nicomachean Ethic*. Edited by David Ross and Lesley Brown. New York: Oxford University Press, 1980.
———. *Physics*. Translated by Hugh Tredennick. Cambridge: Harvard University Press, 1989.
Armbruster, Karl J. *The Vision of Paul Tillich*. New York: Sheed and Ward, 1967.
Armitage, David. *Declaration of Independence: A Global History*. Cambridge: Harvard University Press, 2009.
Augustine. *On the Trinity—De Trinitate*. Translated by Edmund Hill and John E. Rotelle. New York: Oxford University Press, 2015.
Bacon, Francis, et al. *The New Organon*. New York: Cambridge University Press, 2000.
Bagrow, L. "The Origin of Ptolemy's Geographia." *Geografiska Annaler* 27 (1945) 318–87.
Baker, Peter, and Bernard Goldstein. "Theological Foundations of Kepler's Astronomy." *Osiris* 16 (2001) 88–113.
Batchelor, Stephen. *Buddhism without Beliefs*. New York: Bloomsbury, 1998.
Bates, Todd. *Duns Scotus and the Problem of Universals*. London: Continuum Logo, 2012.
Beddall, B. G. "Wallace, Darwin, and the Theory of Natural Selection: A Study in the Development of Ideas and Attitudes". *Journal of the History of Biology* 1 (1968) 261–323.
Bennisi, Elizabeth. "On the Origin of Cooperation." *Science*, September 4, 2009.
Berne, Eric. *Games People Play: The Basic Handbook of Transactional Analysis*. New York: Ballantine, 1964.
Boethius, Anicius Manlius Severinus. *The Theological Tractates and the Consolation of Philosophy*. Translated by H. F. Steward and E. K. Rand. Cambridge: Project Gutenberg, 2004.
Boole, George. *Investigation of the laws of thought, on which are founded the mathematical theories of logic and . . . probabilities*. Summit, PA: Prometheus, 2003.
Bradley, Robert E., and Charles Edward Sandifer. *Cauchy's Cours d'Analyse: An Annotated Translation*. New York: Springer, 2009.
Brett, George Sidney. *The Origin and Goal of History*. New York: Routledge, 1953.
Broughton, John M., et al. *The Cognitive-Developmental Psychology of James Mark Baldwin: Current Theory and Research in Genetic Epistemology*. Norwood, NJ: Ablex, 1982.
Burkeman, Oliver. "How to (Truly) Improve Your Luck." *MentalFloss*, June 2014.

Burnet, John. *Early Greek Philosophy.* 3rd ed. Meridian, ID: Meridian Library, 2014.
Butterworth, G. W. *Clement of Alexandria.* Cambridge: Harvard University Press, 1999.
Cameron, Alexander Thomas. *Radiochemistry.* London: Dent, 1910.
Chen, Ku-Ying. *Lao Tzu Text, Notes, and Comments.* San Francisco: Chinese Materials Center, 1981.
Chidbhavananda, Swami. *The Bhadavad Gita.* Tirupparaitturai, India: Sri Ramakrishna Tapovanam, 1997.
Child, James M., and Carl Immanuel Gerhardt. *The Early Mathematical Manuscripts of Leibniz.* London: Open Court, 1920.
Cicero, Marcus Tullius, and Harry G. Edinger. *On Old Age; On Friendship.* Indianapolis: Bobbs-Merrill, 1967.
Close, Frank. *Antimatter.* New York: Oxford University Press, 2018.
Clutton-Brock, Juliet. *Domesticated Animals from Early Times.* London: British Museum of Natural History, 1987.
Cooper, Dan. *Enrico Fermi and the Revolutions of Modern Physics.* New York: Oxford University Press, 1999.
———. *New Classic American Houses: The Architecture of Albert, Righter & Tittmann.* New York: Vendome, 2009.
Copleston, Frederick C. *A History of Philosophy.* Vol. 4. Tunbridge Wells, UK: Burns & Oates, 1999.
———. *A History of Philosophy.* Vol. 7. Tunbridge Wells, UK: Burns & Oates, 1999.
Curry, Andrew. "Gobekli Tepe: The World's First Temple." Smithsonian.com. November 2008.
Dalley, Stephanie. *The Legacy of Mesopotamia.* New York: Oxford University Press, 2005.
Darwin, Charles, and Gavin De Beer. *Darwin's Notebooks on Transmutation of Species.* London: British Museum of Natural History, 1960.
Dawkins, Richard. *The Selfish Gene.* New York: Ziff-Davis, 1978.
Descartes, Rene. *Discourse on the Method of Rightly Conducting the Reason and Seeking Truth in the Sciences.* Translated by Donald A. Cress. Indianapolis: Hackett, 2008.
Diogenes Laertius, and Charles Duke Yonge. *The Lives and Opinions of Eminent Philosophers.* Andesite, 2015.
Dirac, Paul A. M. "The Quantum Theory of the Electron." *Proceeding of the Royal Society of London* 117 (1928) 610–24.
Dongen, Jeroen van. *Einstein's Unification.* New York: Cambridge University Press, 2018.
Eardley, Peter, and Carl Still. *Aquinas: A Guide for the Perplexed.* London: Continuum, 2010.
Ehrman, Bart D. *How Jesus Became God: The Exaltation of a Jewish Preacher from Galilee.* New York: HarperCollins, 2014.
Einstein, Albert. "How I Constructed the Theory of Relativity." Translated by Masahiro Morikawa from the text recorded in Japanese by Jun Ishiwara. *Association of Asia Pacific Physical Societies Bulletin* 15 (2005) 17–19.
———. "On Science and Religion." *Nature* 146 (1940) 605–7.
European Space Agency. "Planck: Mission Status Summary." March 19, 2013.
Falcon, Andrea. "Four Causes." In "Aristotle on Causality," from the *Stanford Encyclopedia of Philosophy.* Redwood City, CA: Stanford University Press, 2008. https://plato.stanford.edu/entries/aristotle-causality/#FouCau.
Feferman, Solomon, et al. *Kurt Gödel: Essays for His Centennial.* New York: Cambridge University Press, 2010.

Feynman, Richard P. *The Strange Theory of Light and Matter*. Princeton, NJ: Princeton University Press, 1985.

Fisher, Ronald Aylmer. *The Genetical Theory of Natural Selection*. New York: Oxford University Press, 1999.

Fowler, James W. *Stages of Faith: The Psychology of Human Development and the Quest for Meaning*. San Francisco: HarperSanFrancisco, 1981.

Fowler, Thomas B., and Daniel Kuebler. *The Evolution Controversy: A Survey of Competing Theories*. Grand Rapids: Baker Academic, 2007.

Francis, Matthew. "Quantum Entanglement Shows That Reality Can't Be Local." *Ars Technica*, October 30, 2012.

Francis of Assisi. *The Canticle of the Sun of St. Francis of Assisi*. New York: Duffield, 1907.

Froom, Leroy Edwin. *The Prophetic Faith of Our Fathers*. Vol. 4. Washington, DC: Review and Herald, 1954.

Gaiser, Konrad. "Plato's Enigmatic Lecture 'On the Good.'" *Pronesis* 25 (1980) 5–37.

Geddes, Andrew. "Person." *Catholic Encyclopedia*. New York: Appleton, 1911.

Giovino, Mariana. *The Assyrian Sacred Tree: A History of Interpretations*. Göttingen: Vandenhoeck & Ruprecht, 2007.

Glashow, Sheldon L. *From Alchemy to Quarks*. Belmont Grove, CA: Brooks-Cole, 1993.

Gödel, Kurt. *On Formally Undecidable Propositions in Principia Mathematica and Related Systems*. New York: Cambridge University Press, 1931.

Goldstein, Rebecca. *Incompleteness: The Proof and Paradox of Kurt Gödel*. New York: Norton, 2006.

Gould, Stephen Jay, and Niles Eldredge. "Punctuated Equilibria: The Tempo and Mode of Evolution Reconsidered." *Paleobiology* 3 (1977) 115–51.

Greenberg, Robert. *Kant's Theory of A Priori Knowledge*. University Park: Penn State University Press, 2001.

Guth, Alan H. *The Inflationary Universe*. Reading, MA: Perseus, 1997.

Hammond, Richard. *Chien-Shiung Wu: Pioneer Nuclear Physicist*. New York: Chelsea House, 2010.

Harlow, John Martyn. "Recovery from the Passage of an Iron Bar through the Head." Reprint of speech delivered at the Massachusetts Medical Society, 1868. Boston: Clapp, 1869. https://archive.org/details/66210360R.nlm.nih.gov/page/n3/mode/2up.

Harper, Douglas. "Energy." *Online Etymology Dictionary*. https://www.etymonline.com/word/energy.

Hawking, Stephen W. *The Illustrated A Brief History of Time*. New York: Bantam, 1996.

Heath, Thomas Little. *Aristarchus of Samos, the Ancient Copernicus: A History of Greek Astronomy to Aristarchus; Together with Aristarchus's Treatise On the Sizes and Distances of the Sun and the Moon*. Oxford: Clarendon, 1966.

Hick, John. *Evil and the God of Love*. Basingstoke: Palgrave Macmillan, 2010.

Hillman, J., and T. Moore, eds. *A Blue Fire: Selected Writings of James Hillman*. New York: HarperPerennial, 1989.

Hume, David. "Dialogues concerning Natural Religion." In *Essays and Treatises on Several Subjects*. Edinburgh: Kincaid and Donaldson, 1767.

Jaspers, Karl. *The Origin and Goal of History*. Abingdon: Routledge, 2010.

Jolowicz, H. F. *Historical Introduction to the Study of Roman Law*. Cambridge: Cambridge University Press, 1967.

Jung, C. G. *Psychological Types; or, The Psychology of Individuation*. New York: Pantheon, 1964.

Kandel, Eric R., et al. *Principles of Neural Science*. New York: McGraw-Hill Medical, 2013.

Kaplan, I. G. *The Pauli Exclusion Principle: Origin, Verifications and Applications*. 2017. https://doi.org/10.1002/9781118795309.

Kennington, Richard. *Philosophy of Baruch Spinoza*. Washington, DC: Catholic University of America Press, 2018.

Kisner, Matthew J. *Baruch Spinoza*. New York: Oxford University Press, 2016.

Kohlberg, Lawrence. *Essays on Moral Development*. San Francisco: Harper and Row, 2001.

———. "Moral Development." Chapter 12 in *Essays on Moral Development*, vol. 1, *The Philosophy of Moral Development*. San Francisco: Harper & Row, 1982.

Koller, S. M. "Dharma: An Expression of Universal Order." *Philosophy East and West* 22 (1972) 131–44.

Kosmann-Schwarzbach, Yvette. *The Noether Theorems: Invariance and Conservation Laws in the Twentieth Century*. Translated by Bertram Schwarzbach. New York: Springer, 2010.

Koterski, Joseph. *Natural Law and Human Nature: Lecture Transcript and Course Guide Book*. Chantilly, VA: Teaching Company, 2002.

Kübler-Ross, Elisabeth, and Ira Byock. *On Death & Dying: What the Dying Have to Teach Doctors, Nurses, Clergy & Their Own Families*. New York: Quality Paperback Book Club, 2019.

Lao Tzu. *Tao Te Ching*. Translated by Ralph Alan Dale. New York: Barnes & Noble, 2002.

Laplace, Pierre-Simon. *A Philosophical Essay on Probabilities*. New York: Wiley, 1917.

Laplace, Pierre-Simon, and Andrew I. Dale. *Philosophical Essay on Probabilities*. New York: Springer-Verlag, 1995.

LeClercq, Jean. "Influence and Noninfluence of Dionysius in the Western Middle Ages." Introduction to *Pseudo-Dionysius: The Complete Works*, translated by Colm Luibheid. New York: Paulist, 1987.

Lemaître, Georges. "The Beginning of the World from the Point of View of Quantum Theory." *Nature* 127 (1931) 447–53.

Lemaître, Georges, et al. *Learning the Physics of Einstein with Georges Lemaître Before the Big Bang Theory*. Cham, Switzerland: Springer International, 2019.

Lewis, C. S. *The Beloved Works of C. S. Lewis*. New York: Inspirational, 2012.

———. *The Four Loves: The Much Beloved Exploration of the Nature of Love*. New York: Harcourt,1980.

Loevinger, Jane. *Ego Development*. San Francisco: Jossey-Bass, 1976.

Malthus, T. R., and Joyce E. Chaplin. *An Essay on the Principle of Population: Influences on Malthus, Selections from Malthus's Work, Economics, Population, and Ethics after Malthus, Malthus and Global Challenges*. Cambridge: Harvard University Library, 2006.

Malthus, T. R., and Antony Flew. *An Essay on the Principle of Population*. Harmondsworth, UK: Penguin, 1970.

Mason, Stephen F. *A History of the Sciences*. New York: Collier, 1962.

Massimi, Michela. *Pauli's Exclusion Principle: The Origin and Validation of a Scientific Principle*. Cambridge: Cambridge University Press, 2012.

Maxwell, James Clerk. "Dynamical Theory of the Electromagnetic Field." *Philosophical Transactions of the Royal Society of London* 155 (1865) 459–512.

May, Gerhard. *Creatio Ex Nihilo*. London: Bloomsbury, 2014.

Mayr, Ernst. "Change in Genetic Environment and Evolution." In *Evolution in Process*, edited by J. Huxley et al., 158–80. London: Allen and Unwin, 1954.

Michelson, Albert S., and Edward W. Morley. "On the Relative Motion of the Earth and the Luminiferous Ether." *American Journal of Science* 34 (1887) 333–45.

Myers, Isabel Briggs, and Mary H. McCaulley. *Manual: A Guide to the Development and Use of the Myers-Briggs Type Indicator*. Palo Alto, CA: Consulting Psychologists, 1992.

Nave, R. "Quarks." *Hyperphysics* (website). Georgia State University, Department of Physics and Astronomy, 2008. http://hyperphysics.phy-astr.gsu.edu/hbase/Particles/quark.html.

Newman, J. D., and Harris James C. "The Scientific Contributions of Paul D. MacLean." *Journal of Nervous and Mental Disorders* 197 (2009) 3–5.

Nowak, M. A. *Evolutionary Dynamics: Exploring the Equations of Life*. Cambridge: Harvard University Press, 2014.

O'Connor, D. J. *Aquinas and Natural Law*. London: Macmillan, 1968.

Paul, Diane B. "Darwin, Social Darwinism and Eugenics." In *The Cambridge Companion to Darwin*, edited by Jonathan Hodge and Gregory Radick, 219–45. Cambridge: Cambridge University Press, 2003.

Peat, F. David. *Superstrings and the Search for the Theory of Everything*. London: Abacus, 1992.

Peebles, P. J. E., and Bharat Ratra. "The Cosmological Constant and Dark Energy." *Reviews of Modern Physics* 75 (2003) 559–606.

Piaget, Jean. *The Moral Judgment of the Child*. London: Kegan Paul, Trench, Trubner, 1932.

Pope Alexander VI. *Inter caetera*. Papa bull. Rome: Papal Library, 1493.

Pseudo-Dionysius, et al. *Pseudo-Dionysius: The Complete Works*. New York: Paulist, 1987.

Pyle, Andrew. *Hume's Dialogues concerning Natural Religion*. London: Continuum, 2006.

Rabin, Sheila. "Nicolaus Copernicus." In *Stanford Encyclopedia of Philosophy*. Redwood City, CA: Stanford University Press, 2007. https://plato.stanford.edu/entries/copernicus/.

Robinet, Isabelle. *Taoism: Growth of a Religion*. Redwood City, CA: Stanford University Press, 2006.

Roney, John Winter, and Robert Hard. "The Beginning of Maize Agriculture." *Archeology Southwest* 23 (2009) 4–6.

Rutherford, Ernest. *The Scattering of Alpha and Beta Particles by Matter and the Structure of the Atom*. London: Taylor & Francis, 1911.

Ryle, Gilbert. *The Concept of Mind*. Chicago: University of Chicago Press, 1949.

Sachs, Joe. *Aristotle's On the Soul and On Memory and Recollection*. Santa Fe, NM: Green Lion, 2001.

Schaff, Philip, and Henry Wace. *A Select Library of the Nicene and Post-Nicene Fathers of the Christian Church: Second Series*. Grand Rapids: Eerdmans, 1982.

Shubin, Neil. *Your Inner Fish: A Journey into the 3.5 Billion-Year History of the Human Body*. Visalia, CA: Vintage, 2008.

Smith, William, ed. *A Dictionary of Greek and Roman Antiquities*. New York: Cambridge University Press, 2013.

Spinoza, Benedictus de, and E. M. Curley. *A Spinoza Reader: The Ethics and Other Works*. Princeton, NJ: Princeton University Press, 1994.

Stachel, John J. *Einstein from "B" to "Z."* Einstein Studies 9. New York: Springer-Verlag, 2001.

Stahl, William. "Aristarchus of Samos." In *Dictionary of Scientific Biography*, edited by Charles Coulston Gillispie, 1:246–50. New York: Scribner, 1970.

Stolzman, William. *The Pipe and Christ: A Christian-Sioux Dialogue*. Chamberlain, SD: Tipi, 1986.

Swinburne, Richard. *The Evolution of the Soul*. Oxford: Clarendon, 2007.

Tucci, Giuseppe, et al. "Buddhism." *Encyclopaedia Britannica* (online edition). 2009. https://www.britannica.com/topic/Buddhism.

Voltaire. *Philosophic Dictionary*. Boston: Mendum, 1878.

Walsh, James, ed. *The Cloud of Unknowing*. New York: Paulist,1981.

Watson, James D., and Francis Crick. "A Structure for Deoxyribose Nucleic Acid." *Nature* 171 (1953) 740–41.

Weinberg, Sheldon. "High-Energy Behavior in Quantum Field Theory." *Physical Review* 118 (1960) 838–49.

Williams, Rowan. *Arius*. Grand Rapids: Eerdmans, 2002.

Zetta, Edward N., ed. "Aristotle's Metaphysics." In *Stanford Encyclopedia of Philosophy*. Redwood City, CA: Stanford University Press, 2012. https://plato.stanford.edu/entries/aristotle-metaphysics.

Zweig, George. "An SU(3) Model for Strong Interaction Symmetry and Its Breaking." *Developments in the Quark Theory of Hadrons* 1 (1964) 22–101.

Index

Abraham, 77, 80, 84–85, 90, 131, 329, 332, 343, 359
afterlife, 9, 90, 99–100, 105, 320–22, 341–49, 365–70, 378
analogy principle, 34–35
all my relatives, 307–08
appreciative inquiry, 292–93
Aquinas, Thomas. *See* Thomas Aquinas
Archimedes, 285
Aristotle, 14, 52, 52n3, 54–55, 76, 105, 220–28, 226n6, 228n8 253, 262, 299, 344, 355n8
Arius, 122–23
Augustine of Hippo, 52, 70, 82, 121, 267, 288

Bacon, Francis, 253, 253n7
big collapse, 8, 57, 362–65
Becquerel, Henri, 179
Boethius, 300–302
brain, 7, 17, 238–43, 258–59, 344–45
Buddha, 105, 264–65, 343, 358

Cantor, Georg, 107
Cartesian dualism, 338, 346
Cauchy, Augustin-Louis, 154
Caux Round Table, 281
chaos, 128, 214, 362
Chavez, Caesar, 296
Chien-Shiung Wu, 180, 180n2
Cleisthenes, 274
Confucius, 274, 296
cooperation, 121, 202, 216–18, 230, 290–91
Copernicus, Nicolaus, 62, 106, 106n4
Copperrider, David, 292
creatio ex nihilo, 53
Crick, Francis 17

Daoism, 105, 126, 130, 185, 194–95, 265, 317, 331
dark energy, 104, 143, 146–47, 163, 362
dark matter, 104, 143, 161–63

Darwin, Charles, 4–5, 15–17, 15n4, 31–33, 31n1, 61–63, 70, 89, 210–11, 217, 308
Dawkins, Richard, 18
death, 8, 46, 71–73, 83–84, 99–100, 235, 256, 311–13, 342–48, 355–59, 364–70, 378
democratic development, 274–75, 281, 296–97
Democritus, 52, 111, 134
Descartes, Rene, 253, 253n6, 344–45
Dirac, Paul, 171–72, 171n6
Duns Scotus, 328, 328n2

Edison, Thomas, 285
Ehrman, Bart D., 359
Einstein, Albert, 6, 57, 57n7, 62, 63n14,121, 124, 135, 145, 151n2, 157, 157m8, 161, 170n1, 174, 183, 186, 195, 195n4, 215, 349
Eldredge, Niles, 61–62, 210
electromagnetic force, 7, 135, 151, 164–85
Empedocles, 14, 52
entanglement, 104, 124–26, 146, 159, 194, 229
Euclidean, 6, 137–39, 145, 161–62, 170
evolutionary field theory, 7, 150–52, 178, 188
expansion of space, 142–46

Fermi, Enrico, 175, 180, 180n1
Fischer, Ronald, 17
Fowler, James W., 257, 257n5
Fowler, Thomas B., 32, 33n3
Francis of Assisi, 310, 310n6
Freud, Sigmund, 256
Friedman, Milton, 281

Gage, Phineas, 259
Galilei, Galileo, 106, 198, 219, 253
Gandhi, Mahatma, 286
Gell-Mann, Murray, 175
Glashow, Sheldon, 151, 151n4, 180
Gödel's incompleteness theorem, 4, 21–22, 21n1, 61, 99, 151, 232
golden mean, 220–22
golden rule, 122

INDEX

Gould, Stephen Jay, 61–62, 62n12, 210, 210n1
gravitational force, 145, 150–52, 155–69. 174–77, 182–85, 362
Greenberg, Oscar, 175, 233n1
Guth, Alan, 142–44, 142n1

Hawking, Steven, 68, 188 188n5
Hegel, George Wilhelm Fredrick, 201–02
Heisenberg uncertainty principle, 156
Hillman, James, 345, 345n10
hillock, 237–38
Hinduism, 105, 130, 255–56, 311, 331
holocaust, 70, 89, 90
Hubble, Edwin, 57, 106
Hume, David, 69, 69n1, 69n2

interrelational war theory, 285–91
Irenaeus of Lyon, 82

Jasper, Karl, 105, 105n1
Jefferson, Thomas, 228
Jesus of Nazareth, 122, 251, 269–71, 274, 286, 302, 317, 329, 332–33, 358–59
Job, 85–91, 86n17, 86n18
Jung, Carl Gustav, 259, 259n6

Kant, Emmanuel, 7, 233–34, 233n1
King, Martin Luther Jr., 286, 296
Kohlberg, Lawrence, 256, 256n2
Kübler-Ross, Elisabeth, 89, 89n19

Lamarck, Jean-Baptiste, 15
Lao Tzu, 195, 195n2, 277
Laplace, Simon, 187, 187n3, 193
Leibnitz, Gottfried Wilhelm, 55, 127, 152
Lemaitre, Georges, 57, 195
Lewis, C. S., 261–62, 261n9
Loevinger, Jane, 256, 256n4
logistic-soul, 8, 100, 262, 266, 341, 345–49, 364–70, 378
Lovejoy, Arthur O., 18

Malthus, Thomas Robert, 15–16, 16n5, 63
Marshall, George C., 287–91
Marx, Karl, 281
Maxwell, James Clerk, 94, 197n4
Mayr, Ernst, 61–62, 61n11
means and ends, 75–76
Mendel, Gregor, 16–17
mercy, 88, 271–72, 291, 310, 368, 371
Merzenich, Michael, 241
Michelangelo, 226–28, 241

Michelson-Morley experiment, 161, 161n9
Mohammed, 122, 332
Myers-Briggs type indicator, 259, 259n7

Neanderthal, 9, 211, 319, 341
neurons, 235–42
Newton, Isaac, 5, 106, 135–36, 152, 160, 165, 169, 188
Nietzsche, Friedrich, 92
no-soul, 105, 264–66, 343
no-thing, 53, 111, 128, 283, 377
Noether, Emmy, 155, 155n7
Nowak, Martin, 217, 217n3
number systems, 4, 25, 139–42

Occam's razor, 254, 346
original sin, 82–83
ownership, 44–46, 110, 113, 118–20, 202, 212–15, 295–96, 301, 312

panentheistic, 6, 103, 126–32, 148, 199, 213, 226, 306–10, 315–17, 346–47, 376–79
Pauli exclusion principle, 60, 60n10, 175, 175n3, 178, 247
Peale, Norman Vincent, 318
Penzian, Ano, 196
Piaget, Jean, 256, 256n3
Plato, 14, 48, 52, 52n2, 76, 105–06, 246, 258, 342, 344
preferential option for the poor, 276–77
prime matter, 52–53, 57
Ptolemy, Claudius, 105–06, 106n2
punctuated equilibria, 61, 210
purpose of life, 2, 229, 235
Pythagoras, 113, 142

quantized distance, 123, 147, 154–56, 181
quantized time, 111, 134, 186
quidquid recipitur, 30, 46, 247–48, 336, 369–71

Rankine, William, 55
Rutherford, Ernest, 174, 174n1, 179
Ryle, Gilbert, 345, 345n9

second law of thermodynamics, 117–18, 133, 361–62
sacrifice, 27, 85, 121, 133, 186, 217, 262, 274, 312–13, 318, 356
Salam, Abdus, 151, 180
Salinas, Joel, 240
Schrodinger, Ernest, 199
Scotus, Duns. *See* Duns Scotus

Shubin, Neil, 239, 239n3
Smith, Adam, 281
Socrates, 28, 52, 76, 258, 261, 342
Spinoza, Baruch, 63, 63n13
Starr, Kala, 284
straight line, 137–38, 144, 156, 166
subsidiarity, 282, 290, 293–96, 322
Swinburne, Richard, 345, 345n11

taxes, 275–76, 296
Thales of Miletus, 51, 253
Thomas Aquinas, 225, 225n5, 233, 288, 303, 343, 344n6
three-dimension space, 136–37, 173, 181
Tillich, Paul, 256
tree of life, 5, 83–84, 141, 211, 235, 267, 352
Trinity, 119, 121, 122

unified field theory, 6–7, 63, 151, 178, 187, 314
unmoved mover, 53–54
ur-stuff of reality, 4, 51, 55–73, 107, 213, 371, 373

Voltaire, 268, 268n1, 289
Vries, Hugo Marie de, 16

Watson, James, 17, 17n7
wave-particle dualism, 46, 104, 177
weak force, 7, 137, 151–52, 174, 179–81, 185–86
Weinberg, Steven, 151, 151n2, 180
Wilson, E. O., 17
Wilson, Robert, 196
Wiseman, Richard, 284

Young, Thomas, 56

Zweig, George, 175

www.ingramcontent.com/pod-product-compliance
Lightning Source LLC
Chambersburg PA
CBHW060505300426
44112CB00017B/2558